高等院校风景园林类专业"十三五"规划系列教材·应用类

风景园林专业英语

主　编　武　涛　王　霞
副主编　袁　嘉　李晓颖　高　天　赵　炜
　　　　陈　宇　朱晓霞　张齐美晨
主　审　徐　锋

FENGJING YUANLIN
ZHUANYE YINGYU

重庆大学出版社

内容提要

本书是一本适用于风景园林(包括园林、景观、景观建筑)及其他相关专业的英语教材,编写宗旨在于提高学生专业英语的阅读能力、运用能力,同时拓展专业知识。

全书共分为7个部分15单元,第一部分为概述,是有关风景园林专业与职业介绍的总述;第二、三、四部分分别为风景园林史论、近现代风景园林发展、风景园林人物及作品,回顾了风景园林从古至今,直至当前的重要设计思想和作品;第五部分为规划设计理论与方法,是有关场地总体规划、详细景观设计和生态规划设计的方法论;第六部分为工程与技术,从工程角度解析项目实施过程中的专业问题;第七部分为研究与动态,节选专业期刊中热点专题,以开阔读者视野。

本书选材广泛,内容经典,涉及风景园林专业的发展历史、风景园林/景观规划设计的理论与方法、风景园林工程、从业状况以及最新的理论动态。书中诸多珍贵、经典的文献资料有助于读者深入了解该专业在西方的发展历程,同时也有助于将中国园林推向世界,帮助从业人员运用英语这门工具,更多地学习国外风景园林发展情况,更好地与不同背景的专业人士进行交流,进而推动我国风景园林事业的发展。

本书可作为风景园林、城市规划、建筑学、环境艺术、旅游管理等不同学科背景的专业师生及相关建设和管理人员的参考书或教材。

图书在版编目(CIP)数据

风景园林专业英语 / 武涛,王霞主编. -- 重庆:重庆大学出版社,2020.1(2020.9重印)
高等院校风景园林类专业"十三五"规划系列教材. 应用类
ISBN 978-7-5689-1521-2

Ⅰ.①风… Ⅱ.①武… ②王… Ⅲ.①园林设计—英语—高等学校—教材 Ⅳ.①TU986.2

中国版本图书馆 CIP 数据核字(2019)第 059066 号

风景园林专业英语

主 编 武 涛 王 霞
副主编 袁 嘉 李晓颖 高 天 赵 炜
　　　 陈 宇 朱晓霞 张齐美晨
策划编辑:何 明
特约编辑:游奉溢

责任编辑:何 明　版式设计:莫 西 黄俊棚 何 明
责任校对:张红梅　责任印制:赵 晟

*

重庆大学出版社出版发行
出版人:饶帮华
社址:重庆市沙坪坝区大学城西路21号
邮编:401331
电话:(023)88617190　88617185(中小学)
传真:(023)88617186　88617166
网址:http://www.cqup.com.cn
邮箱:fxk@cqup.com.cn(营销中心)
全国新华书店经销
重庆俊蒲印务有限公司印刷

*

开本:787mm×1092mm 1/16 印张:20.25 字数:673千
2020年1月第1版　2020年9月第2次印刷
印数:2 001—5 000
ISBN 978-7-5689-1521-2　定价:48.00元

本书如有印刷、装订等质量问题,本社负责调换
版权所有,请勿擅自翻印和用本书
制作各类出版物及配套用书,违者必究

编委会

陈其兵	陈　宇	戴　洪	杜春兰	段晓鹃	冯志坚
付佳佳	付晓渝	高　天	谷达华	郭　晖	韩玉林
胡长龙	黄　晖	黄　凯	黄磊昌	吉文丽	江世宏
李宝印	李　晖	林墨飞	刘福智	刘　骏	刘　磊
鲁朝辉	马　辉	孙陶泽	申晓辉	唐　建	唐贤巩
王　霞	翁殊斐	武　涛	邢春艳	徐德秀	徐海顺
谢吉容	杨瑞卿	杨学成	余晓曼	袁　嘉	袁兴中
张建林	张　琪	赵九洲	朱　捷	朱晓霞	

·编写人员·

主　编	武　涛	南京农业大学
	王　霞	四川大学
副主编	袁　嘉	重庆大学
	李晓颖	南京林业大学
	高　天	西北农林科技大学
	赵　炜	四川大学
	陈　宇	南京农业大学
	朱晓霞	甘肃农业大学
	张齐美晨	四川大学
参　编	王小文	南京农业大学
	孙陶泽	长江大学
	王倩娜	四川大学
	廖晨阳	四川大学
	母洪娜	长江大学
主　审	徐　锋	剑桥大学

PREFACE／前言

随着近些年风景园林事业的迅速发展，不论是专业发展还是职业需求，都对英语的使用提出了更高的要求。专业教师和学生需要了解国外专业发展动态，专业学者需要运用英语进行国际学术交流，从业者在国际项目中需要运用英语与外籍人士交流合作。可见，提高英语运用水平，加强与国际风景园林界的联系，是推动我国风景园林事业发展的需要。

风景园林专业英语教学的主要目的和任务是帮助专业学生利用英语这门工具，通过阅读获取国外与本专业相关的信息，通过写作训练提高撰写专业性文字的能力，在专业实践中能够综合运用英语进行交流。具体来说，主要解决三个实际问题：一是掌握专业术语和专业常用词汇；二是提高阅读和翻译能力，英文翻译应尽量达到"信、达、雅"的要求，文字表述流畅、通顺，中文翻译成英文应尽量做到用词准确，语法无误；三是提高实际应用能力，包括英文资料查阅、英文论文和设计说明撰写、简单的语言交流等。

正是基于以上的目的，本教材在编写过程中选择难易适中、覆盖面广的文章或论文，以期帮助读者在学习教材的同时，扩大和积累一定量的专业术语和常用词汇。根据国外外语教学专家的研究，生词量在3%以下时，可以不借助词典，自由阅读。此时可通过上下文的联系，把不认识的生词猜出来。目前大学教育阶段，应掌握的词汇量大约在4 000个单词，此时阅读一般性的英文科技文献，生词量大约在6%。那么如何将6%的生词量降到3%以下呢？当然需要增加一些词汇积累。那么增加哪些单词呢？显然是文献中重复出现频率比较高的单词。统计资料表明，在每一个专业的科技文献中，该专业最常用的科技术语大约只有几百个，而且它们在文献中重复出现的频率很高。因此，在掌握4 000个单词的基础上，在专业阅读阶段中，有针对性地通过大量阅读，扩充1 000个左右与本专业密切相关的词汇，便可以逐步达到自由阅读专业文献的目标。基于这样的考虑，本教材设有15个单元，每单元各包含3篇文章，共计45篇，总词汇量约有10万个，涉及生词约1 500个，专业常用词汇不少于1 000个。

在多年的专业英语教学中，我们发现在生词量较低的情况下，学生在阅读和翻译中仍存在许多问题。主要原因在于中英文文法差异较大，转换过程中容易发生理解偏差和错误，翻译不符合中文表达习惯。为了帮助读者理解和学习，每篇文章后附有译文以供参考。

本教材文章的选择主要来源于西方高校风景园林专业本科教材、具有影响力的专业书籍、核心期刊文献、设计大师著述和设计名作赏析。具体内容如下：

第一部分：风景园林概述（第1单元），是有关风景园林专业与职业介绍的总述，文章选自美国风景园林师协会（ASLA）对风景园林的描述与定义，以及美国著名风景园林教育家迈克尔·劳瑞编著，美国和英国大学风景园林专业入门教材《风景园林导论》。

第二部分：风景园林史论（第2—4单元），回顾了世界风景园林发展的过程，根据地区的不同分为东方园林、西方园林和伊斯兰园林，文章选自西方园林史原版论著。

第三部分：近现代风景园林发展（第5—8单元），从风景规划、城市开放空间、公园和广场、庭院和企事业园区几个方面展现了西方近现代风景园林的发展和重要作品，文章选自近现代风景园林教育家、理论家的论著。

第四部分：风景园林人物及作品（第9—10单元），介绍了对现代风景园林影响深远的人物和作品，包括奥姆斯特德、丹·凯利、丘奇、麦克哈格、彼得·沃克和玛萨·舒瓦茨。

第五部分：规划设计理论与方法（第11—13单元），是有关场地总体规划、详细景观设计和生态规划设计的方法论。

第六部分：工程与技术（第14单元），解析了风景园林项目运行过程中有关工程建设的技术问题。

第七部分：研究与动态（第15单元），节选专业期刊论文，内容均为目前热点专题，在开阔视野的同时，帮助学习者了解学术论文的英文撰写方法。

在15个单元中，每单元包括3篇文章，由1篇课文、2篇扩展阅读以及注释、术语表、生词表及思考题组成。在使用本教材过程中，教师可根据侧重点和实际情况，选择不同单元作为授课内容；自学读者也可根据个人爱好进行选择性阅读。

本教材由武涛、王霞担任主编，负责全书的统稿工作。袁嘉、李晓颖、高天、赵炜、陈宇、朱晓霞、张齐美晨担任副主编，王小文、孙陶泽、王倩娜、廖晨阳、母洪娜参编。此外，还要特别感谢重庆大学出版社何明女士为本书付出的大量心血。

由于风景园林专业英语教学内容涉及的专业知识庞杂，英文文法多元，难免会存在不足之处，真诚欢迎各位行业同仁提出宝贵意见，以便日后进一步改进完善。

编　者

2019年11月

CONTENTS 目录

PART I Introduction to Landscape Architecture
第一部分 风景园林概述 ·· 1

    UNIT 1 INTRODUCTION ·· 2

    TEXT What is Landscape Architecture（什么是风景园林） ·················· 2

    Reading Material A A Theory of Landscape Architecture（风景园林的理论） ············· 13

    Reading Material B Garden and Gardening（花园与造园） ·················· 17

PART II History of Garden
第二部分 园林史论 ·· 23

    UNIT 2 EASTERN TRADITIONAL GARDEN ·················· 24

    TEXT Chinese Garden（中国园林） ·· 24

    Reading Material A Two Main Forms of Chinese Garden
    （中国园林的两种主要形式） ·················· 31

    Reading Material B Japanese Garden（日本园林） ·················· 37

    UNIT 3 WESTERN TRADITIONAL GARDEN ·················· 42

    TEXT The Gardens in Ancient Egypt（古埃及的园林） ·················· 42

    Reading Material A André Le Nôtre and French Gardens
    （安德烈·勒·诺特尔与法国园林） ·················· 50

    Reading Material B English Landscape Gardens（英国风景园林） ·················· 59

    UNIT 4 ISLAMIC GARDEN ·················· 70

    TEXT Islamic Garden（伊斯兰园林） ·················· 70

    Reading Material A Islamic Gardens of Spain（西班牙的伊斯兰园林） ·················· 75

    Reading Material B Taj Mahal—Mughal Garden（泰姬陵——莫卧儿园林） ·················· 79

PART III Development of Landscape Architecture in Modern Time
第三部分 近现代风景园林发展 ·· 85

    UNIT 5 LANDSCAPE PLANNING AND PROTECTION ·················· 86

    TEXT Landscape Planning（景观规划） ·················· 86

    Reading Material A Landscape Scenery（自然风景） ·················· 90

    Reading Material B National Parks（国家公园） ·················· 95

    UNIT 6 URBAN SPACE ·················· 101

TEXT　Metropolitan Open Space（都市开放空间）	101
Reading Material A　City Planning（城市规划）	106
Reading Material B　Parkway and Recreational Areas（公园道路及休闲区）	110
UNIT 7　URBAN GREEN SPACE	117
TEXT　Urban Parks（城市公园）	117
Reading Material A　Streetscape, Squares and Plazas（街景和广场）	123
Reading Material B　Waterfronts（滨水区）	129
UNIT 8　RESIDENTIAL, INSTITUTIONAL AND CORPORATE LANDSCAPE	135
TEXT　Gardens（花园）	135
Reading Material A　Housing Environments（居住环境）	141
Reading Material B　Institutional and Corporate Landscapes（社会机构及公司景观）	145

PART Ⅳ　Characters, Ideas and Works of Landscape Architecture in Modern Time

第四部分　近现代风景园林人物、思潮及作品	152
UNIT 9　LANDSCAPE PEOPLE AND THOUGHTS Ⅰ	153
TEXT　Frederick Law Olmsted（弗雷德里克·劳·奥姆斯特德）	153
Reading Material A　Ian McHarg（伊恩·麦克哈格）	158
Reading Material B　Thomas Church（托马斯·丘奇）	162
UNIT 10　LANDSCAPE PEOPLE AND THOUGHTS Ⅱ	168
TEXT　Dan Kiley（丹·凯利）	168
Reading Material A　Peter Walker（彼得·沃克）	174
Reading Material B　Martha Schwartz（玛萨·舒瓦茨）	179

PART Ⅴ　Theories and Methods of Planning and Design

第五部分　规划设计理论与方法	183
UNIT 11　ECOLOGY AND ENVIRONMENT PROTECTION	184
TEXT　Environmental Movement（环境运动）	184
Reading Material A　Ecological Principles for Managing Land Use Ⅰ（土地利用管理的生态原则Ⅰ）	190
Reading Material B　Ecological Principles for Managing Land Use Ⅱ（土地利用管理的生态原则Ⅱ）	197
UNIT 12　METHODS OF PLANNING AND DESIGN	202
TEXT　Site Volume（空间）	202
Reading Material A　The Conceptual Plan（概念规划）	207
Reading Material B　The Planning-Design Process（规划设计过程）	213

UNIT 13　LANDSCAPE DETAIL DESIGN PRINCIPLES ··················· 220
　TEXT　Site Systems（场地系统） ··· 220
　Reading Material A　Significance of Landform（地形的重要性） ················ 226
　Reading Material B　Spatial Characteristics of Plants（植物的空间特性） ········ 232

PART Ⅵ　Engineering and Technology
第六部分　工程与技术 ··· 241
　UNIT 14　LANDSCAPE ENGINEERING ··· 242
　　TEXT　Site Development Guidelines（场地开发导则） ··························· 242
　　Reading Material A　Landscape Engineering Ⅰ（风景园林工程Ⅰ） ············ 251
　　Reading Material B　Landscape Engineering Ⅱ（风景园林工程Ⅱ） ············ 258

PART Ⅶ　Studies and Development of Landscape Architecture
第七部分　研究与动态 ··· 265
　UNIT 15　LANDSCAPE STUDIES AND DEVELOPMENTS ······················ 266
　　TEXT　Creating Character through Sustainable and Successful Landscape Design
　　　　　（以可持续方式营造成功景观设计） ·· 266
　　Reading Material A　Landscape Planning and Design in the Century of the City
　　　　　　　　　　　（城市时代下的风景园林规划与设计） ························ 279
　　Reading Material B　Landscape Infrastructure：in plain view（景观基础设施探析） ········ 290

附表 ··· 302
　附表一　欧美地区设置风景园林专业的院校及网址一览（部分） ···················· 302
　附表二　国际风景园林组织机构一览（部分） ··· 307
　附表三　国外风景园林期刊一览（部分） ··· 308
　附表四　国际风景园林学术会议 ··· 309

PART I

Introduction to Landscape Architecture
第一部分 风景园林概述

UNIT 1 INTRODUCTION

TEXT

What is Landscape Architecture[1]
ASLA[2]
什么是风景园林

[1] Although the term "landscape architecture" was invented by a Scotsman (Gilbert Laing Meason, in 1828), it was an American (<u>Frederick Law Olmsted</u>[3]) who gave birth to the landscape architecture profession. We should therefore pay close attention to the definition of the profession given by the American Society of Landscape Architects (Fig. 1). It was as follows:

Fig. 1 Emblem of American Society of Landscape Architects

A Profession In Demand

[2]　From city council rooms to corporate boardrooms, there is increasing demand today for the professional services of landscape architects. This trend reflects the public's desire for better housing, recreational and commercial facilities, and its expanded concern for environmental protection. Residential and commercial real estate developers, federal and state agencies, city planning commissions, and individual property owners are all among the thousands of people and organizations in America and Canada that will retain the services of a landscape architect every year.

[3]　More than any of the other major environmental design professions, landscape architecture is a profession on the move. It is comprehensive by definition—no less than the art and science of analysis, planning design, management, preservation and rehabilitation of the land. In providing well-managed design and development plans, landscape architects offer an essential array of services and expertise that reduces costs and adds long-term value to a project.

[4]　Clear differences do exist between landscape architecture and the other design professions. Architects primarily design buildings and structures with specific uses, such as homes, offices, schools and factories. Civil engineers apply scientific principles to the design of city infrastructure such as roads, bridges, and public utilities. Urban planners develop a broad overview of development for en-

tire cities and regions. Landscape architects touch on all the above mentioned design professions, integrating elements from each of them. While having a working knowledge of architecture, civil engineering and urban planning, landscape architects take elements from each of these fields to design aesthetic and practical relationships with the land.

A Diverse Profession

[5] Landscape architecture is one of the most diversified of the design professions. Landscape architects design the built environment of neighborhoods, towns and cities while also protecting and managing the natural environment, from its forests and fields to rivers and coasts. Members of the profession have a special commitment to improving the quality of life through the best design of places for people and other living things.

[6] In fact, the work of landscape architects surrounds us. Members of the profession are involved in the planning of such sites as office plazas, public squares and thoroughfares. The attractiveness of parks, highways, housing developments, urban plazas, zoos and campuses reflects the skill of landscape architects in planning and designing the construction of useful and pleasing projects.

[7] From coast to coast, in every region of the world, examples of the landscape architecture profession can be found. Many landscape architects are involved in small projects, such as developing plans for a new city park or site plans for an office building, other members of the profession have contributed their expertise to numerous projects which include:

- Preservation of Yosemite Park and Niagara Falls
- Management plan for the Alaskan Maritime National Wildlife Refuge
- Design of the U. S. Capitol Grounds
- Design of Mount Royal Park in Montreal, Quebec
- Development of Stanford University site
- Creation of Boston's "emerald necklace" of green spaces tying the city to the suburbs
- Plans for Baltimore's park system and Inner Harbor area
- Design of "new towns" such as Columbia, Maryland, and Reston, Virginia
- Landfill reclamation for Fresh Kills in New York and Dyer in Florida
- Plans for Golden Gate National Recreation Area in San Francisco, California
- Sursum Corda Affordable Housing, Washington, D. C.
- Design for water treatment and park facility in Hillsboro, Oregon
- Master plan for King Saud University in Saudi Arabia
- Restoration of the landscape along the Baltimore-Washington Parkway in Maryland

[8] Landscape architects may plan the entire arrangement of a site, including the location of buildings, grading, stormwater management, construction and planting. They may also coordinate teams of design, construction and contracting professionals. Already, federal and state government agencies ranging from the National Park Service to local park planning boards employ a large number of landscape architects. More and more private developers realize that the services of a landscape architect are an integral part of a successful, more profitable project.

Tracing the Profession's Roots

[9] The origin of today's profession of landscape architecture can be traced to the early treatments of outdoor space by successive ancient cultures, from Persia and Egypt through Greece and Rome. During the Renaissance, this interest in outdoor space, which had waned during the Middle Ages, was revived with splendid results in Italy and gave rise to ornate villas, gardens, and great outdoor piazzas. These precedents, in turn, greatly influenced the chateaux and urban gardens of 17th-century France, where landscape architecture and design reached new heights of sophistication and formality. The designers became well-known, with Andre le Notre[4], who designed the gardens at Versailles (Fig. 2) and Vaux-le-Vicomte[5], among the most famous of the early forerunners of today's landscape architects.

Fig. 2 Andre le Notre and his Versailles garden   Fig. 3 "Capability" Brown and his Blenheim Palace

[10] In the 18th Century, most English "landscape gardeners"[6], such as Lancelot "Capability" Brown[7], who remodeled the grounds of Blenheim Palace[8] (Fig. 3), rejected the geometric emphasis of the French in favor of imitating the forms of nature. One important exception was Sir Humphrey Repton[9]. He reintroduced formal structure into landscape design with the creation of the first great public parks Victoria Park[10] in London (1845) and Birkenhead Park[11] in Liverpool (1847). In turn, these two parks would greatly influence the development of landscape architecture in the United States and Canada.

Frederick Law Olmsted: "Father of American Landscape Architecture"

Fig. 4 Frederick Law Olmsted and his Central Park in New York

[11] The history of the profession in North America begins with Frederick Law Olmsted, who rejected the name "landscape gardener" in favor of the title of "landscape architect", which he felt better reflected the scope of the profession.

[12] In 1863, official use of the designation "landscape architect" by New York's park commissioners marked the symbolic genesis of landscape architecture as a modern design profession. Olmsted became a pioneer and visionary for the profession. His projects illustrate his high professional standards, including the design of Central Park in New York (Fig. 4) with Calvert Vaux[12] in the late 1850's and the U.S. Capitol Grounds[13] (Fig. 5) in the 1870's. Olmsted and the Brookline, Mass., firm he founded advanced the concept of parks as well-designed, functional public green spaces amid

the grayness of the urban areas through the well practiced principles of landscape architecture and city planning.

Fig. 5　U. S. Capitol Grounds by Frederick Law Olmsted in the 1870's

Early Developments: Late 1800's

[13]　In the ensuing years, the profession of landscape architecture broadened. It played a major role in fulfilling the growing national need for well-planned and well-designed urban environments. Urban parks, metropolitan park systems, planned suburban residential enclaves and college campuses were planned and developed in large numbers, climaxing with the City Beautiful movement[14] at the turn of the century.

[14]　Although the profession itself grew slowly, its early practitioners, including Olmsted, Vaux and Horace Cleveland[15], were among the first to take part in the town planning movement and to awaken interest in civic design. Olmsted also joined other early landscape architects in working on projects in other urban settings, such as at Yosemite Valley and Niagara Falls.

[15]　In 1899, the American Society of Landscape Architects was founded by 11 people in New York—most of them associated with Olmsted. The Society continued to represent landscape architects throughout the United States. In 1900, Olmsted's son, Frederick Law Olmsted Jr.[16], organized and taught at Harvard University's first course in landscape architecture.

Broadening and Diversifying: The 20th Century

[16]　Landscape architecture continued to influence the city beautification and planning movement well into the 20th century, as growing cities used the services of professionally-trained landscape architects. The L'Enfant Plan[17] for the nation's capitol was revived and expanded by the McMillan Commission of 1901 (Fig. 6). Chicago, Cleveland and other cities also used landscape architects to lay out comprehensive development plans.

Fig. 6　L'Enfant Plan for the nation's capitol by the McMillan Commission (1901)

[17]　By the 1920's, urban planning separated from architecture and landscape architecture as a separate profession with its own degree programs and organizations. Yet, landscape architecture continued to remain a major force in urban planning and urban design. During and after the Depression[18], opportunities to design national and state parks, towns, parkways and new urban park systems broadened the profession. The orientation of American landscape architecture returned to its roots in public projects—a trend which has continued throughout the mid-20th century till today.

The Profession in Practice

[18]　Landscape architecture in the 1990s cannot be described in a few simple terms. The scope of the profession is too broad and the projects too varied. A variety of often interwoven specializations exist within the profession, including the following: Landscape Design, the historical core of the profession, is concerned with detailed outdoor space design for residential, commercial, industrial, institutional, and public spaces. It involves the treatment of a site as art, the balance of hard and soft surfaces in outdoor and indoor spaces, the selection of construction and plant materials, infrastructure such as irrigation, and the preparation of detailed construction plans and documents.

[19]　The scope of professional practice includes:
- Site Planning
- Urban/Town Planning and Urban Design
- Regional Landscape Planning
- Park and Recreation Planning
- Land Development Planning
- Ecological Planning and Design
- Historic Preservation and Reclamation
- Social and Behavioral Aspects of Landscape Design

The Profession of the Future

[20]　The years ahead promise new developments and challenges to the ever-broadening profession. With environmental concerns becoming increasingly important, landscape architects are being called upon to bring their expertise to the table to help solve complex problems. Rural concerns are attracting landscape architects to farmland preservation, small town revitalization, landscape preservation, and energy resource development and conservation. Advances in computer technology have opened the field of computerized design, and land reclamation has become a major area of work for members of the profession. Landscape architects have even begun to use their skill within indoor environments (e.g. atriums) and enclosed pedestrian spaces have been incorporated into commercial development projects. From southern California to the Maine coast, the names of landscape architecture firms appear on signs heralding future developments, as more people seek the expertise and services of the profession.

[21]　Furthermore, the future also promises to increase cooperation among landscape architects and other design professionals. As interest in the profession continues to grow, students are studying of the profession in increasing numbers, nearly 60 universities and colleges in the United States and Canada now offer accredited baccalaureate and post-graduate programs in landscape architecture. Forty-five

states license landscape architects. Today, headquartered in Washington, D. C., the American Society of Landscape Architects has grown to nearly 12,000 members in 47 chapters.

[22] During the past decades, landscape architects have responded to the increased demand and professional responsibilities with new skills and expertise. More and more businesses appreciate the profession and the value that it brings to a project. The public praises the balance achieved between the built and natural environments. According to landscape architectural educator, author and ASLA Fellow, Lane Marshall:"The future of… (the) profession is bright. We are growing in size and stature each day. The profession is expanding its borders constantly and stands at the cutting edges of exciting new practice areas. There are landscape architects who are mortgage bankers, developers, business managers, architects, engineers, and lawyers. Since 1899, the profession has grown steadily and now stands at the threshold of a new period of growth. "

[23] The profession of landscape architecture continues to evolve as it meets the challenges of a society interested in improving the quality of life and the wisdom with which mankind uses the land in many ways, landscape architects are shaping the future.

[NOTES]

[1] Landscape architecture 风景园林/风景园林学:Landscape architecture(LA)是国际社会对现代风景园林学科与职业的统一称谓。对于它的理解与翻译,目前国内学术界尚存有争议。国家科技术语委员会和学术界的主流学者,根据我国的风景园林发展的历史与现状,将其翻译为风景园林以与国际同行相对应。但同时也有一些学者将其称为"景观设计"或"景观设计学"。为不致混淆,本文在涉及职业与学科时,将 Landscape architecture 统一称为"风景园林(学)",Landscape architect 统一称为"风景园林师";在涉及具体案例与项目时,将 landscape planning/design 称为"景观规划/设计"或"风景规划/设计"。本文引自美国风景园林师协会(ASLA)对风景园林的定义与描述。

[2] ASLA 美国风景园林师协会:全称 American Society of Landscape Architects,成立于1899年,是美国职业风景园林师组成的专业团体。它见证了并引领着美国风景园林的发展,已逐渐成为引导世界潮流的重要代表之一。该协会设立的 ASLA 奖是美国最高级别的风景园林奖项,它支持创新、奖励优秀、鼓励业内的思想交流,促进了美国风景园林行业的发展,引导美国乃至世界风景园林向现代化、多元化、可持续的方向发展。

[3] Frederick Law Olmsted 弗雷德里克·劳·奥姆斯特德(1822—1903):美国风景园林之父,近代风景园林发展举足轻重的人物,他大量的设计作品和远见卓识的思想对现代风景园林发展有着重要的影响力。

[4] Andre le Notre 勒·诺特尔(1613—1700):17世纪文艺复兴时期法国最著名的造园家,他开创的法国勒·诺特尔式造园风格是西方古典园林的一种重要风格,它以理性主义思想为基础,采取了严格的几何构图,创造了简洁明朗、庄严华丽的园林风格,在历史上曾经影响了欧洲各国近一个世纪的造园,并且对于欧洲近现代的城市规划和城市设计也产生了深远的影响。

[5] gardens at Versailles and Vaux-le-Vicomte 凡尔赛花园和沃·勒·维贡特府邸花园:法国古典式园林的典范,均为勒·诺特尔设计。

[6] landscape gardener 造园师:1764年由英国作家威廉·申斯通创造的术语,这个词仅存在了短暂的时期,在19世纪末 Landscape architecture 一词被业内普遍采用之后,随即消逝。landscape garden 指自然风景园,尤指盛行于18世纪的英国自然风景园;landscape gardening 指风景造园(学),1933年我国园林界前辈章君瑜先生曾把它翻译为"风致园艺"。

[7] Lancelot "Capability" Brown 兰斯洛特·"可为"布朗(或"万能")布朗(1716—1783):是继肯特之后的英国自然式园林巨匠,他使自然风景园达到理想水平。布朗善于把他人完成的风景园加以"改进",不论在什么地方,都以为"颇有可为",从而博得"可为布朗"(Capability Brown)、"改良者"(Brown the Improver)的称号。但

他的大刀阔斧作风也引起了不必要的破坏，英国过去许多出色的文艺复兴和勒·诺特尔式园林都被平毁而改造成为风景式园林，最引人注目的是1737年对白金汉郡(Buckinghamshire)的斯陀园(Stowe)的改造。

[8]　grounds of Blenheim Palace 布伦海姆宫苑：1764年由布朗改造的宫苑，是布朗的代表作品。

[9]　Humphrey Repton 汉弗莱·雷普顿(1752—1818)：继布朗之后，英国杰出的造家园，他运用效果图演示的方法来对比现状与设计效果，创造富有诗意的浪漫派园林，妥善处理了园林自然化与艺术化的园林。

[10]　Victoria Park 伦敦维多利亚公园：1845年向公众开放，是伦敦最古老的公共绿地之一。

[11]　Birkenhead Park 利物浦伯肯黑德公园：建于1847年，是世界上第一个公共经费资助建设的城市公园。

[12]　Calvert Vaux 卡尔沃特·沃克斯(1824—1895)：美国近代风景园林的重要实践者，与奥姆斯特德共同设计了纽约中央公园(Central Park in New York)。

[13]　U.S. Capitol Grounds 美国国会大厦环境：18世纪70年代由奥姆斯特德设计。

[14]　City Beautiful movement 城市美化运动：是19世纪末20世纪初在美国发起的一场更新城市规划、美化城市环境的一次改革运动。

[15]　Horace Cleveland 霍拉斯·克里夫兰：与奥姆斯特德和沃克斯一起，是最早的城镇规划运动的倡导者与实践者，并唤起人们对城市设计的关注。

[16]　Frederick Law Olmsted Jr. 小奥姆斯特德：奥姆斯特德的儿子，哈佛大学风景园林学科的创始人。

[17]　L'Enfant Plan 朗方规划：朗方规划是1791年由法国建筑师皮埃尔·查尔斯·郎方(Pierre Charles L'Enfant，1754—1825)为美国首都华盛顿所做的城市总体规划。当时朗方的规划方案受到华盛顿总统的肯定，后来的实施却遇到了长达百年的坎坷。19世纪末，针对华盛顿乱建改建的情况，一些有识之士特别是国会议员麦克米伦(Senator James McMillan，1938—1902)，大声疾呼整顿首都建设、恢复朗方规划。于是在国会的同意与支持下，重新从国家档案中查找出了朗方方案的城市规划总图。并于1902年组成了"麦克米伦委员会"(McMillan Commission)，恢复朗方方案，负责整治和建设华盛顿城市的工作，并着重对历史文化风貌特色进行恢复与保护。

[18]　Depression 经济萧条：指1929—1933年美国经济大萧条。

[GLOSSARY]

City Beautiful movement	城市美化运动	landscape preservation	景观保护
civil engineering	市政工程	master plan	总体规划
ecological planning and design	生态规划和设计	metropolitan park system	都市公园系统
energy resource development and conservation	能源开发与保护	park and recreation planning	公园和休闲娱乐规划
environmental protection	环境保护	real estate	地产
farmland preservation	农田保护	regional landscape planning	区域景观规划
grading	地形整理	site planning	场地规划
historic preservation and reclamation	历史保护与改造	small town revitalization	小城镇复兴
L'Enfant Plan	朗方规划	stormwater management	雨洪管理
land development planning	土地开发规划	town planning movement	城镇规划运动
landscape architecture	风景园林	urban design	城市设计
landscape architect	风景园林师	urban park	城市公园
landscape gardener	风景造园师	urban planning/city planning	城市规划

[NEW WORDS]

atrium	n.	中庭	mortgage	n.	抵押
baccalaureate	n.	学士学位	ornate	adj.	装饰的,华丽的
chateaux	n.	城堡(chateau 的复数)	overview	n.	总的看法
enclosed	adj.	围住的,封闭的	pedestrian	n.	步行者
expertise	n.	专业的知识,专家的意见,专门技术	Persia	n.	波斯(现在的伊朗)
forerunner	n.	先驱(者),传令官,预兆	piazza	n.	广场,露天市场,外廊
formal	adj.	规整的,整齐匀称的,形式的	plaza	n.	广场,露天汽车停车场,购物中心
formality	n.	形式,礼节	post-graduate	n.	研究生
genesis	n.	起源	precedent	n.	先例
geometric	adj.	几何的,几何学的	preservation	n.	保护,保存
grading	n.	土方工程	rehabilitation	n.	恢复,复原
headquarter	vi.	以……作总部,设总公司于	remodel	vt.	重新塑造,改造,改变
herald	vi.	预示	Renaissance	n.	文艺复兴,文艺复兴时期
highway	n.	公路,大路	sophistication	n.	复杂,混杂,技巧,完善度
imitate	v.	模仿,效法,冒充	stature	n.	成就,高境界,高水平
incorporate	vt.	合并	thoroughfare	n.	通路,大道
infrastructure	n.	基础设施,下部构造	threshold	n.	开始,开端
integral	adj.	不可或缺的	villa	n.	别墅,乡村庄园,乡村住宅
landscape	n.	风景,山水画,景观,地形,前景	visionary	n.	有远见的人,远见卓识的人
license	vt.	批准,许可,颁发执照	wane	vi.	衰落,变小,消逝
Middle Ages	n.	中世纪,中古时代	waterstorm	n.	排水

[难句]

1. The years ahead promise new developments and challenges to the ever-broadening profession. 对日渐扩大的风景园林职业而言,未来的岁月预示着新的发展和新挑战。

2. Furthermore, the future also promises increase cooperation among landscape architects and other design professionals. 此外,未来还有望增强风景园林师和其他设计专业人员之间的合作。[倒装]

3. The profession of landscape architecture continues to evolve as it meets the challenges of a society interested in improving the quality of life and the wisdom with which mankind uses the land in many ways, landscape architects are

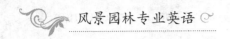

shaping the future. 风景园林在面对社会关注生活条件改善和以多种方式明智地利用土地的挑战中,不断地发展前进,风景园林师正在塑造着未来。[长句]

[参考译文]

什么是风景园林?
美国风景园林师协会

[1] 尽管"风景园林"这个术语是由苏格兰人吉尔伯特·莱恩梅森1828年创造的,但是风景园林职业的诞生却是由美国人弗雷德里克·劳·奥姆斯特德促成的。因此,我们应该更多关注由美国风景园林师协会(ASLA)对风景园林作出的定义。它是这样描述的:

需求量大的职业

[2] 当前,从市政部门到公司企业,对于职业风景园林师的需求越来越多。这个趋势反映了公众对居住环境、娱乐和商业设施更优化的需求,以及由其引起的对环境保护的关注。美国和加拿大每年有数千个居住和商业地产开发商、国家和州机构、城市规划委员会,以及个人地产所有者请风景园林师为他们提供服务。

[3] 比起其他任何专业的环境设计职业,风景园林都是发展最快的职业。从定义来看,它具有复杂性——至少包括艺术、科学分析、规划设计、管理、土地保护和恢复等。风景园林师能够提供至关重要的专业服务,做出良好设计和开发方案,为项目减少投入并获得长期效益。

[4] 风景园林与其他设计职业确实存在明显的不同。建筑师主要设计具有特定用途的建筑和构筑物,比如住宅、办公楼、学校和工厂。市政工程师为城市基础设施做出符合科学原理的设计,比如道路、桥梁和公共设施。城市规划师为整个城市和区域做出总体规划策略。风景园林师要与以上所有提及的设计职业相接触,综合每个职业的要素。风景园林师要具备建筑、市政工程和城市规划的知识,并利用这些领域的知识从审美和实践的角度对土地进行规划设计。

多样性高的职业

[5] 风景园林是设计行业中多样性最高的职业之一。风景园林师设计街区、城镇和城市的外环境,同时也保护和管理从森林、农田到河流、海岸等自然环境。这个职业的成员通过为人类和其他生物的生活环境提供最好的设计,而承担着改善生活质量的特殊义务。

[6] 实际上,风景园林师的工作无处不在。他们致力于场地规划,比如公司广场、公共广场和道路环境等。公园、道路、居住环境、城市广场、动物园和校园的空间质量及吸引力能够反映出风景园林师在规划和设计这些实用且令人愉悦的项目工程方面所具备的能力。

[7] 从西海岸到东海岸,在世界的每个区域都有风景园林的案例。多数风景园林师从事小型项目的设计,例如为新的城市公园制定开发规划或为办公建筑进行场地规划,还有一些风景园林师则致力于大量的大尺度规划项目,其中有:

- 约瑟米蒂国家公园和尼亚加拉大瀑布的保护
- 阿拉斯加国家海上野生动植物避难所管理规划
- 美国国会大厦环境设计
- 加拿大魁北克蒙特利尔皇家山公园设计
- 斯坦福大学校园规划
- 将城市与郊区相连的波士顿"翡翠项链"绿色空间的创举
- 巴尔的摩公园系统和内海湾规划
- "新镇"设计,例如哥伦比亚、马里兰,以及弗吉尼亚莱斯顿
- 纽约清泉垃圾填埋场以及佛罗里达州染厂景观改造
- 加州旧金山金门国家休闲区规划

- 华盛顿特区什科达经济适用房规划
- 俄勒冈州西斯博罗水处理和公园设施设计
- 沙特阿拉伯国王大学总体规划
- 马里兰州巴尔的摩-华盛顿公园道路景观恢复

[8]　风景园林师有能力规划场地中所有的内容,包括建筑选址、地形整理、雨水管理、施工及种植。他们还能够协调组织包括设计、施工和工程承包在内的专业团队。从国家公园管理局到地方公园规划委员会的联邦和州政府机构,已经雇用了大量风景园林师。越来越多的私人开发商也意识到风景园林师的工作是项目获得成功和盈利的不可缺少的一部分。

职业溯源

[9]　今天的风景园林职业起源可以追溯到从波斯和埃及直至希腊和罗马的古代文明中,人们对室外环境的不断改善。文艺复兴期间,人们对室外空间的热爱,经过中世纪的消退,又在意大利以辉煌的成就获得重生,产生了华丽的别墅、花园和壮观的户外广场。这些先例,进而又大大地影响了17世纪法国城堡园林和城市花园,法国的风景园林和设计在复杂程度和形式上达到了新的高度。设计师也因此扬名,就像勒·诺特尔——当今风景园林设计师最有名的早期先驱者之一,因设计了凡尔赛花园和沃·勒·维贡特府邸花园而闻名于世。

[10]　18世纪,大多数英国"风景造园师",抛弃了法国重视几何的布局而采用模仿自然的形式,比如兰斯洛特·"可为"布朗,他曾对布伦海姆宫苑进行改造。但其中有一个例外——汉弗莱·雷普顿。他在第一个大型公园——伦敦维多利亚公园(1845年)和利物浦伯肯黑德公园(1847年)中,再次将规则结构引入风景设计中。之后,这两个公园极大地影响了美国和加拿大风景园林的发展。

弗雷德里克·劳·奥姆斯特德:"美国风景园林之父"

[11]　北美风景园林职业的历史开始于弗雷德里克·劳·奥姆斯特德,他抛弃了"风景造园师"的称谓,以"风景园林师"的称呼取代之,他认为这样更能反映这个职业的范畴。

[12]　1863年,纽约公园委员会正式采用"风景园林师"称谓,标志着风景园林成为一个现代设计职业的起始。奥姆斯特德成为这个职业的先驱者和具有远见卓识的人物。他的设计项目说明了他高超的专业水平,其中包括他和卡尔沃特·沃克斯在19世纪50年代晚期设计的纽约中央公园,以及18世纪70年代美国国会大厦环境设计。奥姆斯特德和他在麻省布鲁克林的公司,进一步发展了公园的概念——在城市灰色区域中,符合风景园林和城市规划实践原理的,经过良好设计、功能合理的公共绿地空间。

早期的发展:19世纪后期

[13]　在这之后的几年,风景园林职业的范围更加宽泛。它在满足国家对良好规划和设计的城市环境日益增加的需要中发挥着重要的作用。大量城市公园、都市公园系统、郊区居住区和大学校园都进行了规划和开发,并且在世纪之交时,随着城市美化运动达到高潮。

[14]　尽管这个职业本身发展很慢,它早期的从业者,包括奥姆斯特德、沃克斯和克里夫兰,是最早参加城镇规划运动,并唤起人们对城市设计关注的人群。奥姆斯特德也加入到早期从事其他城市环境项目的风景园林师中,比如他的约瑟米蒂河谷和尼亚加拉大瀑布规划。

[15]　1899年,美国风景园林师协会由纽约11人发起,他们中大多数都与奥姆斯特德合作过。该协会一直代表着全美国的风景园林。1900年,奥姆斯特德的儿子,小奥姆斯特德,在哈佛大学开设了第一门风景园林专业课程。

宽领域和多样性:20世纪

[16]　由于城市发展需要受到过专业训练的风景园林师参与,风景园林对城市美化和规划运动的影响一直持续到20世纪。1901年,国家首都的郎方规划由麦克米伦委员会进行再次规划和扩大。芝加哥、克里夫兰和其他城市都请风景园林师对城市进行综合布局和开发规划。

[17]　到20世纪20年代,城市规划从建筑和风景园林分离成为一个独立的职业,并拥有自己的学位课程和组

织机构。然而,风景园林师在城市规划和城市设计中仍然保持着主导力量。在经济大萧条期间和之后,国家和州公园、城镇、公园道路和新的城市公园系统规划设计的机会使风景园林行业得到壮大。美国风景园林的定位重新回到它的起源公共项目——一个贯穿20世纪中期直至今天的趋势中来。

实践中的职业

[18]　20世纪90年代的风景园林很难用几个简单的词语来描述。风景园林的职业范畴极其广阔,项目类型极其多样。职业内的多样性还伴随着专业化的发展,其中包括:景观设计,历来都被看作职业的核心,它关注居住、商业、工业、公共机构和公共空间的外部空间细部设计。它包括场地的艺术化处理,室内外空间硬景与软景的平衡,施工与植物材料的选择、基础设施(比如排水),以及细部施工图纸与文件的准备等。

[19]风景园林的职业实践范畴包括:
- 场地规划
- 城市/城镇规划和城市设计
- 区域景观规划
- 公园和休闲娱乐规划
- 土地开发规划
- 生态规划和设计
- 历史保护与改造
- 景观设计的社会和行为研究

未来的职业

[20]　今后,越来越宽广的风景园林职业将面临着新的发展和挑战。随着人们对环境越来越多的关注,风景园林师需要运用专业知识帮助解决复杂的问题。人们对农村的关注吸引风景园林师从事农田保护、小城镇复兴、景观保护和能源开发与保护。计算机技术的发展开创了计算机化设计领域,同时,土地改造成为风景园林师的主要工作领域。风景园林师甚至开始将他们的技术运用到室内环境(比如中庭)中,封闭的步行空间环境与商业开发项目结合成一个整体。从南加州到缅因州海滨,越来越多的人对风景园林职业技术和服务有所需求,预示着风景园林公司在未来将有更大发展。

[21]　此外,未来的风景园林将会与其他设计职业有越来越多的合作。随着人们对这个职业兴趣的增加,学习这个专业的学生也大量增加,美国和加拿大有近60个大学和学院开设了风景园林方面的学士学位和研究生学位课程。有45个州制订了风景园林从业许可制度。今天,设于华盛顿的美国风景园林师协会总部已经拥有47个分会近12 000个会员。

[22]　在过去的几十年,风景园林师为满足日益增加的需求和职业责任,发展了新的技术和专业技能。越来越多的商业开发受益于这个职业,以及这个职业为项目带来的价值。公众也赞誉它为人工与自然环境之间取得平衡所做的工作。根据风景园林教育家、作家和ASLA成员雷恩·马歇尔的观点:"这个职业的未来是光明的……我们的规模和地位每天都有所增加。这个职业正在不断地拓宽它的边界,并处于令人兴奋的新的实践领域的最前沿。风景园林师来自房地产抵押银行家、开发商、商业经理、建筑师、工程师,以及律师。自从1899年,这个职业就稳健地发展着,目前正处于一个新发展的开端。"

[23]　风景园林在面对社会关注生活条件改善和以多种方式明智地利用土地的挑战中,不断地发展前进,风景园林师正在塑造着未来。

Reading Material A

A Theory of Landscape Architecture[1]
Michael Laurie[2]

风景园林的理论

[1] Five major components of a theory are natural process, human factors, methodology, technology, and values. Whatever the scale or emphasis of operation, these five components are consistently relevant. Social and natural factors clearly permeate every facet of a profession that is concerned with people and land. Problem solving, planning, and design methods apply at all scales. Good judgment is consistently required.

[2] Consider how natural factors data are relevant to both planning and design. At the regional scale, in a responsible society, the impact of development or change in use on a landscape must be known and evaluated before a policy to allow such action is set. An inventory of the natural factors, including geology, soils, hydrology, topography, climate, vegetation and wildlife, and the ecological relationships between them is fundamental to an understanding of the ecosystem to which change is contemplated. Equally important is an analysis of visual quality which is the sum of the components. Land use policy can thus be made on the basis of the known vulnerability or resistance of the landscape. In other circumstances the natural processes which add up to a given landscape at a given moment in its evolution may, as at Grand Canyon[3] and other unique places, be considered resources to be preserved, protected, and managed as a public trust. On a smaller scale, soil and geological conditions may be critical in the determination of the cost and the form of building foundations: where it is most suitable to build and where it is not. Sun, wind, and rain are important factors of design where the development of comfort zones for human activity or the growth of plants is a primary objective. Thus, natural factors influence land use, site planning, and detailed design.

[3] Similarly, human factors apply equally at all scales. In site planning and landscape design, cultural variation in the use and appreciation of open space and parks and the physical and social needs of the young and old are some of the many variables to be considered in a design process that aims to be responsive to social values and human needs. In decisions related to appropriation of landscape for recreation and aesthetic value people's perception of the environment and the behavioral patterns and tendencies of people in the out-of-doors are clearly relevant. It is also important that designers understand the impact of environment on behavior and also appreciate the basic human need to manipulate and control the environment. The value of community participation in urban planning and design is now widely recognized.

[4] Technology is the means by which a design is implemented or on which a policy depends. Some of it changes year by year as new materials, machinery, and techniques are developed. Its role in the three types of landscape architecture is clear. Specific areas of technology include plants, planting and ecological succession, soil science, hydrology and sewage treatment, microclimate control, sur-

face drainage, erosion control, hard surfaces, and maintenance. Other techniques of importance in landscape architecture relate to communications, community participation, development economics, and political process.

[5] Design and planning methodology involve systems where by landscape problems are defined, all relevant factors and variables are assembled, given values are incorporated in the solutions. Computer graphics, analytical techniques, and notation systems aid in this process. As a device to modify the bias of planners in the creative process, Halprin[4] suggests scoring techniques, as in music or choreography. These open up the design process, allowing more people to participate in decision making and facilitate the generations of more humanistic ways to plan and design large scale complex environments.

[6] Finally, landscape architecture must be based on a set of values; this is perhaps the most difficult part of the theory to deal with. Natural and social science, methodology, and technique can be learned; values have to be lived and felt. Experience and good sense tell us that we need to develop a set of priorities and subscribe to a land ethic related to our belief in the "alternative for survival," in which short term profit at the expense of long term regeneration and conservation of resources would be unthinkable. Environmental impact must be seen in a regional context. Quantity must be equated with quality. We must learn to make judgments in terms of what is considered best for the common good and the future of mankind. Even Third World countries, who have exploitation without much benefit, must see the importance of this. The professional must present such considered judgments to the investment banker, government agency chiefs, and others in whose hands lie the ultimate decisions—even though his recommendations may be at variance with their programs.

[7] The objective of combining these components is the development of a basis from which landscape planning and detailed design can be made responsive to human behavioral patterns (people) and specific situation characteristics (the setting). Since both will vary in terms of culture, region, and neighborhood there can be no panacea and no preconceived solutions.

[8] Little has been so far about aesthetics or visual quality. This is because it is possible to stress these aspects of landscape architecture with a resulting lack of attention to the other sources of form already mentioned. But of course, how things look is important and proportion and relative size are particularly critical where new work is fitted into an existing framework. Color and form influence human comfort and often have symbolic meaning. It can be seen, then, that aesthetic principles are essentially techniques associated with human factors in the production of meaningful form. Thus a design which fully satisfies all criteria is likely to be aesthetically pleasing. It has also been said that a landscape in ecological repose will be beautiful. When all is said and done, responsible, effective, and pleasing landscape design results from clear and objective thinking and a synoptic view from start to finish.

[9] The process of design, the aim of which is the evolution of forms and relationships suited to the need of people, may be compared to the fundamental formgiving processes which have created the geomorphology of the great natural landscapes of the world. Here the visual form of the land's surface— valleys and ridges, water-filled basins, and jagged peaks—represents an evolutionary stage in the in-

teraction between the geological structure and the agents of erosion. The forms we see result from the response of inorganic material to a set of imposed conditions of weathering. The variations of vegetative cover from north to south slope, from meadow to subarctic plateau, from river valley to rocky talus consistute exact responses to the range of environmental conditions created by the physiographic differentiation of the landscape. In turn, the wildlife distribution is dictated by the type and extent of the vegetation. No one aspect of the pattern is without cause or consequence. All merge irrevocably into a self-sustaining and evolving ecological system, representing the resolution of the natural forces and processes up to a specific moment in time.

[10] This model for the creation of form is similar to the one of the design process which also produces form(s) representing the resolution of forces with a build-in capacity for adaptation. Such forms are the goal of the landscape planner and designer.

[NOTES]

[1] 本文节选自 *An Introduction to Landscape architecture*（第二版）中第一章"The Human Environment: Landscape Architecture"。该书 1986 年由 Elsevier 出版社出版，已经成为美国和英国大学风景园林专业入门的著名教材。

[2] Michael Laurie 米歇尔·劳瑞（1932—2002）：美国加利福尼亚大学伯克利分校（University of California, Berkeley）风景园林系教授，曾三次出任该系主任，是美国著名的风景园林教育家和学者。

[3] Grand Canyon 科罗拉多大峡谷：位于美国亚利桑那州西北部，由科罗拉多河数百万年冲蚀而成，1979 年被列入世界遗产。

[4] Halprin 哈普林（1916—2009）：全名劳伦斯·哈普林（1916—2009, Lawrence Halprin），美国现代风景园林的重要代表人物。哈普林的设计重视自然和乡土性，提倡从自然环境中获取创作灵感，为激发人们的行为活动提供一个具有艺术感召力的背景环境，如著名的海滨农场住宅开发项目。哈普林的作品还强调空间参与性，体现现代城市景观的开放性、公共性和大众化，同时带有强烈的"隐喻"（metaphor）意境，他的代表作有罗斯福总统纪念园、伊拉·凯勒水景广场（即演讲堂前庭广场）、爱悦广场、柏蒂格罗夫公园等。他还是最早的生态规划的实践者，提出广泛适用于各个领域的创作过程"RSVP Cycle"。

[GLOSSARY]

ecological succession	生态演替	detailed design	细节设计	site planning	场地规划
out-of-doors	室外，户外	sewage treatment	污水处理	surface drainage	地表排水

[NEW WORDS]

aesthetic	adj.	美的，审美的，具有美感的	facet	n.	（事物的）一个方面，多面体的面
	n.	美学，美感，审美观	geomorphology	n.	地形学，地貌学
bias	n.	倾向，偏见	inorganic	adj.	无机的
consistently	adv.	一致地，和谐地，始终如一地	inventory	n.	详细目录，财产清单，总量
criteria	n.	标准	irrevocably	adv.	不能取消地，不能撤回地
dictate	vt.	指示，命令，规定	jagged	adj.	参差不齐的，锯齿状的

Continued

manipulate	vt.	（熟练地）使用，操作处理，控制	proportion	n.	均衡，相称，协调
methodology	n.	方法论，方法	repose	n.	静止，休息，睡眠
misleading	adj.	误导的，使人误解的	resistance	n.	抵制，抵抗；抵制力，抵抗力
panacea	n.	万能药，灵丹妙药	subarctic	adj.	亚北极区的，靠近北极的
peak	n.	山顶，顶点，最高峰	symbolic	adj.	象征的，象征性的，符号的
permeate	v.	渗透，弥漫，使充满	synoptic	adj.	概要的
physiographic	adj.	地形学的，地文学的	talus	n.	斜面，山麓
plateau	n.	高原，高地，台地	variation	n.	变化，变动，变化程度
preconceive	vt.	事先形成（意见、看法），预想，预见	vulnerability	n.	易受伤，易受攻击，脆弱

[参考译文]

风景园林的理论

［1］（风景园林）理论的五个主要组成部分包括自然过程、人为因素、方法论、技术和价值观。不论什么尺度或侧重点，这五个组成部分都相互关联。社会和自然因素很明显地渗透到有关人和土地的职业的方方面面。解决问题、规划和设计的方法适用于各个尺度。良好的判断力也是始终需要的。

［2］要考虑自然要素数据是如何与规划和设计相关。在区域尺度上，在一个负责任的社会，一处景观的开发或用途改变所带来的影响，在政策许可之前，必须要了解和评估。自然要素主要包括地理、土壤、水文、地形、气候、植被和野生动物，以及这些要素之间的生态关系，是理解生态系统将会发生何种变化的基础。同样重要的是视觉质量分析，这是各要素的汇总分析。进而，土地利用政策可以在这些已知的景观脆弱性和韧性基础上得出。在另一种情况下，特定景观在特定时刻的自然演化过程，例如大峡谷和其他独特的地方，可以被视为值得保留、保护和管理的公共资源。在一个较小的尺度上，土壤和地理条件对确定造价和建筑地基的形式可能是关键性的：哪里最适宜建设，哪里不适宜。当适宜人类活动的舒适区或者植物生长是开发建设的主要目标时，太阳、风和雨水是重要的设计要素。因此，自然要素影响土地利用、场地规划和细节设计。

［3］同样地，人为要素也一样适用所有尺度。在场地规划和景观设计中，对开放空间和公园在使用和欣赏上的文化差异，以及年轻人和老年人生理和社会需求，是旨在响应社会价值观和人类需求的设计过程中，需要考虑的许多变量中的一部分。在与休闲景观适宜性和美学价值相关的决策中，人们的环境感知、行为模式和户外活动的偏好是与其明显相关的变量。设计师理解环境对行为的影响以及重视人类对环境控制的基本需求，也是同样重要的。社区参与在城市规划和设计中的价值如今也被广泛认可。

［4］技术是设计实现的方法或政策所依赖的载体。一些技术随着材料、机械和技术的发展而逐年变化。它的作用在风景园林的三种类型中显而易见。具体的技术领域包括植物、种植和生态演替、土壤科学、水文和污水处理、微气候控制、地表排水、侵蚀控制、硬质地面和维护。风景园林中其他重要的技术与通信、社区参与、经济发展和政治过程相关。

［5］设计和规划方法论包括景观问题的定义系统，所有相关要素和变量的汇总，既定价值观和解决方案的整合。计算机图形、分析技术和符号系统辅助这个过程。为了调整规划师创作过程中的偏差，哈普林建议采用评分技术，如同音乐和舞蹈的评分系统。这些开放了设计过程，允许更多的人参与决策，并促使更多代人以人文主义方法来规划和设计大尺度综合环境。

［6］最后，风景园林必须基于一系列的价值观；这也许是该理论要处理的最棘手的部分。自然和社会科学、方法论和技术是可以学习的；价值观则需要亲身体会和感知。经验和理智告诉我们，我们需要建立一组优先

级,并遵守与我们"生存选择"信仰相关的土地伦理,以牺牲长期资源再生和保护为代价的短期利益是不被接受的。环境影响必须在区域背景下审视,数量与质量必须相互匹配。我们必须学会做出符合普遍利益和人类未来发展的最佳判断。即使是利益已被耗竭的第三世界国家,也必须看到这个问题的重要性。专业人士必须将这些深思熟虑的意见展示给投资银行家、政府机构负责人和其他有最终决定权的人——即使建议可能与他们的项目有分歧。

[7] 结合这些要素的目标是使景观规划和细部设计适应人类行为方式和特定环境特征的发展基础。由于它们都因文化、区域和邻里而变化,因此没有万全之策和既定的解决方案。

[8] 到目前为止关于美学或视觉质量的研究都很少。这是因为强调风景园林的这些方面可能会导致缺乏对已经提及的那些方面的关注。但当然,当新的设计放置于一个现有环境中,事物看起来的面貌很重要,比例和相对尺度也尤为关键。颜色和形式影响人们的舒适感,并且常常具有象征意义。可以看出来,美学原则在创造有意义的形式中,是与人类因素相关的重要技术。因此,一个完全满足所有标准的设计有可能是美观的。据称一个生态静止的景观也会是美的。当所有提及的都能做到,负责的、高效的、令人愉快的景观设计会从清晰客观的思考和自始至终的全局观中产生。

[9] 设计的过程,目标是形式的演化和创造适合人们需要的关系,可以同自然世界景观中地形地貌的形态创造过程相比较。大地表面的视觉形式——山谷和山脊、充水盆地和参差的山峰——代表了地质结构和侵蚀力交互作用的演化过程。我们看到的形式来自于无机物质受到一系列风化作用后的结果。从北坡到南坡,从草原到亚北极高原,从河流山谷到岩屑坡地,植被覆盖的变化正是适应不同地形差异所创造的不同环境条件的结果。继而,野生动物的分布又由植被的类型和分布所定义。这个模式里没有哪个要素是没有原因或结果的。所有要素都融入不可逆的自我维持和演变的生态系统中,并代表着某一特定时刻自然力量和自然过程的一个部分。

[10] 这个形式创造的模式与设计过程相似,设计过程也创造代表内在适应能力的形式。这些形式正是景观规划师和设计师追求的目标。

Reading Material B

Garden and Gardening
花园与造园

Garden

[1] A garden is a planned space, usually outdoors, set aside for the display, cultivation, and enjoyment of plants and other forms of nature. The garden can incorporate both natural and man-made materials. The most common form is known as a residential garden. Western gardens are almost universally based around plants.

[2] See traditional types of eastern gardens, such as Zen gardens[1] (Fig. 1), use plants such as parsley. Xeriscape gardens use local native plants that do not require irrigation or extensive use of other resources while still providing the benefits of a garden environment. Gardens may exhibit structural enhancements, sometimes called fol-

Fig. 1 The representative of Zen gardens RyoanJi-Dry garden, in Japan

lies, including water features such as fountains, ponds (with or without fish), waterfalls or creeks, dry creek beds, statuary, arbors, trellises and more.

[3] Some gardens are for ornamental purposes only, while some gardens also produce food crops, sometimes in separate areas, or sometimes intermixed with the ornamental plants. Food-producing gardens are distinguished from farms by their smaller scale, more labor-intensive methods, and their purpose (enjoyment of a hobby rather than produce for sale).

[4] Gardening is the activity of growing and maintaining the garden. This work is done by an amateur or professional gardener. A gardener might also work in a non-garden setting, such as a park, a roadside embankment, or other public space. Landscape architecture is a related professional activity with landscape architects tending to specialize in design for public and corporate clients.

[5] The term "garden" in British English refers to an enclosed area of land, usually adjoining a building. This would be referred to as a yard in American English. Flower gardens combine plants of different heights, colors, textures, and fragrances to create interest and delight the senses.

Garden design

[6] Garden design is the creation of plans for layout and planting of gardens and landscapes. Garden design may be done by the garden owner themselves, or by professionals. Most professional garden designers are trained in principles of design and in horticulture, and have an expert knowledge and experience of using plants. Some professional garden designers are also landscape architects, a more formal level of training that usually requires an advanced degree and often a state license. Elements of garden design include the layout of hard landscape, such as paths, rockeries, walls, water features, sitting areas and decking, as well as the plants themselves, with consideration for their horticultural requirements, their season-to-season appearance, lifespan, growth habit, size, speed of growth, and combinations with other plants and landscape features. Consideration is also given to the maintenance needs of the garden, including the time or funds available for regular maintenance, which can affect the choices of plants regarding speed of growth, spreading or self-seeding of the plants, whether annual or perennial, and bloom-time, and many other characteristics.

Gardening history

[7] The history of gardening extends across at least 4,000 years of human civilization. Egyptian tomb paintings of the 1500s BC are some of the earliest physical evidence of ornamental horticulture and landscape design; they depict lotus ponds surrounded by symmetrical rows of acacias and palms. Another ancient gardening tradition is of Persia: Darius the Great [2] was said to have had a "paradise garden[3]" and the Hanging Gardens[4] of Babylon (Fig. 2) were renowned as a Wonder of the World[5]. Persian gardens were also organized symmetrically, along a center line known as an axis.

[8] Persian influences extended to post-Alexander's Greece: around 350 BC there were gardens at the Academy of Athens, and Theophrastus[6], who wrote on botany, was supposed to have inherited a garden from Aristotle[7]. Epicurus[8] also had a garden where he walked and taught, and bequeathed it to Hermarchus[9] of Mytilene[10]. Alciphron[11] also mentions private gardens.

Fig. 2 The illustration of Hanging Gardens of Babylon depicted by the 16th-century Dutch artist Martin Heemskerck

Fig. 3 Canopus at Hadrian's Villa

[9] The most influential ancient gardens in the western world were the Ptolemy's[12] gardens at Alexandria[13] and the gardening tradition brought to Rome by Lucullus[14]. Wall paintings in Pompeii[15] attest to elaborate development later. The wealthiest Romans built extensive villa gardens with water features, topiary and cultivated roses and shaded arcades. Archeological evidence survives at sites such as Hadrian's Villa[16] (Fig. 3).

[10] Byzantium and Moorish Spain kept garden traditions alive after the 4th century AD and the fall of Rome. By this time a separate gardening tradition had arisen in China, which was transmitted to Japan, where it developed into aristocratic miniature landscapes centered on ponds and separately into the severe Zen gardens of temples.

[11] In Europe, gardening revived in Languedoc[17] and the Ile-de-France[18] in the 13th century, and in the Italian villa gardens of the early Renaissance. French parterres developed at the end of the 16th century and reached high development under Andre le Notre. English landscape gardens opened a new perspective in the 18th century. The 19th century saw a welter of historical revivals and Romantic cottage-inspired gardening. In England, William Robinson[19] and Gertrude Jekyll[20] were strong proponents of the wild garden and the perennial garden respectively. Andrew Jackson Downing and Frederick Law Olmsted adapted European styles for North America, especially influencing public parks, campuses and suburban landscapes. Olmsted's influence extended well into the 20th century.

[12] The 20th century also saw the influence of modernism in the garden: from Thomas Church's[21] kidney-shaped swimming pool to the bold colors and forms of Roberto Burle-Marx[22]. A strong environmental consciousness is driving new considerations in gardening today.

[NOTES]

[1] Zen gardens 日本禅宗花园:是中国传统园林艺术传到日本之后,受佛教思想和日本本土文化影响而形成的具有日本民族特色的园林形式。

[2] Darius the Great 大流士大帝(公元前549—前486):公元前521年至前485年波斯帝国阿契美尼德王朝君主。

[3] paradise garden 乐园:来自于古波斯语,最初paradise本身的含义就是指由墙围合成的花园,是理想化的生活净土,该词传入欧洲之后在英语和其他语言中常与heaven(天堂)一词互用。

〔4〕 Hanging Gardens 空中花园:又称悬空花园、悬园,古代世界七大奇迹之一,公元前6世纪由新巴比伦国王尼布甲尼撒二世(Nebuchadnezzar),在巴比伦城为其王妃安美依迪丝(Amyitis)所建。

〔5〕 Wonder of the World 世界奇观。

〔6〕 Theophrastus 泰奥弗拉斯托斯(公元前372—前287):公元前4世纪的古希腊哲学家和科学家,先后受教于柏拉图和亚里士多德。

〔7〕 Aristotle 亚里士多德(公元前384—前322):古希腊著名哲学家和科学家。

〔8〕 Epicurus 伊比鸠鲁(公元前341—前270):古希腊著名哲学家,伊比鸠鲁学派的创始人。

〔9〕 Hermarchus 赫马库斯:古希腊哲学家,伊比鸠鲁学派的创始人之一。

〔10〕 Mytilene 米蒂利尼:希腊莱斯博斯岛(Lesbons)的首府。

〔11〕 Alciphron 阿尔茨弗隆:古希腊最著名的诡辩家之一,著名的书信体作家。

〔12〕 Ptolemy 托勒密(公元90—168):古希腊天文学家、地理学家和光学家。

〔13〕 Alexandria 亚历山大港:埃及第二大城市、亚历山大省省会,地中海岸的港口,按其奠基人亚历山大大帝命名,它是托勒密王朝的首都,后成为古希腊文化中最大的城市。

〔14〕 Lucullus 卢卡库鲁斯(公元前118—前57):古罗马共和时期著名的政治家和执政官。

〔15〕 Pompeii 庞贝城:古罗马城市之一,位于那波利湾的岸边,公元79年被维苏威火山爆发时的火山灰所淹埋,1748年被挖掘出来。

〔16〕 Hadrian's Villa 哈德良宫苑:是公元2世纪由罗马皇帝哈德良在提沃里所建造,它融合古代埃及、希腊、罗马的建筑风格。该遗址于1999年被联合国教育科学文化组织列入世界文化遗产名录。

〔17〕 Languedoc 朗格多克:法国南部一地区。

〔18〕 Ile-de-France 法兰西岛:法国的一个行政区域,于巴黎盆地中部,现以巴黎为中心,俗称为大巴黎地区。

〔19〕 William Robinson 威廉·罗宾逊(1838—1935):爱尔兰著名的园林家和记者,野趣造园活动的倡导者,他倡导应用乡土植物以俭朴和简洁的形式来营造花园,以反对当时盛行的维多利亚式华丽而呆板的花园风格。罗宾逊留有大量的文章和著作,其中有著名的《野趣花园》(*The Wild Garden*)和《英式花卉园》(*The English Flower Garden*)。

〔20〕 Gertrude Jekyll 格特鲁德·杰基尔(1843—1932):英国现代非常有影响力的园林设计家、作家和艺术家。她在英国、欧洲和美国所设计的花园超过400个,她为《乡村生活》(*Country Life*)、《花园》(*The Garden*)和其他杂志所撰写的文章达1 000多篇。

〔21〕 Thomas Church 托马斯·丘奇(1902—1978):全名Thomas Dolliver Church,美国现代著名的风景园林师。丘奇1922年毕业于加州大学伯克利分校,后在哈佛大学获得硕士学位,曾在俄亥俄州立大学执教1年,随后在旧金山成立了自己的设计所,设计作品多达2 000项,直到1977年退休。他被认为是现代风景园林运动的奠基人之一,其设计风格被誉为"加利福尼亚风格"(California Style)。丘奇等加州现代园林设计师群体被称为加利福尼亚学派,其设计思想和手法对今天美国和世界的风景园林设计有深远的影响。在他的著作《园林是为人的》(*Garden are For People*)中,丘奇提出了设计的"四项原则",即统一(unity)、功能(function)、简洁(simplicity)和尺度(scale)。

〔22〕 Burle-Marx 布雷·马科斯(1909—1994):全名为Roberto Burle-Marx,巴西著名的现代风景园林师、画家、生态学家和自然主义者。他早年在德国学习绘画期间开始接触景观,1930年回到巴西后,开始热心于收集植物,他在自己住宅周围种植了所收集的3 500多个植物品种。马科斯于1955年建立了自己的设计公司,从事了大量富有独创性的设计工作。同时,他还花费了很多时间在巴西的林业上,在新植物发现等方面都做出巨大的贡献。

[GLOSSARY]

English landscape garden	英国风景园	gardening	造园	water feature	水景
bloom-time	花期	perennial garden	多年生宿根草木园	wild garden	野趣花园

[NEW WORDS]

acacia	n.	洋槐,阿拉伯橡胶树	lotus	n.	莲,荷花,莲属植物,莲花图案
adjoining	n.	贴近,毗邻,邻近	miniature	adj.	小型的,缩小的,小规模的
amateur	adj.	业余的,非职业的	Moorish	adj.	摩尔人的
annual	adj.	一年生的,每年的,一年一次的	ornamental	adj.	装饰,装饰性的
arbor	n.	藤架,棚架,凉亭	palm	n.	棕榈,棕榈叶
arcade	n.	连拱廊,拱形走道,拱形建筑物	parsley	n.	欧芹,香芹菜
archeological	adj.	考古学的	parterre	n.	花坛,花圃
aristocratic	adj.	贵族的,上层社会的	perennial	adj.	多年生的,终年的,多年的,长期的
attest	v.	作证,证明,证实	perspective	n.	远景,景;透视,透视图;观点
bequeath	vt.	将……遗赠给,传给后人	proponent	n.	提倡者,支持者,拥护者
botany	n.	植物,植物学	revive	v.	恢复,复兴,再生
Byzantium	adj.	拜占庭的,拜占庭风格的	rockery	n.	假山,假山工程
consciousness	n.	意识,知觉,觉悟;察觉,感觉	statuary	n.	雕塑,塑像
creek	n.	(海岸或河岸边的)小湾,小溪,小港	symmetrical	adj.	对称的,匀称的
depict	vt.	描绘;描画,描写;描述	texture	n.	质感,质地,肌理,纹理
elaborate	v.	详尽说明,详细制定	topiary	adj.	(灌木、树枝等)修剪成形的
embankment	n.	防护堤,护坡,路堤	trellis	n.	格子棚架,格子结构,格架
folly	n.	装饰性建筑物	welter	n.	混乱,起伏,汹涌
kidney	n.	肾,肾脏	xeriscape	n.	(指自然环境)干旱景观,旱生植物景观

[参考译文]

花园与造园

花园

[1] 花园是规划的空间,通常位于室外,被用作展示、栽培、植物欣赏和其他的自然形式。花园可以采用或自然或人工的材料。最普通的花园形式是住宅花园。西方花园基本上以植物为主。

[2] 传统的东方园林,例如禅宗花园,使用像香芹这样的植物。旱生花园运用不需要浇灌的乡土植物,也不过多使用其他材料,仍然能够提供花园般的环境。花园可以展示景观构筑物,有时也称装饰性建筑,包括像喷泉这样的水景、水池(有鱼或没有鱼)、瀑布或小溪、旱溪、雕塑、凉亭、花架等。

[3] 一些花园仅用来装饰,一些花园还生产食物,有时在单独分开的区域,有时同装饰性植物混杂在一起。与农场不同,生产食物的花园由于尺度小,需要更费力的养护和特定的用途(兴趣爱好多过生产销售)。

[4] 造园是种植和养护花园的活动。这项工作可以由业余爱好者或专业造园师完成。造园师也可能在一个不是花园的环境中工作,比如公园、道路堤岸,或其他公共空间。风景园林是由风景园林师为公共或公司客户提供的专业设计活动。

[5] 在英国英语中"花园"这个词是指一处围合的区域,通常与建筑相邻。这个词在美国英语中可能指一个庭院。花园种植有不同高度、色彩、质感和香气的植物,来创造趣味和愉悦感。

花园设计

[6] 花园设计是创造花园布局和植物种植方案的过程。花园设计可以由园主自己完成,也可以由专业人士完成。大部分专业的花园设计师受到过设计方面和园艺学的培训,拥有运用植物的专门知识和经验。一些专业的花园设计师还是风景园林师,受到过更正式的要求高学历和州执照的培训。花园设计要素包括硬质景观的布局,例如道路、岩石、墙、水景、座椅区和木平台,以及植物本身,植物要考虑他们的园艺要求,每个季节的面貌、生命周期、生长习性、大小、生长速度,以及与其他植物和景观的搭配。此外,还要考虑花园的维护需求,包括常规养护的时间和资金,它们会影响植物的选择,要根据生长速度、自播繁衍的能力、一年生还是多年生、开花时间以及许多其他特点来选择植物。

造园历史

[7] 造园史至少跨越了4 000年人类文明。公元前1500年前的埃及墓葬绘画是最早的观赏园艺和景观设计实物证据之一,它们描绘了由对称种植的刺槐和棕榈树列包围的荷花池塘。另一个古代造园传统是在波斯:据称大流士大帝有一处"天堂园"和被称为世界奇迹的巴比伦空中花园。波斯花园也是沿着中轴线对称式布局。

[8] 波斯影响还延伸到亚历山大之后的希腊:大约公元前350年,在雅典学院的一些花园,泰奥弗拉斯托斯写到的植物园,应该是继承了亚里士多德的花园。伊壁鸠鲁也有一处他散步和教书的花园,后来把它赠与了米蒂利尼的赫马库斯。阿尔茨弗隆也提到过私人花园。

[9] 西方最有影响力的古代花园是亚历山大港的托勒密花园和由卢卡库鲁斯带到罗马的造园传统。庞贝的墙面绘画证明了后期发展的精致化。最富有的罗马人也建造了大量有水景、整形灌木、栽培玫瑰和凉阴廊架的别墅花园。哈德良宫苑的遗址留存着他们的考古证据。

[10] 拜占庭和摩尔人的西班牙花园传统一直活跃至公元4世纪和罗马衰落以后。此时,一个独立的造园传统在中国出现,后来传播到日本,在日本发展为中心是水池的贵族缩景园,然后又从中分离出严谨的禅宗花园。

[11] 在13世纪欧洲,造园又在朗格多克和法兰西岛复兴,出现了文艺复兴早期的意大利别墅花园。法国花坛在16世纪末发展起来,并在安德烈·勒·诺特尔园林中达到顶峰。英国风景园在18世纪打开了新的视角。19世纪出现了历史复兴和浪漫村舍花园的混杂,在英国,威廉·罗宾逊和格特鲁德·杰基尔分别是野趣花园和多年生观赏花卉园的有力支持者。安德鲁·杰克逊·唐宁和弗雷德里克·劳·奥姆斯特德调整了欧洲风格以适应北美,尤其影响了公园、校园和郊区景观。奥姆斯特德的影响一直延伸到20世纪。

[12] 20世纪见证了现代主义对花园的影响:从托马斯·丘奇的肾形泳池到布雷·马科斯的大胆色彩和形式。今天强烈的环境意识正在推动着对造园新的思考。

[思考]

(1)何谓风景园林?
(2)风景园林一词是谁发明的?
(3)风景园林包括哪些学科和知识领域?
(4)风景园林师的职业可以划分为哪几种类型?
(5)简述世界风景园林的造园历史。

PART II

History of Garden
第二部分 园林史论

UNIT 2　EASTERN TRADITIONAL GARDEN

TEXT

Chinese Garden

中国园林

[1]　Garden (Fig. 1) is an artifact, made by someone. At the same time its elements are independent of man (they have a life of their own, which may take a different course beyond the designers' intention, if not attended). The garden is created and received within a framework of conventions, but it can exist, although in deteriorated way, even if there is no gardener to keep it.

Fig. 1　Traditional Chinese garden

Fig. 2　The Jichang Garden borrowing scenery of HuiShan

[2]　Many Chinese gardens were in a state of deterioration when Siren[1] visited them (between 1922-1935). The garden is designed to be perceived aesthetically, for the "production of aesthetic pleasure". This pleasure is the joint result of all the values of the object, of the connoisseur's contemplation and communion with the mysterious forces present in the garden and its design. Even very big (imperial) gardens were actually limited, but even small gardens were designed so that one cannot really tell their actual size. Actual size was in this sense less real than the experience, which really counted. One of the favorite design strategies was "borrowing scenery"[2], actually delimiting the fixed, limited space of the room, or garden, toward the outer environment, borrowing part of its scenery for the space and the onlooker inside the wall (Fig. 2).

[3]　The shape of the opening served as a frame for the borrowed scenery, which was now a ready-made painting. By framing it, the designer borrowed the exterior scenery of the outside environment for the interior. Whatever was suitable outside the frame of the garden, could become a part of the aesthetic experience relevant for the space inside the frame.

[4]　Chinese gardens(and their designers) had names, like any other work of art. In big gardens, which had separate parts, each part had a separate name. In one of the classic Chinese novels, The Dream of the Red Chamber (The Story of the Stone)[3], a process of naming is described in details. After finishing the designing, the owner gathered guests, relatives, and friends, and each name was proposed and chosen, in a friendly contest, after careful consideration. In considering various names, they had in mind the original intentions of the designer, and the literary, philosophical, and art tradition.

[5]　Certain gardens included miniature replicas/simulacrums of famous sites (famous environments), whose purpose was similar to landscape paintings to depict a favorable environment in three-dimensional setup, or installation. These designs were not considered as inferior to designs that presented something yet unseen, something new. In this case it was important to catch and recreate the spirit of the particular environment. Also, sometimes the designer would attempt to recreate the landscape (or part of it), already presented in some landscape painting(in many cases the designers were landscape painters). In Chinese art(between 12-13th century) landscape, or its elements(rocks, bamboo, waters, birds, etc.) become more important than human figures(face, in particular), even when the main subject was an emotion, or sentiment.

[6]　To define valuable art, Chinese art criticism differentiated two pairs of concepts: substance (chil), and ornament (wen); skill (kung-fu), and spontaneity (tzu-jan). In the Chinese tradition the obsession was to be close to nature; the ideals were spontaneity, and sincerity, or authenticity. One of the most important elements of natural spontaneity in landscape design were garden rocks[4]. From the middle of the Tang dynasty(8th century), important element in landscape design became rocks eroded by water[5], whose creation was absolutely spontaneous, and therefore they embodied naturalness (Fig. 3). These rocks were, perhaps, the first ready-mades in the history of world art.

Fig. 3　The Cloud Capped Peak in the Lingering Garden

[7]　Here were two types of rocks: recumbent, and standing. The first were piled up to make hills/mountains. Vertical stones, larger and with more interesting shapes, were treated as monuments, sometime set up on pedestals, and solitary. Their decorative function in the Chinese gardens is often the same as that of the statues, obelisks, and urnfound in European gardens, only with the difference that they merge so much more naturally in the picturesque play of light and shade of their surroundings. The function of standing stone was not mimetic, on purpose, although they were usually named according to associations they provoked. They were like abstract sculptures in modern art, at the time when European gardens and parks(in the Renaissance and Baroque[6]) were full of naturalistic sculptures. However, associative and mimetic function of rocks was also utilized, and many visitors would associate upon their resemblance, "seeing" in them goblins, lions or dragons.

[8] Designers of Chinese landscapes introduced amorphous rocks that could not be explained away, and could provoke the experience of pure existence: inexplicable, unnamable, astounding, mysterious, like ultimate Tao, or Buddhist ultimate suchness. The chinese saw the "abstract" nature force displaying in the "concreteness" of gnarled branches or roots, or water eroded rocks. "Faith in nature's self-disclosure was expressed aesthetically by an intense, empathic interest in natural forms, old trees, and odd-shaped rocks, pitted an cut through by natural forces". Rocks pitted and cut by water brought to mind lines of <u>Tao Te Ching</u>[7]: "Nothing under heaven is softer or more yielding than water. But when it attacks things hard and resistant there is not one of them that can prevail".

[9] The process of choosing and putting rocks in the right setting, was usually very meticulous and slow, as can be also depicted from recorded episodes in designing the Japanese gardens. Chinese landscape designers also had to solve the antagonism of naturalness and style. To follow nature and discard artificiality can mean identification of style with spontaneity, or seen the way round-absence of style, negation, or abandon of style. In Chinese painting this was achieved by utilizing chance (<u>splashing, spattering, or dripping ink</u>[8], or color), by lack of ostentation, or deliberate clumsiness, changing the learned right, for the unlearned left hand, getting drunk, etc.

[10] In certain Chinese gardens (or in certain parts of the garden) the point of observation was determined, the observer was guided by the design (pathways, corridor, bridge, tunnel, pavilion, or tower) to move to certain points of observation. In other he was free to choose the point of observation. The Chinese garden can never be completely surveyed from a certain point. It consists of more or less isolated sections which must be discovered gradually and enjoyed as the beholder continues his stroll: he must follow the paths, wander through tunnels, ponder the water, reach a pavilion from which a fascinating view unfolds (Fig. 4). He is led on into a composition that is never completely revealed. We see that, although Chinese gardens belong to <u>plastic arts</u>[9] broadly defined, they also have marks of <u>temporal arts</u>[10].

Fig. 4 The winding path leads to a secluded quiet place in the Jiangnan gardens

[11] They are observed and contemplated gradually, in time, through a succession of scenes, designed to unfold one after another. Siren compares this to unfolding of a Chinese landscape painting in a form of scroll. Mirroras a metaphor of art played a considerable role in European tradition, but in this context we could say that the garden was a place where nature as a mirror was held up to mind (to paraphrase: "art is a mirror held up to nature").

[12] Aesthetic judgment of environmental works of art is challenging, because environments offer a

broader perspective for philosophy of art, and aesthetics, than standard works of art. Certain types of environment are particularly valuable because of the view they offer, but some because of fragrance, tactile qualities, etc. Some are poor in view, but rich in "whispers" of nature(sounds of water, wind, birds, frogs, or other animals).

[13]　Describing the range of perception in Chinese gardens, Johnston says: "Chinese gardens are marking a direct appeal to the emotions and devoted exclusively to serving all the senses: visually unfolding a succession of pleasing surprises; introducing textures which seek to be touched; mingling the perfumes of blossoms and bark; capturing whispers of moving leaves and water; exploiting the ever changing character of the trees whose varying beauties enhance each season." Aesthetic contemplation of the environment can be either general, or related to particular(visual, audible, tactile, or olfactory) aspects of the environment.

[14]　Listening to various sounds(of water, rain, wind, birds), or watching particular objects, or sight, sometimes develops as a separate affinity in relation to the overall contemplation of the environment. For example, bird-watching developed as a particular pastime and chapter in environmental aesthetics in England, at the beginning of this century. Some Chinese paintings from the 12th century—like the masterpiece ("<u>Birds among plum treesand bamboo</u>"[11] (Fig. 5) seem to prove that the pastime was probably known in China at the time, as well as listening to the "whispers" of nature, recorded in the painting by <u>Ma Lin</u>[12], "<u>Listening to the wind in pines</u>"[13] (Fig. 6).

Fig. 5　*Birds among plum trees and bamboo*

Fig. 6　*Listening to the wind in pines*

[15]　Most damaging for perception is "speeding-up" of perception, forced by contemporary film and TV(especially by commercials, and spots), high-powered speakers, and other gadgets invented to "hit you into guts", to force an information into you. People are conditioned to be targets of machine-gun fire of chaotic short-duration percepts. This damages the sense of time, and replaces meaning and depth with speed, force, and distortion.

[16]　Sensitivity to sounds in nature seems now almost lost for man in the West and East. Sounds of the natural environment are drowned in aggressive sounds of the urban environment, or they are already bellow the damaged faculty of hearing, or just absent, because nature is dead. However, some contemporary poets try to keep in our memories particular sounds, like rain drops hitting leaves, sounds of crickets, birds, or frogs, or sound of the wind. Perhaps the tactile qualities of the environment are

less understood or recognized than other percepts. Some people have tactile experiences, whether they touch the texture, or just watch it. They spontaneously "translate" part of the visual experience into tactile(synaesthesia) sometimes just because it is not possible to touch it.

[NOTES]

[1] Siren 塞伦:奥斯伍尔德·塞伦(Osvald Siren),瑞典美术史家、哲学博士。曾于 1920、1921、1930、1934、1935、1954 和 1956 年先后五次访问中国,对中国古代艺术十分热爱,并进行了深入研究。这方面主要著作有:《北京的城墙和城门》(The Walls and Gates of Peking,1924 年),《中国雕刻》(Chinese Sculpture,1925 年),《北京故宫》(The Imperial Palace of Peking,1926 年);《中国绘画史》(Histoire de art anciens,1929—1930 年);《中国花园》(Gardens of China,1949 年)等。这些著作在向世界人民介绍中国灿烂的古代文化方面起了很大作用。1966 年 6 月 26 日逝世,享年八十七岁。

[2] borrowing scenery 借景:借景是古典园林建筑中常用的构景手段之一。在视力所及的范围内,将好的景色组织到园林视线中的手法。古典园林中常用的构景手段还有障景、框景、漏景等。

[3] The Dream of the Red Chamber (The Story of the Stone)《红楼梦》(《石头记》)。

[4] garden rocks 园林假山:园林中以造景为目的,用土、石等材料构筑的山。

[5] rocks eroded by water 水冲奇石:江河中的石头在自然水流不断地冲刷和自身相互碰撞摩擦中,石上形成各种形态的凹槽、孔穴,显得光洁温润。其肌理丰富,以质、色、形、纹和独特的神韵和外观而著称,变化多端让人品味无穷。

[6] Renaissance and Baroque 文艺复兴和巴洛克:前者是盛行于 14 世纪到 17 世纪的一场欧洲思想文化运动;后者是一种代表欧洲文化的典型艺术风格,是在 16 世纪下半叶在意大利发起的,在 17 世纪的欧洲普遍盛行,是背离了文艺复兴艺术精神的一种艺术形式。

[7] Tao Te Ching《道德经》:是春秋时期老子(李耳)的哲学作品,又称《道德真经》《五千言》等,是中国古代先秦诸子分家前的一部著作,为其时诸子所共仰,是道家哲学思想的重要来源。

[8] splashing, spattering, or dripping ink 泼墨、溅墨、滴墨:泼墨是国画的一种画法,用笔蘸墨汁大片地洒在纸上或绢上,画出物体形象,像把墨汁泼上去一样,如泼墨山水;而后两者则为墨汁在纸上或绢上的形态,如无规律的大小不等的墨点,则称为溅墨。

[9] plastic arts 造型艺术:指以一定物质材料(如绘画用颜料、墨、绢、布、纸、木板等,雕塑、工艺用木、石、泥、玻璃、金属等,建筑用多种建筑材料等)和手段创造的可视静态空间形象来反映社会生活与表现艺术家的思想情感。主要包括绘画、雕塑、摄影艺术、书法艺术等。

[10] temporal arts 时间艺术:当物体一经活动的时候就具有了时间性,这里指的是中国园林中有变化,在心理感受中有节奏的体验艺术。

[11] Birds among plum trees and bamboo《梅竹雀图》:此图体现了南宋院体花鸟画的典型特征。图中梅竹相依,由左上向下斜势而出。枝头黄莺宛然欲起,饶有生趣。精巧的构图,细腻的勾勒,淡雅的设色,与画院名家林椿的画风相近。

[12] Ma Lin 马麟:南宋画家,画承家学,擅画人物、山水、花鸟,用笔圆劲,轩昂洒落,主要画作有《层叠冰绡图》《橘绿图》《静听松风图》《芳雨春霁图》等。

[13] Listening to the wind in pines《静听松风图》:是中国南宋画家马麟创作的国画作品,内容描绘的是树木山水和人物景象,作品由台北"故宫博物院"收藏。

[GLOSSARY]

amorphous rocks	怪石、奇石	famous sites	风景名胜	pavilion	亭子
baroque	巴洛克式的	garden rocks	园林假山	plastic arts	造型艺术
borrowing scenery	借景	imperial gardens	皇家园林	rocks eroded by water	水冲奇石
corridor	游廊	monuments	纪念碑	temporal arts	时间艺术
exterior scenery	外景	obelisks	方尖碑		

[NEW WORDS]

aesthetically	adv.	审美地	fragrance	n.	芳香
affinity	n.	吸引力	gadget	n.	小机械,小器具
artifact	n.	人工制品	gnarled	adj.	扭曲的
beholder	n.	观看者,旁观者	goblin	n.	小妖怪
chaotic	adj.	混沌的、无秩序的	gut	n.	内脏
concreteness	n.	具体性	inferior	adj.	次要的
connoisseur	n.	鉴定家、鉴赏家、行家	metaphor	n.	隐喻
contemplation	n.	沉思、思考	mimetic	adj.	模仿的
convention	n.	常规、惯例	olfactory	n.	嗅觉
cricket	n.	蟋蟀	perception	n.	感知、知觉
delimit	vt.	限制、定……的界	picturesque	adj.	如画的、美丽的
depict	vt.	描绘	recumbent	adj.	躺着的、斜靠的
deteriorate	vi.	变坏、恶化	relevant	adj.	相宜的
disclosure	n.	揭露、揭示	resemblance	n.	形状、外观
distortion	n.	扭曲、变形	scroll	n.	卷轴
drowned	vt.	淹没	synaesthesia	n.	通感、联觉
episode	n.	插曲、片段	tactile	adj.	触觉的、可触摸的
exclusively	adv.	唯一地、专门地	urn	n.	瓮,罐

[参考译文]

中国园林

[1] 园林是由人创造的工艺品。同时它的组成元素又是独立于人的意志之外(它们有自己的生命,如果没有受到照管,它们会超出设计者的意图,按另外的方式发展)。园林的建造和欣赏都是在一个常规框架内进行,但是即便在最糟糕的情况下,哪怕是没有园丁来管护,它也能够存在。

[2] 当塞伦探访中国园林时(1922—1935年),许多园林正处于衰败状态。为了成为审美愉悦的产物,当时的园林是以审美感知来设计的。这种愉悦是多种价值共同作用的结果,包括:所有景观实物的价值,以及鉴赏家

对园林所呈现的神秘力量和设计本身的关注与交流。即使是大型皇家园林实际上范围都是有限的,但哪怕是小型私家园林的设计都令人难以真正判断其实际大小。从这种意识上来说,实际测量的尺度倒不如体验真实。中国园林最常用的设计手法之一是"借景",实际上就是通过房屋或园林中有限的固定空间,朝向外部环境,将外部景色的一部分借入到围墙内的空间中来,供游人观赏。

[3]　开口的形状充当了借景的边框,从现在来理解就是一幅现成的画面。通过框景,设计者将园外环境借入园内。景框之外任何适宜的景观,均可成为与墙内空间相关联的审美体验的一部分。

[4]　如同其他艺术品一样,中国园林(及其设计者)都是有名字的。大型园林可分为独立的几部分,每部分都有各自的命名。在中国古典小说《红楼梦》(《石头记》)中详细描述一个命名的过程。园子设计完成之后,主人遍邀亲朋好友,通过友好的争论,仔细的斟酌,提出并选定了每一处的名称。在推敲各个命名时,他们考虑了设计者的创意,并且还综合了文学、哲学和艺术传统。

[5]　某些园林,对风景名胜微缩复制或模仿,其目的如同风景画一样,通过三维立体的方式组织或安排,从而描绘美景。这样的设计并未被认为不如那些呈现前所未有的新设计。这种情况下,领悟并再现独特景观的神韵是很重要的。有时,设计师也会试图再现那些已呈现在风景画上的景观(或其一部分),那时的设计师也往往是风景画家。在12世纪到13世纪的中国风景画艺术中,哪怕是要表现情感或情绪的主题时,风景画或是其部分元素(如石、竹、水、鸟等)都变得比人物(特别是肖像)更为重要。

[6]　为了精确地解释这种珍贵的艺术,中国艺术评论家区分了两对概念:质地(质)与修饰(纹);技巧(功夫)与自然性(自然)。中国传统的固有思想是接近自然,其理想是自然、真诚或真实。表现自然的最重要的景观设计元素是园林假山。从唐代中期(8世纪)开始,水冲奇石成为了景观设计的重要元素,其创造过程绝对天然,因此,能体现自然。这些山石或许是世界艺术史上最早的现成艺术品。

[7]　奇石有两种类型:卧式和立式,前者被堆叠成假山。立石中,体量大,造型有趣的,像纪念碑一样,有时是被独立安放在基座上。在中国园林内,它们的装饰性功能通常与西方园林中常见的雕塑、方尖碑、花坛的作用一样。它们之间仅有的区别在于,奇石在光影之间,非常自然地融入了周围环境。尽管通常以其激发的联想命名,立石的设计功能却不是模仿,而是像现代艺术中的抽象雕塑,如同欧洲园林里一段时期内(文艺复兴和巴洛克时期)充满着的自然主义雕塑一样。当然,奇石的联想与模仿的功能也被利用,一些游客会通过其象形之处进行联想,从中"看"出一些妖怪、狮子、龙之类的形象来。

[8]　中国的园林设计师引入了奇怪形状的石头,其内涵很难解释得清,但却能唤起纯粹存在的体验,包括无法解释的、难以言状的、令人惊骇的、神秘莫测的,就像是道家终极的道或是佛家终极的禅。中国人通过形象的扭曲的树干或树根以及水冲奇石来理解其展示的抽象深奥的自然力。凭着对自然的自我揭示的信念,一种强烈的共情的审美情趣通过自然形成被表达。一段老树,一块奇怪形状的石头,都被自然的力量刻上了疤痕。水对于岩石的冲刻被古人写入了《道德经》的至理名言:"天下莫柔弱于水,而攻坚强者,莫之能胜。"

[9]　挑选石块及安放石块在适宜的位置的过程,通常来说是非常谨慎而缓慢的,以致于在日本庭园的设计中,也可以被记录描述。中国园林设计师还不得不解决自然性与造型之间的对立问题。追求自然,放弃使用人工制造就意味着认同一种自然的风格,或是从另一方面看,缺乏风格,或拒绝、放弃风格。中国绘画为实现这种追求,采取利用偶然性的手法(如泼墨、溅墨、滴墨或泼彩),不事张扬,有意的笨拙,如将熟练的右手换为笨拙的左手,喝醉等。

[10]　在一些中国园林(或园林的特定部分)中,从某一观景点开始,观者在规划设计的引导下(通过园路、游廊、桥、隧洞、亭子或塔等)走到特定的观赏点。园林的其他部分,游人则自由选择观景点。中国园林绝对不能在一个特定地点一览无余。它或多或少地包含一些独立的部分,观赏者只有持续游赏才能发现和享受其美景,如必须顺着园路,穿过山洞,流连于水畔,然后才能到达亭子,在那里展开了一幅迷人的风景。他被引入了一个绝不能完整展露的布局构成。我们了解到,尽管中国园林属于广义上的造型艺术,它也有时间艺术的痕迹。

[11]　设计上,通过逐个展开的一系列的场景,使观赏者逐步地观察揣摩它。赛伦将这种方式比作以卷轴的形式逐步展开中国风景画。"映射"作为艺术隐喻的手法在欧洲传统文化中起着相当重要的作用。但在中国园林的背景下,我们可以认为在园林中,自然作为一种映射来表现人的思想(以此来解释:艺术是一种对自然的

映射)。

[12]　对环境艺术品作出审美评价是有挑战性的,因为和标准的艺术品相比,环境为艺术哲学和美学提供了更加宽泛的视角。一定类型的环境价值的特别,取决于它们展示的风景,但另一些则是由于它们的芳香和可触摸的质感等。一些环境本身景观单一,却富有天籁之音(水声、风声、鸟啼、蛙鸣及其他动物的鸣叫声)。

[13]　在描述中国园林的感知类别时,约翰斯顿说过:"中国园林是在营造一种对情感的直接吸引力,专一地致力于为所有感官服务:视觉上展开一连串的惊喜;引入促使人触摸的质感;糅合了花草树木的芬芳;捕捉流水和树叶摇动的声音;利用树木四季不断变化的美来增添每个季节的特点。"对于环境的美学思考既可以是概括的,也可以是具体的环境因素(视觉的、听觉的、触觉的、嗅觉的)。

[14]　聆听各种声音(水声、雨声、风声、鸟鸣),观看特别的景物或景观,有时会发展为与整体环境设计因素相关的单独吸引力。例如在19世纪初的英国,将观鸟作为一种独特的休闲方式,并且作为环境美学的一部分。一些12世纪的中国画,比如名画《梅竹雀图》——似乎是证明了在那时中国或许就有了这种消遣方式,同时也有了对天籁之音的倾听,这一点被记录在马麟的名作《静听松风图》(1246年)中。

[15]　对感知最严重的损害是"加速"感知而被迫接受信息。如在当代电影、电视(特别是商业节目和现场直播),大功率的扬声器,以及其他发明出来用以"撞击你的五脏六腑"的玩意儿。人们习惯于混乱且短时间的感知,如同成为机枪火力的目标一样。它损害了时间的感觉,速度、强迫和扭曲代替了深度和意义。

[16]　无论是东方还是西方人,几乎都失去了对自然声音的敏感。自然的声音被城市环境的具有侵略性的声响所淹没,或者他们已经对破损的听力发出咆哮,亦或是充耳不闻了,因为自然已经死去。然而,一些当代诗人却尝试让我们记住那些独特的声音,比如雨滴敲打树叶、蟋蟀、鸟、青蛙的鸣叫,或是风声。与其他感知对象相比,人们对环境中触觉的理解和认识较少。一些人有过触觉的体验,无论是感触过材料的质地,或是仅仅观看。人们本能地将部分视觉体验译为触觉(即通感),有时这仅仅因为不可能去触摸。

Reading Material A

Two Main Forms of Chinese Garden
中国园林的两种主要形式

[1]　There is considerable variety in Chinese gardens depending on the date and the place, but for our purposes (in search of prototypes), oversimplification leads to two main forms. First, imperial gardens, built as settings for the summer palaces of the imperial families, were large estates. Consists of mountains, forests, streams, lakes, and islands, often stocked with exotic plants and animals from distant countries, they were frequently furnished with bright pavilions and bridges for the amusement of the royal party, removed temporarily, as they were, from social and political life. The pavilions in the large gardens were often built for special purposes, e. g, viewing the moon or lotus flowers. The second major form derives from the smaller private gardens attached to town dwellings and suburban villas. These belonged to landlords, rich merchants and bureaucrats. But, regardless of whether it was a large estate or a small town garden, the objectives were the same: to create a symbolic landscape in which the contrasting forces of nature were harmoniously arranged as a setting for the individual in contemplation or for a release from the conformity of social life.

The Summer Palace

[2]　Situated in the western outskirts of Haidian District, the Summer Palace is 15 kilometers from

central Beijing. Constructed in the Jin Dynasty (1115-1234), it was extended continuously. By the time of the Qing Dynasty (1644-1911), it had become a luxurious royal garden providing royal families with rest and entertainment. Originally called "Qingyi Garden[1]" (Garden of Clear Ripples), it was known as one of the famous "three hills and five gardens[2]". In 1888, Empress Dowager Cixi[3] embezzled navy funds to reconstruct it for her own benefit, changing its name to Summer Palace.

[3] Composed mainly of Longevity Hill and Kunming Lake[4] (Figs. 1, Figs. 2), The Summer Palace occupies an area of 294 hectares, three quarters of which is water. It can be divided into four parts: the imperial court area, front-hill area, front-lake area, and rear-hill and back-lake area.

Fig. 1 Longevity Hill

Fig. 2 Kunming Lake

[4] The imperial court area: this is where Empress Dowager Cixi and Emperor Guangxu met officials, dealt with state affairs and rested. Entering the East Palace Gate, one immediately faces the architectural group of Hall of Benevolence and Longevity[5]. This hall was the place where emperors held court. Surrounding the Hall of Benevolence and Longevity, there are several groups of quadrangle courtyards—the Yi Yun House, Yu Lan Hall and the Happiness and Longevity Hail[6]. These are the imperial living quarters that belong to the "back residence" in the convention of imperial court.

Fig. 3 Tower of Buddhist Incense

[5] Front-Hill Area: this area is the most magnificent area in the Summer Palace with the most constructions. Its layout is quite distinctive because of the central axis from the yard of Kunming Lake to the hilltop, on which important buildings are positioned including Gate of Dispelling Clouds, Hall of Dispelling Clouds, Hall of Moral Glory, Tower of Buddhist Incense (Fig. 3), the Hall of the Sea of Wisdom[7], etc.

[6] Front Lake Area: The designers of Kunming Lake scenic area adopted the layout of "three islets in one lake", of those the most important being the South Lake Islet[8]. There stands the Guang Run Ancestral Temple[9] in the islet. The 17-arch Bridge[10] connects the east side of the islet with the east bank of the lake. The West Dyke[11] of Kunming Lake was built in the same layout of the Su Dyke[12] of the West Lake in Hangzhou. Like the Su Dyke, six bridges were built.

[7] Rear-Hill and Back-Lake Area: The landmark in the rear hill area, the Hou Xi River traveling from west to east, was dug on the north piedmont of the Longevity Hill. A special scenic spot is the shopping street lying at the middle of the Hou Xi River that imitates the style of business streets in

Suzhou and Nanjing. The Xu Mi Ling Jing Temple[13] in the center of the back hill is the largest Buddhist architecture in the imperial garden.

[8] At the northeast foot of the Longevity Hill, there is a small and relatively independent "garden in a garden[14]" that imitates the Ji Chang Garden[15] in Wuxi. It is the Harmonious Interests Garden[16].

[9] The Summer Palace is a melting pot of natural landscape and man-made wonders, and the magnificence of the imperial garden and the delicacy of the garden in the south.

The Humble Administrator's Garden[17]

[10] The garden is situated at the north-eastern part of the city of Suzhou, and was first built in the Zhengde Period (1506-1521) of the Ming Dynasty. The garden is divided into eastern, central and western parts, which altogether add up to 4.1 hectaresin area, which is quite large for private gardens.

[11] The central part of the Humble Administrator's Garden is the main part of the garden, and in terms of overall planning, this part can be divided into the northern water area and the southern land area. In terms of scenery area arrangements, this part can be divided into three areas from east to west, with the central part being the most important. The Yuanxiang (Distant Fragrance) Hall[18] (Fig. 4) built facing the water at center of the garden is the largest hall type architecture of the entire garden. The eastern area is made up of a group of architecture including Haitang Chun Wu (Spring Crabapple Flower Dock)[19], the Linglong House (Exquisite Garden)[20], etc. The western area holds the Yulan Tang (Magnolia Hall)[21], Dezhen (Attainment of Truth) Pavilion[22] and Xiangzhou (Fragrant Islet)[23] Stone Bridge inside an open courtyard space formed with corridors and artificial hills.

Fig. 4 Distant Fragrance Hall

Fig. 5 Little Canglang

[12] The central part of the Humble Administrator's Garden is mainly a scenic area of open waters, with complimentary architectural scenes. The northern area of the central part of the garden is the water scene area, and built within the water are two islets. On the western islet are built a Fragrant Snow and Clouds Pavilion[24] and a pavilion named "Lotus Breeze from all Sides"[25]. On the eastern islet there is the Pavilion of the Northern Hills[26]. Flat bridges are built between the two islets and between each islet and the water bank. Northwest of the pond is built a "Tower of Viewing the Mountain"[27], which is a major scenic spot of the northern part of the garden. The Little Flying Rain-

bow[28] and the Little Canglang[29] (Fig. 5) are two covered bridges across the water. The southern area of the central part, although an area concentrated with architecture, has a variety of architectural forms of halls, mansions, pavilions and stone bridges interspersed with corridors, bridges and artificial hills, and decorated with crabapple flower trees and loquat trees.

[13] The western part of the Humble Administrator's Garden is also an area of mainly water scenery. The pond runs from north to south, and at the central part where the water widens there is also an islet built within. The surface of the water is shaped like a narrow carpenter's square. The main sceneries are concentrated at the northern part. At the north of the pond there is a Tower of Water Reflection[30]. Opposite at the south of the pond is built the House of 36 Mandarin Ducks[31].

[14] The current Humble Administrator's Garden, in comparison with the original one in the Zhengde Period of the Ming Dynasty, has more buildings and the islets added to it. Although the scenes somewhat lack the natural, open and lofty feel, the garden is still amasterpiece of meticulous work.

[**NOTES**]

[1]　Qingyi Garden (Garden of Clear Ripples) 清漪园：即颐和园。
[2]　three hills and five gardens 三山五园：北京西北郊一带皇家行宫苑囿的总称，是从康熙至乾隆年间陆续修建起来的。三山是指香山、玉泉山和万寿山，五园是指香山静宜园、玉泉山静明园、万寿山清漪园（颐和园）、圆明园和畅春园。
[3]　Empress Dowager Cixi 慈禧太后。
[4]　Longevity Hill and Kunming Lake 万寿山和昆明湖。
[5]　Hall of Benevolence and Longevity 仁寿殿。
[6]　Yi Yun House, Yu Lan Hall and the Happiness and Longevity Hail 宜芸馆、玉澜堂和乐寿堂。
[7]　Gate of Dispelling Clouds, Hall of Dispelling Clouds, Hall of Moral Glory, Tower of Buddhist Incense, the Hall of the Sea of Wisdom 排云门、排云殿、德辉殿、佛香阁、智慧海。
[8]　South Lake Islet 南湖岛。
[9]　Guang Run Ancestral Temple 广润祠。
[10]　17-arch Bridge 十七孔桥。
[11]　West Dyke 西堤。
[12]　Su Dyke 苏堤。
[13]　Xu Mi Ling Jing Temple 须弥灵境。
[14]　garden in a garden 园中园：是指在大园中套有不同内容的小园，从而丰富全园的景致，使空间层次多变，产生不同的空间艺术效果。园中园为中国古典园林中的一种造园手法，也是藏景的一种，多见于皇家园林中，也见于少数私家园林。
[15]　Ji Chang Garden 寄畅园。
[16]　Harmonious Interests Garden 谐趣园。
[17]　Humble Administrator's Garden 拙政园。
[18]　Yuanxiang(Distant Fragrance) Hall 远香堂。
[19]　Haitang Chun Wu(Spring Crabapple Flower Dock)海棠春坞。
[20]　Linglong House (Exquisite Garden)玲珑馆。
[21]　Yulan Tang (Magnolia Hall)玉兰堂。

[22] Dezhen (Attainment of Truth) Pavilion 得真亭。
[23] Xiangzhou (Fragrant Islet) 香洲。
[24] Fragrant Snow and Clouds Pavilion 雪香云蔚亭。
[25] Lotus Breeze from all Sides 荷风四面亭。
[26] Pavilion of the Northern Hills 北山亭。
[27] Tower of Viewing the Mountain 见山楼。
[28] Little Flying Rainbow 小飞虹。
[29] Little Canglang 小沧浪。
[30] Tower of Water Reflection 倒影楼。
[31] House of 36 Mandarin Ducks 三十六鸳鸯馆。

[GLOSSARY]

artificial hill	假山	hall	厅,堂	stone bridge	石桥
corridor	廊	mansion	堂,馆	tower	楼,塔
covered bridge	廊桥	pavilion	亭,阁		

[NEW WORDS]

bureaucrat	n.	官僚	loquat	n.	枇杷
complimentary	adj.	免费的;赠送的;赞美的	magnificent	adj.	壮丽的,宏伟的,华丽的,高贵的
conformity	n.	依照,遵从;符合,一致	masterpiece	n.	杰作,代表作,名著
contemplation	n.	意图,期望;沉思,冥想;凝视	merchant	n.	商人
convention	n.	惯例,习俗,会议	meticulous	adj.	极仔细的,一丝不苟的
corridor	n.	走廊,通道	outskirt	n.	市郊,郊区
crabapple	n.	山楂	oversimplification	n.	过分简化
delicacy	n.	精美,精致,微妙	pavilion	n.	亭,阁,临时展出馆,华美建筑
dyke	n.	堤,坝,堰	piedmont	n.	山麓地带
embezzle	vt.	挪用,盗用	prototype	n.	原型,蓝本
exotic	adj.	由外国引进的,非本地的,奇异的	quadrangle	n.	四边形,方庭,四合院
furnish	vt.	陈设,布置,提供	quarter	n.	住处
harmoniously	adv.	和谐地,调和地	rear	adj.	后面的,后部的,背面
lofty	adj.	高耸的,极高的;高尚的,崇高的	stock	vt.	储备,保持……的供应

[难句]

1. Consisting of mountains, forests, streams, lakes, and islands, often stocked with exotic plants and animals from distant countries, they were frequently furnished with bright pavilions and bridges for the amusement of the royal party, removed temporarily, as they were, from social and political life. 它们由山脉、森林、溪流、湖泊和岛屿组成,

园中通常遍植奇花异草,蓄养珍禽异兽,建有绚丽的亭桥,以供皇族暂时从社会和政治生活中抽身而出,来此处娱乐休息之用。[长句分译](often stocked with...bridges 因语句过长,分译成三个独立分句。)

2. But, regardless of whether it was a large estate or a small town garden, the objectives were the same: to create a symbolic landscape in which the contrasting forces of nature were harmoniously arranged as a setting for the individual in contemplation or for a release from the conformity of social life. 但是,无论是大型庄园还是小型城市花园,他们的目标都是相似的:创造一个与外界迥异的具有自然面貌的写意式园林,其中布置和谐的自然景观是他们表述个人志向抱负、远离世俗的境地。[长句综合译法]

[参考译文]

中国园林的两种主要形式

[1]　基于时代的变迁与地域性的差异,中国园林有许多不同的类型。但依据我们的目的(为寻求原型),过度简化后主要存在两种主要形式。一类是皇家园林,为皇室避暑而建造的大型宫殿。在这类园林中有许多的群山、溪流、湖泊、岛屿,还有各色来自外国,充满异域风情的植物和动物。园林中常常点缀有敞亮的亭廊与各式的桥,以供皇室能够短暂地忘记宫中烦劳的政务,享受山水之乐。在这类大型园林中所修建亭子往往带有一定的目的性,比如赏月或观荷。另一类则是较小的私家园林,往往依附于城郊的住宅或乡野间的私院。这类园林为地主、富商、官员所有。但无论是规模宏大的皇家园林还是小巧精致的私家园林,它们的目的都是相同的:创造一种象征性的景观,具有对比性的自然力量被协调组织起来,给人们以沉思或者能从千篇一律的社会生活中得以解脱的地方。

颐和园

[2]　位于海淀区的西郊,颐和园距北京城中心有 15 千米。它始建于金代(1115—1234),历经之后朝代的不断扩建。直至清代(1644—1911),它成为了一座豪华的皇家园林,以供皇室的休闲娱乐。颐和园最初起名清漪园(清澈的涟漪之园),它是著名的"三山五园"之一。1888 年,慈溪太后私自挪用海军军费进行重建,并改名为颐和园。

[3]　颐和园总占地 294 公顷,万寿山与昆明湖占据了大部分面积,整个园区的四分之三都是水域。颐和园被分为四个部分:皇宫区、前山区、前湖区和后山后湖区。

[4]　皇宫区:皇宫区是慈溪太后与光绪皇帝接见朝臣,处理国事与休息的地方。一迈进东宫的大门,映入眼帘的便是仁寿殿建筑群。这个宫殿是皇上处理政务的地方。仁寿殿周围有几组四合院——宜芸馆、玉澜堂和乐寿堂。这些都是皇帝的居所,遵循"背居"朝堂的传统习俗。

[5]　前山区:前山区是整个颐和园中最壮丽,建筑最多的区域。它的布局非常具有特色,因为许多重要的建筑物都布置在由昆明湖到山顶的这条中轴线上,包括排云门、排云殿、德辉殿、佛香阁、智慧海等。

[6]　前湖区:昆明湖的设计工匠采用了"一池三岛"的传统布局,其中最重要的是南湖岛。广润祠就建在这座小岛上。十七孔桥将小岛的东边与湖东岸相连接。昆明湖的西堤是仿造杭州西湖苏堤所建造。如同苏堤,在这里修建了六座桥。

[7]　后山后湖区:后溪河是后山区的地标,起源于万寿山的北麓,自西向东。具有特色的一个景点是一条坐落于后溪河中段的购物街,仿造苏州与南京的商业街道建造而成。位于后山区中部的须弥灵境是这个皇家园林中最大的佛教建筑。

[8]　在万寿山的东北脚下,有一座小巧且相对独立的"园中园",仿造无锡的寄畅园而建造。这便是谐趣园。

[9]　颐和园不仅融合了自然景观与人造奇观,还结合了皇家园林的宏伟壮丽与南方私家园林的精致小巧。

拙政园

[10]　拙政园坐落于苏州的东北部,始建于明代正德初年(1506—1521)。花园被分为东、中、西三部分,总占地达 4.1 公顷,对私家园林而言已是相当大规模了。

[11]　拙政园的中部是全园精髓所在,从总体的规划而言,中部可被分为北部的水域和南面的陆地区域。从景观的布局来看,可以分为自东向西三部分,中心区域是最重要的部分。修建在园林中心且面向水面的远香堂是全园中最大的厅堂式建筑。东部由许多建筑组成,包括海棠春坞、玲珑馆等等。西部有玉兰堂、得真亭、香洲石舫,都在由回廊和假山所组成的开敞庭院中。

[12]　拙政园的中部主要是一片开阔的水景与优美的建筑。中部的北面是水景,在其中有两座小岛。在西边的小岛上建有雪香云蔚亭和荷风四面亭。东边的岛屿上建有北山亭。在两座岛屿之间、岛屿与水岸之间均建有平板桥。见山楼在水池的西北岸,是北园区主要的景点。小飞虹与小沧浪是横跨过水面的两座廊桥。中部的南面区域建筑密集,有各式各样的厅堂、馆、亭阁、石舫,点缀着廊、桥、假山,并以海棠花和枇杷树装饰。

[13]　拙政园的西部区域同样是以水景为主。池中流水由北向南,在池中水面变宽的地方也建有一座小岛。水面的形状如同窄的木工尺。主要的景点集中在北面。水池的北岸有一座倒影楼。与之相对的南岸上建有三十六鸳鸯馆。

[14]　如今我们所看到的拙政园与明代正德年间最初建成的相比,增加了更多的建筑物与岛屿。虽然如今的景色某种程度上缺少自然、开阔和大气的感觉,这依旧是一件精致的园林杰作。

Reading Material B

Japanese Garden

日本园林

[1]　The historical development of the Japanese garden is exceedingly complicated and the early garden forms which were adopted by nobles, were heavily influenced by religious beliefs, symbolism, and Chinese influence in varying proportions. Meditation was the gardens chief purpose, through which the meaning and purpose of life was revealed. Gardens of the Nara period[1] (645-784), often built by craftsmen from Korea and China, included lakes and rocks arranged to resemble nature based on the Chinese model. Subsequent periods, especially those associated with the location of the capital at Kyoto, saw the refinement of this garden type as a pleasure ground representing paradise and within which imperial courtiers amused themselves, boating on the lakes, writing poetry, and discussing aesthetics. The gardens, too, contained symbols of longevity and purity, as well as allusions to specific place in Japan. To the initiated, then, the garden could be read and enjoyed like a book. The importance of Zen Buddhism[2] in the Kamakura period[3] (1185-1392) brought new concepts to life. Zen contrasted to the more formal Buddhist symbol-laden and elaborate doctrine. The garden was seen more strictly as an aid to meditation. For this purpose it was enclosed with a wall, and the relationship of the viewer to the garden was fixed. Later, during the Muromachi period[4] (1393-1568), the dry garden[5], of which Ryoan-ji[6] (Fig. 1) is a prime example, was produced in a time of nostalgia for the eleventh and twelfth centuries. The dry garden was the ultimate Zen aesthetic. Temples contained dry gardens as places to find spiritual peace in turbulent times.

[2]　The Zen dry garden is a perfect reflection of the monk's life, imbued with simplicity and austerity, leading to spiritual enlightenment.

[3]　In the Edo period[7] (1620-1645) political power moved to Tokyo(Edo). The Emperor, whose power was merely symbolic, remained in Kyoto and spent time embellishing court and family life with

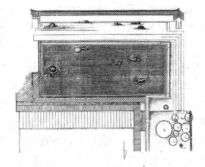

Fig. 1　The dry garden in Ryoan-ji

the arts. Many of the earlier garden concepts were reexamined and brought together in what is usually called the stroll garden[8]. This is the second type of garden most commonly associated with Japan. Several good examples exist in Kyoto.

[4]　The idea of the stroll garden was the creation of a series of views and experiences in the garden. In this sense it resembled the gardens of China, although the route to be followed was more clearly prescribed. Ideally it followed a clockwise route around an irregularly shaped lake and was laid out with bends and turns in relation to planting and topography such that the whole garden could not be seen at any one time. Each view was carefully composed and framed, the villa, the teahouse, the temple, and bridges and other garden structures featured unobtrusively in these views as did rocks, pebble beaches, and planting. The path, itself, would consist of various materials and forms: gravel, cobbles, stepping stones. Symbolism and allusion pervaded the elements and their composition: rock, streams, and plants that were carefully pruned to emphasize their essence.

[5]　The illusion of space and landscape continuity was a major goal of the Japanese garden. The concept of borrowed landscape, i.e., opening up a view of a distant valley or mountain while concealing the garden boundary, was frequently applied. To aid the illusion, the trees inside the garden were the same species as those in the distant view. Other techniques such as planting large trees in the foreground and smaller ones in the background, or placing a large hill in front of a smaller one, also contributed to a sense of distance and space within the garden.

Fig. 2　Tea house in Katsura Imperial Villa

[6]　At the Katsura Imperial Villa[9] (Fig. 2) near Kyoto we find all of these features and concepts combined into a magnificent stroll garden. The villa, which also remains intact, is a fine example of Japanese architecture. Together they illustrate the idea of integration which typifies the national Japanese house and garden. It also demonstrates the adaptable architectural form. Built in the Edo period over several years starting in 1620, it is thought to be the conception of Prince Toshihito, a brother of Emperor Goyozei. Many of the views are based on an eleventh century novel and the garden is imbued with literary allusions to the initiated.

[7]　The villa would be approached through a series of gates and along paths following an indirect route to the main door at the side of the villa. The door itself is not seen until the last moment. From the building, raised for protection from floods and for ventilation, specific views of the garden can be controlled by sliding screens or the entire front can be opened up to the garden which lies to the south. The villa consists of a series of units, built on the tatami mat module, added together and forming an informal edge on the garden side. The orientation and design of the buildings permits winter sunshine to enter but summer sun is shaded. The central feature of the garden is a lake with five islands.

[8]　Along the pathway, the lake is always present. The stroller encounters landscape views composed with trees, shrubs, rocks, a roofed bench or stone basin, lanterns, stone and wooden bridges, a waterfall and a pebble beach reminiscent of a famous scenic coastline in Japan. A rustic pavilion on top of a hill provides a place to view the moon reflected in the lake. There are several tea pavilions and a temple. This complicated landscape and its revealed views are filled with aesthetic experience and opportunities for meditation.

[NOTES]

[1]　Nara period 奈良时代。
[2]　Zen Buddhism 禅宗佛教。
[3]　Kamakura period 镰仓时代。
[4]　Muromachi period 室町时代。
[5]　dry garden 枯山水庭园：是源于日本本土的缩微式园林景观，多见于小巧、静谧、深邃的禅宗寺院。在其特有的环境气氛中，以耙出水纹模样的白砂象征溪流和海洋，布置石组象征山岳、洞壑、落瀑及海中岛屿，配以自然式造型的树木和苔藓，将山河海洋缩于很小的庭园之中。
[6]　Ryoan-ji 龙安寺：是日本枯山水庭园的代表作，长方形平庭长28米，宽12米，一面临厅堂，其余三面围以土墙。庭院地面上全部铺白砂，除了15块石头外，再没有任何树木花草。其中白砂象征水面，以15块石头的组合、比例搭配、向背的安排来体现岛屿山川，与咫尺之地幻化出千七万壑的气势。石头放置奇特，不管从哪个角度看，均有一块石头隐身不见，是寺内传授知足常乐的禅教。这种庭园纯属观赏的对象，人不能在里面活动。
[7]　Edo period 江户时代。
[8]　stroll garden 回游式庭园：宅邸、别墅中可供绕池漫步或乘船游览的环游式园林，茶庭的内容和风格也扩大到其中，茶室、书院、亭、轩、涉石、水滨、水滨卵石滩、绿篱、草坪等陆续出现。
[9]　Katsura Imperial Villa 桂离宫：建于1620年，是日本皇室的行宫，位于京都西南郊的桂川岸边，因是后阳成天皇（Emperor Goyosei）的弟弟智仁亲王（Prince Toshihito）的离宫，故称为桂离宫。桂离宫是典型的回游式园林风格，庭园中心为水池，池心有大小岛屿，岛建有桥相连，池苑周围主要园路环回引导到茶庭洼地以及亭轩院屋等建筑。

[GLOSSARY]

borrowed landscape	借景	stroll garden	回游式庭园
dry garden	枯山水庭园	tea garden	茶庭

[NEW WORDS]

abbot	n.	寺院住持,方丈	module	n.	单元,单位;模数,模块
allusion	n.	提及;暗示	monk	n.	僧侣,修道士
austerity	n.	严厉,严格;简朴;苦行	moss	n.	苔藓,地衣
coarse	adj.	粗的;粗糙的	nostalgia	n.	乡愁;留恋过去
cobble	n.	鹅卵石,圆石,中砾石	pebble	n.	卵石,小圆石,鹅卵石,砾石
complicated	adj.	复杂的,结构复杂的	pervade	vt.	遍及,渗透,弥漫
conducive	adj.	导致……的;有助于……的	prescribe	vt.	指示,规定
continuity	n.	连续(性),持续(性)	priest	n.	僧侣,神父,牧师
courtier	n.	侍臣,朝臣	proportion	n.	比例,面积,规模;均衡,协调
doctrine	n.	教义,学说	prune	v.	剪除,修剪
eave	n.	屋檐	rake	vt.	用耙子……耙
elaborate	adj.	精心制作的,细心完成的	reminiscent	adj.	回忆往事的
enlightenment	n.	启发,开导;开明,文明	resemble	vt.	效仿,类似于
exceedingly	adv.	非常,极其,过分地,极为	reveal	vt.	显示,透露
gravel	n.	沙砾,砾石,石子	rustic	adj.	乡村的
illusion	n.	错觉,幻想	strictly	adv.	严厉地,严格地;完全地,绝对地
imbue	vt.	使充满,灌输,浸透;激发	stroller	n.	散步者,闲逛者
initiate	vt.	开始,着手	subsequent	adj.	后来的,随后的
intact	adj.	完整无缺的,未受损伤的	tatami	n.	榻榻米,草席子,日式垫子
integration	n.	结合,整合,综合,一体化	tile	n.	瓦片,瓷砖
Kyoto	n.	京都(曾是日本古都)	turbulent	adj.	狂暴的,吵闹的
laden	adj.	装满的,负载的,负重的	ultimate	adj.	最后的;主要的
lantern	n.	灯笼,灯塔	uneven	adj.	不规则的,不一致的,有差异的
lavish	adj.	太多的,过分丰富的;无节制的	unobtrusively	adv.	不引人注目地,不容易看到地
longevity	n.	长寿;寿命	ventilation	n.	通风,流通空气
mat	n.	席子,垫子	veranda	n.	阳台,走廊
meditation	n.	默想,沉思,冥想	Zen	n.	禅,禅宗

[参考译文]

日本园林

[1] 日本园林的发展历程十分复杂,早期为贵族所采用的园林形式受到了宗教信仰、象征主义与中国园林不同程度的影响。而冥想是这个时期园林的主要功能,通过冥想,人们探寻到生命的目的与意义。奈良时代(645—784)的日本园林大多由韩国与中国的工匠建造,其中不乏运用到中国的造园模式,以造湖与叠石模拟自

然。随后的时期,尤其是在首都京都周边,园林演化成为如天堂般游玩的地方,皇权贵族们取乐于此,划船、写诗、讨论美学。这些园林同样包含了长寿与纯洁的象征,也包含了日本特定地方的寓意。然后对于那些先驱者,日本园林才能像一本书一样地被人们阅读享受。镰仓时代(1185—1392)尤为重要的禅宗佛教赋予了生命新的理念。禅宗与更加正式的传统的佛教充满符号和精致教义有所不同。此时的园林更被视作帮助冥想的一种途径。为了达到这个目的,园林被墙体围起来,观者与园林的关系也被固定。随后,在室町时代(1393—1568),枯山水庭院产生于对十一、十二世纪的怀旧思潮中,龙安寺便是这类庭院的典型代表。枯山水庭院是禅宗的终极美学。拥有枯山水庭院的寺庙便成为了人们在这个动荡的时代寻求心灵平和的地方。

[2]　禅宗的枯山水庭院是僧侣生活的完美体现,充满了简约与简朴,直至心灵上的启示。

[3]　江户时代(1620—1645),政治中心转移到东京,此时的天皇的权利仅仅是象征性的,他依然居住在京都,将大量时间用于美化庭院与家庭生活。早前的很多庭院理念被重新审视,并组合在一起,成为了我们所谓的回游式庭院。这也是第二类常见的与日本相关的园林。至今在京都也保有些许好的案例。

[4]　回游式庭院的理念在于在庭院中创造一系列的风景与体验。在这个意义上,它类似于中国园林,虽然游园的路线有着更为明确的规划。路线顺着形状不规则的湖泊顺时针方向曲折延伸,加以植物和地形的处理,使得整个花园不会一览无余。每一处的景色都经过了精心的组合与框定,别墅、茶亭、寺庙、桥梁,其他不显眼的特色园林构筑物,如岩石、鹅卵石滩、植被也是同样的。园路本身就有很多种材料和组成形式:砾石、鹅卵石、汀步。各要素以及它们的构图都弥漫着象征主义和寓意,如岩石、溪流和精心修剪的植物组合,以突显它们的本质。

[5]　空间的错觉与景观的连续性是日本园林的一个主要目标。借景的手法被频繁地运用,例如,隐藏庭院的边界而取远处山谷或高山的景象。为了增强空间的错觉感,庭院中栽植了与远处相同的树种。其他的技术手法,例如在前景种植大树而在背景处栽种较小的树,或在小山丘前放置一座大的假山,同样也在庭院中增强了空间感与距离感。

[6]　坐落于京都附近的桂离宫,是一座瑰丽宏伟的回游式庭院,集成了日本庭院所有的特点与理念。如今,这座离宫依旧保持完好无损,是日本建筑一个非常好的代表。它们很好地阐释了日本建筑与庭院相互融合的典型。桂离宫也展示了多样的建筑风格。从江户时代1620年开始建造历经多年完工,是后阳成天皇的弟弟智仁亲王的离宫。许多风景都基于十一世纪的小说,这个庭院充满了文学典故。

[7]　桂离宫通过一系列大门进入,其通道到主门之间由一条非直接的路线相连。若非走到尽头,则看不到这扇门。为了防洪与通风,建筑整体被抬升起来,通过推拉窗户,或者打开整个朝南面的窗户可以看见特定的庭院景观。桂离宫由一系列的单元构成,均基于榻榻米的模数,被组合起来构成一个庭院一侧不规则的边缘。桂离宫的朝向与设计使得冬季阳光能够照入,而夏日的太阳却被遮挡了。庭院的中部由一个湖泊与五座岛屿组成。

[8]　小路旁的湖水四季都是充盈的,散步者所见之景观有大树、灌木、石头、石凳、石盆、灯笼、石桥、木桥、瀑布、铺满卵石的河滩让人联想起日本著名而优美的海岸线。位于山顶的质朴的凉亭为人们提供了一个观赏水中月影的好去处。桂离宫中还有几处茶亭与一座寺庙。这里复杂的景观以及它所展现的景色都充满了美学体验与冥想的机会。

[思考]
(1)中国古典园林的设计要素有哪些?
(2)中国传统园林的两种主要形式是什么?
(3)日本最有代表的传统园林类型有哪些?这些类型的特点和代表作是什么?

UNIT 3 WESTERN TRADITIONAL GARDEN

TEXT

The Gardens in Ancient Egypt
Jimmy Dunn
古埃及的园林

[1] One really hears very little about gardens, and yet, they were an essential element to the ancient Egyptian people those who could afford to do so laid out gardens in front of both their houses and tomb chapels. The gods were even thought to enjoy gardens and so almost every temple was surrounded by lush greenery. Gardens seem to have been particularly important during the New Kingdom[1]. It should also be noted that certain types of gardens had religious symbolism.

[2] We know that gardens often consisted of both trees and other plants. Popular trees included the sycamore fig, pomegranate, nut trees and jujube. However, willows, acacia and tamarisk were also found. In all, there were about eighteen varieties of trees grown by the Egyptians. Flowers were also abundant, and included daisies, cornflowers, mandrakes, roses, irises, myrtle, jasmine, mignonettes, convolvulus, celosia, narcissus, ivy, lychnis, sweet marjoram, henna, bay laurel, small yellow chrysanthemums and poppies. Of course, there were also papyrus, lotus and grapes.

[3] Gardens were not simply for pleasant environs to the Ancient Egyptians. There were many symbolisms associated with trees, including to specific gods such as Osiris[2], Nut[3], Isis[4] and Hathor[5]. They also had creation overtones, as well as funerary. The Papyrus and Lotus plants were symbolic of the two regions of Lower and Upper Egypt (respectively). Of course, gardens also provided food including vegetables and wine, and in the final analysis, we might know much less about ancient Egypt if it were not for the papyrus paper used through most of Egyptian history.

[4] Regrettably, we know of very few depictions of gardens that surround normal houses, but several literary descriptions of a country estate mention the lush cultivated grounds around a villa of the New Kingdom. One owner, who obviously enjoyed his garden tells us that, "You sit in their shades and eat their fruit. Wreaths are made for you of their twigs, and you are drunk with their wines." There were even models of gardens made and placed within tombs (Fig. 1).

[5] Houses, palaces, temples and chapels, whether funerary or private, when in the paintings of the tombs nearly always have a garden connected to the building. We even very often find a whole layout of an elaborate nature detailed, and thus an adequate picture of the various types of gardens

during the New Kingdom can be reconstructed from this pictorial evidence. Until the end of the Middle Kingdom, gardens had to be watered from jars carried at the end of a pole slung on the shoulders of water carriers. The primitive counterpoised sweep for elevating water (Arabic shaduf[6]), which is connected by Winlock[7] with the invasion of the Hyksos[8], enabled a much easier irrigation of cultivated land.

House Gardens

[6]   Even in ancient Egypt, the value of land was almost prohibitive in the cities and we have today no real evidence of any gardens in these locations. Occasionally, a few trees were planted along the sides of the house, usually date palms alternating with another species, which can al-

Fig. 1   The model of ancient Egyptian garden

so be grown in brickwork containers. In the harem of Pharaoh Ay[9], a large court surrounding the structure is planted with a row of trees in mud copings, and on the farthermost side a kiosk on columns supports a vine.

[7]   However, in the country where the land was much less expensive, the houses and palaces were set in a large garden surrounded by a wall. Numerous depictions in tombs show what might be considered to be the standard type of garden. Typically, a symmetrical layout was used with a rectangular or T-shaped pond in front of the house on the main longitudinal axis. This garden would then be surroun-

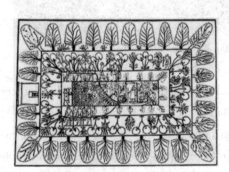

Fig. 2   The typical illustration of ancient Egyptian garden

ded by rows of trees of various species, possibly alternating in the same rows. It was not uncommon to find a pergola bordering the main alley along the axis or surrounding the pond. It should be noted that many times these ponds were stocked with fish, and at times included exotic examples. Fruit trees have their leaves or branches supported on the trelliswork of the pergolas. The shortest species of trees are planted nearest the pond, while the tallest, such as doum palms and date palms, are in the outside rows (Fig. 2). This arrangement provided a graded perspective about the center of the garden.

[8]   Sometimes, there was more than one pond. In the formal garden of the Temple of Amenhotep II[10] and the attached house of its attendant Sennufer[11] at Thebes[12] as depicted in his Sennufer's tomb, the layout is symmetrical about an axis perpendicular to the river and running from the entrance along an alleyway flanked with two pergolas and leading to the small temple with three shrines (Fig. 3). Each half of this garden, on either side of the alley, is divided transversely into three areas. The front section, which has a rectangular pond parallel to the river has water plants, and there is also date palms and sycamores. A second section in the middle area is enclosed within a wall and planted with

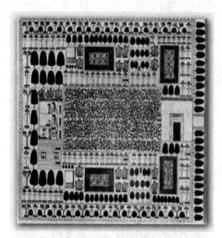

Fig. 3　The illustration of garden in the Temple of Amenhotep Ⅱ

light green trees that are perhaps a rare species. Finally, a rear section is the largest area and again has a rectangular pond bordered on one side by date palms and on the other by sycamores. Near this rear section is a small open kiosk of the type we find at Amarna[13]. On either long side of the whole garden an enclosed path is planted with trees of alternating species, while tall trees form an effective screen at the back of the estate.

[9]　A formal layout is also followed in the large palace gardens. Usually the approach is symmetrical, usually with a pond on either side of the axis, bordered with rows of trees. At Amarna, where the ground is not arable, trees were planted in pits filled with humus and bordered with a round coping. At the rear of the various groups of buildings a large area is laid out as an independent garden around a square pound with sloping sides. In one of the pond's corners, a stairway descends to its bottom. A deeper basin opening in the bottom is probably filled with infiltration water. Interestingly, the distribution of the trees seems particularly informal and may have been another aspect of the Amarna trend toward freedom and naturalism in art.

Sacred Gardens

[10]　Gardens on processional approaches to pylons, or in front of the temple quay along the river, are also represented in tombs. In the temple of Hatshepsut[14] at Deir el-Bahari[15] (Fig. 4), a garden with four ponds, papyrus, flowers and vegetables is represented schematically. There were exotic trees

Fig. 4　Trees and bushes from the tomb of Sennedjem at Deir el-Medina

that were brought from the new countries subdued during the New Kingdom and planted in the gardens of Amun[16]. Such rare species are represented at Deir el-Bahari, Medinet Habu[17] and Karnak[18], but the representations of these "botanical gardens", though fascinating due to their innumerable exotic species, do not offer any clue regarding their layout. Private chapels were erected by rich people in their gardens at Amarna or on the bank of a river or canal, and formed an important element in the layout, being situated at the crossing of two axes or at the end of the main axis. Often the chapel stands at the rear of the enclosed garden on a higher terrace, with a rectangular pond flanked by two rows of sycamore trees, or what seems to be two rows of tall jars surrounded by climbing growth. The formal layout of the Persian garden, where an artificial pond mirrored the glittering splendor of a rich facade beyond it, had already been carried out to perfection in Egypt, at least as early as the New Kingdom.

Funerary Gardens

[11]　Most of the depictions of funerary gardens are schematic in nature. They are usually reduced to

a T-shaped basin shown in plan on a background of a few date palms. Here, the origin of the peculiar plan of the basin may be investigated. It is certain that the dead end of a canal, when shaped as a transverse rectangular basin, would facilitate the mooring and circulation of boats. On the other hand, the offering table for the presentation of funerary offerings often assumed the shape of a T-slab, in the middle of which is a deep basin. Whether there is any real relation between the funerary T-shaped pond and the offering table is uncertain. What is certain, however, is that even in the beginning of the New Kingdom, the T-shaped plan had a symbolic implication. There were two T-shaped ponds flanking the central alley at the bottom of the lower stairway in Hatshepsut's temple.

[12]　At Memphis[19], there are at least two paintings depicting a funerary ceremony where the mummy is conveyed by boat to a rectangular island in the middle of a rectangular pond. In one of these, the pond is bordered on tree of its outer sides by a double row of funerary structures in the shape of light awnings containing a stand which alternate with date palms and trees planted in brickwork containers. A quay protrudes into the water from one small side of the pond, and in one painting it is accessible by a stairway. In this latter representation there is a quay that is set at both smaller ends of the island. This could be a symbolic representation of the Osireion[20] at Abydos[21] (Fig. 5).

[13]　The location of the funerary garden has been the subject of controversy but it can be safely assumed that some kind of small garden was occasionally laid out in front of the tomb itself and that more often a larger garden was laid out below on the riverbank, and probably also near the portal of the tomb complex.

Fig. 5　Scene from the Book of the Dead papyrus of Nakht showing he and his wife approaching Osiris and Ma'at in their garden

[14]　Today, and throughout history really, gardens have played a big part in the lives of Egyptians. Gardens seem to have become a part of their being doubtless as much because of the nearby barren desert and the need to see life everywhere within that tiny strip of land which fosters life.

[NOTES]

[1]　New Kingdom 新王国：古埃及的历史大体上可以划分为前王国时期(公元前 4500—前 3100)、早期王国时期(公元前 3100—前 2686)、旧王国时期(公元前 2686—前 2134)、中王国时期(公元前 2030—前 1640)、新王国时期(公元前 1648—前 1549)和后王国时期(公元前 664—前 332)。

[2]　Osiris 欧西里斯：也拼作 Usiris 乌西里斯，埃及神话中的冥王，即掌管阴间的神，同时也是生育之神和农业之神。欧西里斯是九柱神之一，也是古埃及最重要的神祇之一。

[3]　Nut 努特：也拼作 Nuit，埃及神话中的天空之神，是一位女神，九柱神之一。

[4]　Isis 艾西斯：古埃及的母性与生育之神，九柱神之一。

[5]　Hathor 哈索尔：全称哈索尔·迪特拉(Hathor Ditera)，古埃及女神，是爱神、富裕之神、舞蹈之神、音乐之神。

[6]　Arabic shaduf 阿拉伯式吊桶取水装置：俗称"桔槔"，在近东地区特别是在古埃及使用，一种在支架一端系有重物，另一端吊有水桶的取水灌溉装置。

[7]　Winlock 温洛克(1884—1950)：全名为 Herbert Eustis Winlock，美国 20 世纪上半叶最杰出的埃及学(Egyp-

tology)研究专家。

［8］　Hyksos 喜克索斯人：古代亚洲西部的一个混合民族，于公元前 17 世纪进入埃及东部并在那里建立了第 15 和第 16 王朝（公元前 1674—前 1548）。

［9］　Pharaoh Ay 法老阿伊：古埃及第 18 王朝的法老，在位期间大约在公元前 1323—前 1319 年，或为公元前 1327—前 1323 年。

［10］　Amenhotep Ⅱ 阿蒙霍特普二世：古埃及第 18 王朝第七位法老，在位约 26 年（公元前 1427—前 1401）。

［11］　Sennufer 赛努佛：古埃及贵族，法老阿蒙霍特普二世统治时期的底比斯城（Thebes）的市长和阿蒙（Amun）的谷仓与田地、花园与城堡的总监。

［12］　Thebes 底比斯城：上埃及的一座古城，有"百门之都"之称，濒临尼罗河，位于今埃及中部，即今天的卢克索城（Luxor）。作为皇室的居住地和宗教膜拜中心，它从公元前 22 世纪中期到公元前 18 世纪繁荣一时。

［13］　Amarna 阿玛纳：古埃及城市，位于埃及中部尼罗河的东岸，现位于明亚省（Minya）首府明亚市。

［14］　Hatshepsut 哈特谢普苏特：古埃及第 18 王朝法老，在位时间为公元前 1479—前 1458 年，是古埃及历史上唯一的女法老。

［15］　Deir el-Bahari 代尔巴赫里：埃及一地名，其字面意思为"北方的修道院"（the Northern Monastery），位于尼罗河的西岸，与卢克索城相对。

［16］　Amun 阿蒙：也拼作"Amon"，在古埃及语中意为"隐形"，是古埃及底比斯的主神，因底比斯的兴起而成为国家的主神，因而成为新王国时期的国神。阿蒙与太阳神拉组合成为阿蒙·拉神，以头戴双翎冠男人身形象出现。

［17］　Medinet Habu 美迪奈特哈布：埃及一地名，是埃及考古和旅游的重要地点，位于卢克索城的西岸。

［18］　Karnak 卡纳克：卡纳克神庙建筑群（Karnak temple complex）的简称，其神庙是底比斯最为古老的庙宇，内有 134 根巨大石柱，分为 16 列。其中 122 根高度为 10 m，另外 12 根高为 21 m，大厅面积约 5 000 m²，总占地约为 30 ha（公顷）。

［19］　Memphis 孟斐斯：古代下埃及的一个城市，位于开罗以南，相传是埃及首位法老美尼斯所建。自古王国时期开始，除第 11 和第 12 王朝以外孟斐斯一直为埃及的首都，直至新王国时期才被底比斯取代。

［20］　Osireion 奥西里昂：也拼作"Osirion"，是位于埃及阿拜多斯（Abydos）的塞提一世（SetiⅠ，古埃及第 19 王朝法老）庙宇与陵墓的遗址。

［21］　Abydos 阿拜多斯：埃及一地名。

［GLOSSARY］

house garden		府邸花园	funerary garden		墓园	sacred garden		神庙园林

［NEW WORDS］

acacia	n.	刺槐	bay laurel	n.	月桂
alley	n.	胡同,小巷,小径	celosia	n.	鸡冠花
alleyway	n.	小巷,窄街,通道	ceremony	n.	典礼,仪式
arable	adj.	可耕的,适于耕种的	chapel	n.	小教堂,礼拜堂,礼拜仪式
attendant	n.	服务员,侍者	chrysanthemum	n.	菊花
awning	n.	凉篷,遮篷,雨篷,天篷	circulation	n.	循环,流通,环流,发行
barren	adj.	贫瘠的,不孕的,不结果的	controversy	n.	争论,辩论,争吵

continued

convolvulus	n.	旋花,旋花属攀缘植物	jujube	n.	枣子,枣树
coping	n.	顶部,顶盖,墙的顶部	kiosk	n.	小摊棚,售货亭,电话亭,凉亭
cornflower	n.	矢车菊	longitudinal	adj.	长度的,纵向的,经度的
counterpoise	vt.	平衡,使平衡,平均	lotus	n.	莲,莲花,睡莲
daisy	n.	雏菊	lush	adj.	茂盛的,葱翠的,豪华的,奢华的
date palm	n.	海枣,枣椰,椰枣	lychnis	n.	剪秋罗属植物
depiction	n.	描写,描述,描画	mandrake	n.	曼德拉草
descend	vi.	下来,下降,下落,降下,下斜	mignonette	n.	木犀草
doum palms	n.	埃及姜果棕	moor	vi.	使(船、浮标等)固定,停泊
elaborate	adj.	复杂的,精巧的,详尽的,详述的	mummy	n.	木乃伊,干尸
elevate	vt.	举起,提高,提升	myrtle	n.	桃金娘科植物,香桃木属植物
environ	vt.	包围,围住	narcissus	n.	水仙,水仙花
erect	vt.	建立,使直立	offering	n.	(宗教)供品,祭品;捐赠物
exotic	adj.	外国的,奇异的,古怪的	overtone	n.	暗示,含义,弦外之音
façade	n.	(房屋的)正面,外观,假象	papyrus	n.	纸莎草,(古埃及的)莎草纸
facilitate	vt.	使变得容易,使便利,促进	parallel	adj.	相平行的,相同的,类似的
farthermost	adj.	最远的	peculiar	adj.	奇怪的,古怪的,特有的,独特的
flank	vt.	位于……的侧面,保卫……的侧翼	pergola	n.	藤架,棚架,廊架
			perpendicular	adj.	垂直的,成直角的,直立的
foster	vt.	养育,照料,培养,促进	perspective	n.	景色,景观,透视图,观点
funerary	adj.	殡葬的,殡葬用的,葬礼上用的	pharaoh	n.	法老,古埃及统治者的称号
glitter	vi.	闪闪发光,闪光,闪烁,炫耀	pictorial	adj.	以图画表示的,图示的
greenery	n.	绿叶,绿色植物	pomegranate	n.	石榴,石榴树
harem	n.	(伊斯兰教徒的)女眷,内房,闺房	poppy	n.	罂粟,罂粟属植物
			portal	n.	门,壮观的大门,豪华的入口
henna	n.	指甲花,散沫花属植物	processional	adj.	列队行进的,与列队行进有关的
humus	n.	腐殖质	processional	n.	供(宗教)游行用的,祭神大典
implication	n.	含义,暗示,暗指	prohibitive	adj.	禁止使用或购买的,禁止性的
infiltration	n.	渗入,渗透,透过	protrude	vi.	伸出,突出
iris	n.	鸢尾	pylon	n.	(古埃及庙宇的)塔式门楼,电缆塔
ivy	n.	常春藤			
jasmine	n.	茉莉	quay	n.	码头

continued

regrettably	adv.	令人遗憾地,可惜地,不应该地	symmetrical	adj.	对称的,匀称的
schematically	adv.	图表地,图式地,略图地	tamarisk	n.	柽柳
shrine	n.	圣地,圣坛,神龛,神殿	terrace	n.	梯田,平台,露台
slab	n.	(尤指石质的)厚板,平板,停尸桌	transverse	adj.	横的,横向的
sling	vt.	投,掷,吊挂,悬挂	transversely	adv.	横断地,横切地
stairway	n.	楼梯	trelliswork	n.	格子棚架,格子结构
subdue	vt.	征服(敌人或自然等),抑制	twig	n.	(树木或灌木的)细枝,嫩枝
sweet marjoram		马约兰	willow	n.	柳树
sycamore fig	n.	桑叶无花果,西克莫无花果	wreath	n.	花冠,花环,花圈
symbolism	n.	象征主义,象征手法			

【难句】

However, in the country where the land was much less expensive, the houses and palaces were set in a large garden surrounded by a wall... Typically, a symmetrical layout was used with a rectangular or T-shaped pond in front of the house on the main longitudinal axis. 不过,在地价大为低廉的乡间,府邸和宫殿设于由围墙围合起来的大花园中……尤为典型的是,他们常常采用在建筑前主要长轴线上布置长方形或 T 形水池的对称布局。【被动句 was used 译为主动句,增加主语"他们"】

[参考译文]

古埃及的园林
吉米·杜恩

[1] 有些人确实很少听说园林,然而,对于那些能够负担得起在他们府邸和墓室前建造园林的古埃及人而言,园林是一个必不可少的元素。众神们甚至更享受园林,以至于几乎每个庙宇都被繁茂的植物所环绕。园林在新王国时期似乎显得特别的重要。需要指出的是某些类型的花园具有宗教象征性。

[2] 我们知道花园通常是由树木和其他植物组成。常用的树木包括桑叶无花果、石榴、坚果树和枣树。不过,也会见到柳树、刺槐和柽柳。总体上,古埃及人所栽种的树木约有 18 种。他们所使用的花卉品种也是相当丰富的,包括雏菊、矢车菊、曼德拉草、玫瑰、鸢尾、桃金娘、茉莉、木犀草、旋花、鸡冠花、水仙、常春藤、剪秋罗、马约兰、指甲花、月桂、小黄菊和罂粟。当然,有时还有纸莎草、莲花和葡萄。

[3] 对古埃及人而言,园林不仅仅是个愉悦的空间。园林树木具有许多象征意义,例如欧西里斯、努特、艾西斯、哈索尔等所意寓着的特定神明。它们还暗指生命的创造与消亡。纸莎草和莲花分别是下埃及和上埃及(各自的)两个地区的象征。园林还为人们提供了像蔬菜和葡萄酒在内的食物,如果莎草纸没有在大部分的埃及历史时期里被使用的话,那么我们对古埃及的了解可能会大大减少。

[4] 令人遗憾的是,就我们所知极少有对那些环绕普通住宅花园的描述,不过在几篇记述乡村庄园的文学作品中提到了新王朝时期的一处乡村庄园周围所环绕着的浓密植被。一位古代明显陶醉于其花园的主人告诉我们:"你坐在它们的树荫下,品尝着它们的果实。头戴嫩枝条编成的花环,而由它们酿制而成的葡萄酒又使你醉意朦胧。"甚至还有花园模型被制作并放置在陵墓里。

[5] 府邸、王宫、神庙和教堂,不论是用于殡葬的或是私家的,墓穴里的绘画几乎总是显示出有一个和建筑相连的花园。我们甚至能够经常从绘画中见到带有详尽自然细节的花园完整布局,这样使得我们可以根据这些

图示证据重建新王国时期各种类型的花园。直到中王国时期末,花园的浇水仍是由挑水工用吊桶肩挑取水后再灌溉。最初的提水平衡装置(阿拉伯桔槔),是由温洛克从喜克索斯人的入侵受到启迪所想到的,它使耕地灌溉变得容易多了。

府邸花园

[6] 即便是在古埃及,城中的土地买卖也几乎是禁止的,所以今天人们在这些城市中没有任何关于花园的真正证据。偶尔只会看到,在住宅的两侧栽种有为数不多的树木,通常是由海枣树和另一树种交替种植,并且它们有时也会被种植在砖砌的树池内。在埃亚法老的墓室内,环绕墓室结构的大庭院中摆放了一行由泥做成的树木模型,在最远的一侧是一个亭子,亭柱上缠绕着葡萄藤。

[7] 不过,在地价大为低廉的乡间,府邸和宫殿设于由围墙围合起来的大花园中。墓室中数量众多的描绘显示出了可能被公认的花园标准类型。尤为典型的是,他们常常采用在建筑前主要长轴线上布置长方形或T形水池的对称布局。而其花园被一排排不同种类的树木所环绕,同一排树木中也有可能交替种植不同品种。在沿花园轴线的主园路上或水池四周布置藤架是非常常见的做法。值得注意的是,这些水池中还经常养鱼,甚至有时还有外来的品种。果树的叶子或枝干由棚架上的木格子支撑着。越矮小的树木,越靠近水池。而越高大的树木,如埃及姜果棕和海枣树,则被种植在花园的最外侧。这种栽种方式营造出了以花园为中心具有不同层次的视觉效果。

[8] 有时花园中还有多处水池。阿蒙霍特普二世神庙及其侍者赛努佛位于底比斯与之相连的房子,其规整的花园布局如同赛努佛墓室中所描绘的一样,花园的轴线从主入口沿园路入内后与河流相垂直,其两侧各有一个廊架,并最终通向具有三个神龛的小神庙。园路把花园分成两半,每一半都被分为横向的三个区域。在最前面的区域中有一个与河流相平行的长方形水池,其中生长有水生植物,此外该区域中还有海枣树和桑叶无花果树。第二个区域位于中间位置,是一个由墙围合起来的封闭空间,栽种有可能是稀有品种的浅绿色树木。最后,是面积最大的后部区域,这里也同样有一个长方形水池,池塘的一侧为海枣树,而另一侧则为桑叶无花果树。靠近这一后部区域有一个小型开敞式亭子,其形式我们曾经在阿玛纳见到过。在花园的每一侧长边,都有一个由不同品种树木交替种植所围合起来的园路,而在府邸后面则由高大的树木组成了有效的屏障。

[9] 在大型的宫殿苑园中,也采用规则式布局。常用手法是对称式,一般在轴线的每一侧都有一个有成排树木围绕的水池。在土地不适宜耕种的阿玛纳,树木就被栽植在填有腐殖质的坑内,坑上还罩有圆形覆盖物。在各式各样的建筑群后面,有一块大场地被作为一处独立的花园,花园环绕在一个四周有坡度的方形水池周边。在池塘的一角上,都有一个能下到池底的阶梯。而在更深处的一个底部有开口的水池,可能是由渗透而入的水所注满。有趣的是,树木的布局方式看上去特别不规整,也许这是阿玛纳花园的另一个特质,即在艺术上的自由和自然主义倾向。

神庙园林

[10] 花园无论是位于祭祀队伍通往神庙塔楼路上,还是沿河位于在神庙码头前,都在墓穴中有所反映。在代尔巴赫里的哈特谢普苏特神庙里,有一个有着四个水池的花园、纸莎草、花卉和蔬菜作为不同主题被展示出来。花园中有在新王国时期从新征服的国家带来的,并栽种于阿蒙神庙花园里的外来树种。这些稀有品种在代尔巴赫里、美迪奈特哈布和卡纳克等地也被展示过,尽管难以计数的奇异品种很是迷人,但这些"植物园"式的表现方式却没能提供任何有关其布局的线索。在阿玛纳,私人教堂建在富人们的花园里或河流及运河旁,成为花园整体布局的重要因素,它们往往位于两条轴线的交会处,或主轴线的末端。通常教堂位于花园后部较高的台地上,园中有长方形水池,其两侧各有两排桑叶无花果树,或两排仿佛是有攀爬植物缠绕的高坛。在波斯花园的规则式布局中,用人工水池波光粼粼地如镜子般映照出周围壮丽的景致,其手法在埃及就已经到了炉火纯青的地步,至少早在新王国时期就是如此。

墓园

[11] 大部分对于墓园的描述都是以自然为主题的。它们在平面布局上通常被简化为一个"T"形水池并以一

些海枣树为背景。这里,关于水池奇特平面布局的起源仍有待探究。可以确定的是,运河的尽头,其横向的长方形水池,便于船只的停泊与调头。从另一方面来看,为葬礼祭品摆放之用的祭品桌,通常被呈现为"T"形石板,而在其中部有一条深深的水道。墓园的"T"形水池与祭品桌是否真的有任何关联,现在还不确定。不过,能够确定的是早在新王国初期,"T"形平面布局就具有了象征性的暗示。在哈特谢普苏特女王神庙下层台阶底部的中间通道的两侧,就分布有两个"T"形水池。

[12]　在孟斐斯,至少有两处壁画描绘了葬礼仪式,木乃伊被船运送到一个长方形池塘中央的长方形小岛上。在其中的一幅画中,池塘周围有树相围,其外面有两倍行距宽的葬礼构筑物,海枣与栽在砖砌树池内的树木交替种植,为构筑物提供了遮挡阳光的庇荫之地。码头从池塘稍短的一侧延伸至水中,一幅画中显示可以通过台阶登上码头。在后面这幅绘画中,码头设在小岛的两个短边上。这有可能是阿拜多斯的奥西里昂的象征性代表。

[13]　墓园的位置一直是存有争议的话题,但可以比较有把握地假设在陵墓自身的前端偶尔设置有某种类型的小型花园,而更为常见的则是在河岸下方布置有大型的花园,可能也靠近陵墓群的入口处。

[14]　现在,乃至整个历史,园林在埃及人的生活中扮演了重要的角色。由于靠近贫瘠的沙漠以及在这片滋养生命的狭小土地上能处处见到生命的需要,园林似乎已将变成他们毋庸置疑的一部分。

Reading Material A

André Le Nôtre and French Gardens[1]

Autum Matysek Snyder

安德烈·勒·诺特尔与法国园林

Fig. 1　André Le Nôtre

[1]　The French landscape designer André Le Nôtre (Fig. 1), best known for his work and designs for Louis the XIV palace Versailles, utilized classical elements in every nook and cranny of each and every landscape he designed. He employed the classical style garden, French architectural or Formal style garden, where man was dominant over nature. "The classical garden took form in an age when man imposed precise patterns of his own making on mankind as well as on the landscape…" The gardens were strict and contrived, it is obvious that the gardens were altered and groomed by man. Le Nôtre, born into a royal French landscape family in 1613, was destined to be a huge success; he was a third generation landscapist and was initiated into the practice of landscape design at an early age by his grandfather (Pierre) and his father (Jean). Le Notre's education and training was the epitome of a classical and academic training, he followed the "proper training of a gardener set by Boyceau[2] and studied painting and architecture with Simon Vouet[3] and Francois Mansart[4]". Le Nôtre also learned about optical illusion, which he utilized throughout his successful career as a royal garden designer.

[2]　The monumental gardens that Le Nôtre designed possessed a typical and standard form, the gardens were very geometric and mathematical and utilized the technique of optical illusion to create the desired effect and play on the eye of a vast, sprawling landscape. Le Nôtre's gardens were manip-

ulated and "constructed garden, Paradise itself... controlling every inch of space to the visible horizon and, by Cartesian[5] projection, 'indefinitely' beyond it. Le Nôtre created tiers, walls, waterways and pools to create optical illusions of infinite space. Usually Le Nôtre was not restricted to a limited space, but had unlimited space at his disposal, even though he had unlimited space he would still utilize optical illusions to create his masterpieces. Le Nôtre had a standard plan that he followed when creating a design. The classical scheme that he employed for his designs were used over and over again in each garden, the gardens all relied on the same basic standard plan. He would draw the chateau or main house into the landscape and use it as a focal point for the design.

[3] In the 17th century landscape and garden designs "house and garden were indivisible and complemented each other." The designs laid out by Le Nôtre always had a main axis leading towards the facade of the chateau, and the "central axis was bisected at intervals by transverse axes in the form of walks, pools and canals." The main axis would be the "spine" of the plan, which would extend along the main avenue, even though there were many deviations off the main path.

[4] "His creations... are characterized by great axes opening onto vast perspectives reaching to the horizon. The layout of his plans is very geometric, using octagonal, tree-lined allies to define wooded spaces, at the entrance to which are groves-chambers of green where the gardener, by contrast, gives rein to his fantasy. Order, clarity, simplicity, symmetry, amplitude are the rules applied by Le Nôtre in the creation of his gardens..." This type of strict plan of harmony and balance between buildings and landscape was a key element in Le Nôtre's plans as well as a leading aspect of classical landscape design.

[5] Le Nôtre was credited with creating a set of rules for gardens, as written down by one of his pupils Alexander Le Blond[6] in *La theorie et la pratique du jardinage*[7] 1739. The book laid out the rules and "grammar" of gardens. The rules of Le Nôtre's gardens became to be known as the classical scheme. The rules are as follows:

[6] "The garden must be in harmony with the conformation of the land, that is, in accord with the situation whether it is in the mountains, on the seashore or in a desert such as the American Southwest; it must be planned for the climate, in terms of whether it is hot or cold, damp or dry, or combinations of these; it must form a unit with the house, harmonizing with its proportions as well as its size and style... The garden should be one-third longer than wide, and that one feature should always be balanced by another, such as wood by parterre. There should be variety, and no designs in the beds should be repeated. Statues should look inevitable, canals should be used in low places, and woods should be planted on the sides and tat the back of the house. Moreover, planting should be done in terms of how it would look twenty years later..." From these rules stemmed Le Nôtre's unique, yet very ordered and structured gardens. This set of rules was later applied to gardens to explain when something went wrong and was consulted to create the ideal garden landscape.

[7] The classical idea of beauty and creating the perfect image by complying every "perfect" aspect of nature was used full heartedly by André Le Nôtre in his landscape designs. Every detail was in harmony and perfect proportion to one another, and everything was on a monumental scale. The landscapes were a visual reference of Man's dominations over nature and the creation of the ideal. In Le

Nôtre's landscapes, buildings accentuated the beauty of the landscape, although contrived and manipulated, and vise versa the landscape accentuated the buildings. Although the landscape was very important, "architectural features such as pavilions, stairways, balustrades and statues were considered the most important ornaments in the gardens, and living plants were treated as subsidiary furniture."
[8] The shrubs, flowers and trees were "pruned to follow geometric shapes, and flowers were trimmed to provide lines or sections of color... outlines of ponds were carried out from designs first drawn on paper with a ruler and compass." In all of Le Nôtre's plans he incorporated and focused on statues and architectural devices to create his classical beauty' with nature. These mathematical and geometric designs that Le Nôtre designed stemmed from his academic training and the interest of the French Academies[8] emphasis on line as the key element to an artistic career. He incorporates the use of line into every aspect of his planning and relies heavily on mathematical proportions and line to create his desired effect of infinite space. Now we will look at how Le Nôtre applied and adapted his standard plan to each of the above-mentioned chateaus.

[9] André Le Nôtre's first major work, Chateau Vaux-le-Vicomte[9] (Fig. 2), was in collaboration with Louis le Vau[10], and Charles Le Brun[11], shows his excellent skills as a classical landscape designer. The geographical landscape of Vaux-le-Vicomte is flat with expansive green fields (Fig. 3). To overcome the flatness of the geographical site, Le Nôtre chose to create optical illusions to create the ideal landscape.

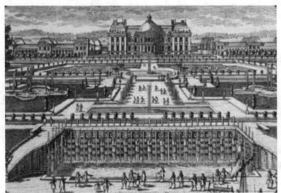
Fig. 2 A view of Chateau Vaux-le-Vicomte from the end of garden to Chateau

Fig. 3 The master plan of Chateau Vaux-le-Vicomte

[10] The flat fields had been geometricized to fit into the landscape; the shrubs trimmed and planted in patterns to show the "rescued landscape from nature's wild". The actual geometric layout of the gardens shows Le Nôtre's goal of harmony, clarity and balance. He strived very hard to balance the gardens to fit the harmony and grandeur of the chateau. Le Nôtre employed optical illusions to make the vast gardens actually seem smaller and more inviting than they were. He encouraged the visitors to explore and veer off of the main path by having "openings and exits from one garden to the next..." A characteristic of Le Nôtre's garden designs was the leveled terraces. These terraces created the optical illusion that the chateau was not diminishing in size, even though the distance between viewer and chateau was growing, but that it remained the main focus of the grand estate. The optical illusions used at Vaux-le-Vicomte were not only to enhance the beauty of the garden and chateau, but also cre-

ated for practical endeavors; the terraces were gradually "sloped for the sake of drainage though they appeared to be level".

[11] Incorporated into the gardens at Vaux-le-Vicomte, Le Nôtre situated a grotto and canal as the final climax of the garden. The grotto was the balance to the chateau, it could be seen from anywhere in the garden, it was a beckoning point for exploration. The grotto's large niches were filled with monumental reclining classical sculptures. The canal and cascades could not be seen from the chateau, but could be heard throughout the garden, this was an encouraging device for the walker to keep exploring the gardens until they reached the grand finale. From the terrace above the grotto, as well as from the chateau's terrace, the whole park could be seen and the order and harmony of the design could be appreciated (Fig. 4). The carefully sculpted parterre beds, pools and avenues showed Le Nôtre's control over nature and the geometric design that he had created. Every aspect of the park was perfect and in complete harmony with every other element. Nothing was over shadowed by another, but complemented one another (Fig. 5).

Fig. 4 A view of Chateau Vaux-le-Vicomte from Chateau to garden

Fig. 5 An aerial view of Chateau Vaux-le-Vicomte

[12] Versailles was André Le Nôtre's grand master piece, he had unlimited space to create the most massive, yet controlled park of the 17th century and for decades to come (Fig. 6). At Versailles Le Nôtre, again utilized his standard plan of a central axis with the palace as the focal point and diverging side avenues, to create the ideal and perfect park. Versailles optical illusions were so dramatic that from the main terrace of the palace you could not see most of the park, the park is multi-leveled with the palace raised above the park itself. Each level the promenader descends brings a new and magnificent small park. "... The promenader realizes that the garden, which had appeared relatively flat when viewed from the building, is in reality marked by

Fig. 6 Plan view of the gardens of Versailles

a series of descending levels, which, once discovered, reveal hitherto concealed embellishments in the form of stairways, pools and sculptures." The main avenue is an optical illusion itself, when riding up the avenue the viewer is struck by an impressive view of the palace, but to get to the palace they actually have to ride through the many small parks even though it appears that the main avenue runs straight into the facade of the palace. In actuality the avenue splits around pools, terraces and parterre. The parterre of Versailles is grand feats of controlled nature. The shrubs are sculpted into geometric designs with every aspect of the parterre's depicting the classical idea of beauty.

[13]　The gardens are controlled and ruled by man, the best parts of nature were improved on to create the ideal. The parterre is incorporated with pools and fountains to create a more impressive display of nature at its ultimate beauty. Versailles was in its best, an impressive conquering of chaotic nature and civilizing it. Every aspect is controlled and carefully manicured, the trees were cut to focus on the palace, the sculpted parterre emphasized the intricate design of the facade and the fountains and pools tied all elements together to create a unity of parts and a magnificent complex (Fig. 7).

Fig. 7　View of Versailles from the Avenue de Paris, ca. 1662 by Pierre Patel (left) and View of the château de Versailles as seen from the Place d'Armes, 1722, ca. 1722, by Pierre-Denis Martin (right)

[14]　Versailles was no small undertaking, it took Le Nôtre 33 years, hundreds of workers and thousands of hours to complete. Transforming the harsh landscape of the original hunting cabin into the classical idea for beauty was nothing short of a miracle. The original land was described as a, "unhealthy, swampy site, which had no view and almost not water..." This unfit land was manipulated and crafted into a magnificent example of classical beauty; it was man over nature, where man was the ultimate victor. Le Nôtre stuck to his rules for creating the perfect "classical landscape"; he situated the park to fit in with the landscape to best catch all the beauty that nature had to offer. The lie of the land sloped west and he created "board terraces to catch the sun during damp, cool winter months; shaded woods with splashing fountains for the sake of coolness on hot summer days; and out of door ballrooms and concert halls to provide a setting for entertainment, while dramatic stairways furnished backgrounds for the pageant like parties." Throughout the park Le Nôtre incorporated classical architecture, sculpture, vases and pedestals to draw connections between the glory of the Sun King[12] and the magnificent of the classical Greek and Roman rulers and gods, such as with the fountain of Apollo[13]. The fountain is a representation of Louis XIV and his connection to the gods and

glory of the ancients.

[15]　Le Nôtre's classical schemes, and desire to create gardens were harmony and order prevailed over nature, cannot be better displayed then in his grand works of Vaux-le-Vicomte and Versailles. In these garden parks Le Nôtre employed a standard set plan. Le Nôtre reviled in the fact of knowing that he had transformed an uninhabitable and harsh landscape into a spectacular and "flawless landscape". His gardens could be compared to a virtual heaven on earth, everything was perfect, ordered, controlled and in harmony with everything else. Nothing was out of place; his designs epitomized the classical ideal of beauty.

[NOTES]

[1]　本文节选自 *The Supreme Architect of Gardens：André Le Nôtre's classical beauties* 一文。
[2]　Boyceau 博伊索(1560—1633)：全名为 Jacques Boyceau, sieur de la Barauderie,法国造园师,路易十三时期的皇家园林总监。
[3]　Simon Vouet 西蒙·武埃(1590—1649)：法国画家、工艺师,路易十三时期的宫廷首席画师,他将意大利巴洛克风格传入了法国。
[4]　Francois Mansart 弗朗西斯·芒萨特(1598—1666)：法国17世纪建筑师,他因将古典主义引入法国巴洛克式建筑而闻名。
[5]　Cartesian 笛卡儿的/笛卡儿哲学的：笛卡儿(1596—1650),全名勒内·德斯卡特斯(René Descartes),法国哲学家、数学家、物理学家。他对现代数学的发展做出了重要的贡献,因将几何坐标体系公式化而被认为是解析几何之父。他还是西方现代哲学思想的奠基人。
[6]　Alexander Le Blond 亚历山大·勒·布隆(1679—1719)：全名为 Jean-Baptiste Alexander Le Blond,法国园林设计师、建筑师。
[7]　*La theorie et la pratique du jardinage*《造园的理论与实践》。
[8]　French Academies 法兰西学院：法国最高的学术机构,由红衣主教黎塞留创办于1635年,当选法兰西学院院士是极高的荣誉。
[9]　Chateau Vaux-le-Vicomte 沃·勒·维贡特府邸庄园：勒·诺特尔最重要的作品之一,是为当时路易十四的财政大臣尼古拉斯·福凯(Nicolas Fouquet,1615—1680)建造的。花园面积为 70 hm², 工程于 1656 年开始,历时 5 年建成。富丽广袤的府邸和花园招致了路易十四的妒忌和愤怒,福凯很快就沦为了阶下囚,之后路易十四仿造维贡特庄园,建造了规模更加宏大的凡尔赛宫苑。沃·勒·维贡特花园可以说是勒·诺特尔式园林的一个原型,中轴明显,起主导作用,各造园要素布置得合理有序,在轴线上依次展开；花园美观大方,处处显得宽阔,却又不是巨大无垠。地形处理精致细腻,与原有地形关系和谐。
[10]　Louis le Vau 路易·勒·沃(1612—1670)：是法国著名的古典主义建筑师,沃·勒·维贡特府邸及凡尔赛宫的建筑师。
[11]　Charles Le Brun 查尔斯·勒·布伦(1619—1690)是法国17世纪具有影响力的画家和艺术理论家。
[12]　Sun King"太阳王"：即路易十四,因为他自喻为太阳,所以也被人称为"太阳王"。
[13]　Apollo 阿波罗神。

[NEW WORDS]

accentuate	vt.	强调,使……突出	intricate	adj.	错综复杂的
amplitude	n.	广阔,宽阔,丰富	landscapist	n.	风景画家
balustrade	n.	栏杆,扶手	magnificent	adj.	宏伟的,华丽的,豪华的
beckoning	adj.	引诱人的;令人心动的	manicure	vt.	修剪(草坪等),给(某人)修剪指甲
bisect	vt.	把……一分为二,把……二等分	manipulate	vt.	(熟练地)使用,操作,处理
cabin	n.	小木屋	niche	n.	壁龛
cascade	n.	小瀑布(尤指大瀑布的分支)	nook	n.	角落,隐匿处
chaotic	adj.	混乱的,杂乱的,一团糟的	octagonal	adj.	八角形的,八边形的
chateau	n.	(法国封建时代的)城堡,庄园	optical	adj.	视觉的,视力的
collaboration	n.	合作,协作	ornament	n.	装饰品,点缀品
complemented	adj.	有补助物的	pageant	n.	盛装的游行,壮丽的场面,伟观
comply	vi.	遵从,依从,服从	pedestal	n.	底座,基座
conformation	n.	一致,符合;结构,形态	prevail	vi.	战胜,压倒;盛行,流行
contrived	adj.	不自然的,勉强的	projection	n.	投影,投射;推断,设想
cranny	n.	裂缝,裂隙	promenader	n.	散步者
damp	adj.	潮湿的	pruned	vt.	修剪(树木等)
depict	vt.	描画,描述,描写	recline	vi.	斜倚,倚靠,躺卧
disposal	n.	排列,布置	rein	n.	缰绳,驾驭;控制,统治
diverge	vi.	分歧,分叉,偏离	rescue	vt.	解救,援救,营救,搭救
embellishment	n.	美化,装饰,渲染,润色	revile in	vi.	得意于,纵情于;着迷
endeavor	n.	努力,尽力	scheme	n.	计划,方案
enhance	vt.	增强,提高,改进,改善	spine	n.	(尤指山坡或山脉的)脊,脊柱
epitome	n.	代表,典型,象征,缩影	splash	vi.	溅,泼,溅湿,溅污
epitomize	vt.	为……的典型,为……的代表	split	vi.	裂开,劈开,分离
feat	n.	功绩,业绩,技艺	sprawling	adj.	蔓延的,蔓生的,不规则地伸展的
finale	n.	结局,终曲,尾声	stem from	vi.	来自,起源于
flawless	adj.	无裂痕的,无瑕的,完美的	struck	adj.	目瞪口呆,被某人/某物打动
geometricize	vt.	(设计)成几何形状	subsidiary	adj.	辅助的,附带的,从属的
groom	vt.	整饰,照料	swampy	n.	沼泽,湿地
harsh	adj.	粗糙的,刺目的,严厉的,严酷的	tat	vi.	梭编,梭织
hitherto	adv.	到目前为止,迄今	tier	n.	层级,等级;行,列
illusion	n.	错觉,幻想,假象	transverse	adj.	横向的,横断的,横切的
indefinitely	adv.	无限定地,不确定地	trim	vt.	修剪,整修,整理;使整齐;修饰
indivisible	adj.	不可分的	unlimited	adj.	无限的,无边的
inevitable	adj.	不可避免的,必然发生的	veer	vi.	改变方向,转向

[参考译文]

安德烈·勒·诺特尔与法国园林
奥特姆·马蒂塞克·斯奈德

[1] 法国景观设计师安德烈·勒·诺特尔,因其为路易十四凡尔赛宫苑所做的工作和设计而闻名,他在每一个角落和缝隙,每一处设计的景观都运用传统要素。他采用古典风格的花园,法国建筑式或规则式花园,即人工主导自然的设计。"古典园林的形成于人们将其自身的精确模式施加于人类也同样施加于景观的时代……"花园是严整和刻意的,明显是由人改造和修饰过的。勒·诺特尔1613年出生于一个法国皇家景观世家,注定要获得巨大的成功。他是第三代风景画家,很小受祖父(皮埃尔)和父亲(让)影响开始练习景观设计。勒·诺特尔的教育和培训是古典和学院培训的缩影,他遵循博伊索设立的"造园师适合的培训"并向西蒙·武埃和弗朗西斯·芒萨特学习绘画和建筑。勒·诺特尔还学习了有关视错觉知识,并将其运用在他整个成功的皇家园林设计师职业生涯中。

[2] 勒·诺特尔设计的宏伟园林都有一个典型的标准形式,园林非常几何和精确,运用视错觉技术创造想要达到的效果,呈现出一个宏大、延展的景观。勒·诺特尔式园林是被操控的"人工建造的花园,天堂一般……控制每一寸空间直至视平线,通过笛卡尔投影,无限延伸至远方。"勒·诺特尔创造了层级台地、墙、水道和水池来创造无限深远的视错觉。通常勒·诺特尔不局限于有限的空间,反而自由支配那些无限空间,尽管他有无限空间,他仍然利用视错觉来创造他的杰作。当进行一个设计时,勒·诺特尔遵循一个标准平面。他的设计所运用的古典方案在每一个园林中都反复使用,所有园林都基于同样的标准平面。他会在景观图中画出城堡或主要建筑,并将其作为设计的焦点。

[3] 在17世纪景观和园林设计中,"房屋和花园是不可分割并相辅相成的"。勒·诺特尔的设计布局总有一个主轴线,导向城堡的正立面,并且"中轴线不时地被道路、水池或水渠形成的交叉轴所切分。"主要的轴线会成为平面的"脊骨",沿着主林荫道延伸,尽管还有许多偏离主干道的小道。

[4] "他所创造的特色是宏大的轴线向广阔的远景延伸,直至地平线。他的规划布局非常几何化,运用八边形,以成排的树木来界定植物空间,从入口进入,可以看到由树丛形成的绿色房间,通过对比,造园师在这里可以充分发挥想象力。秩序、清晰、简洁、对称、广阔是勒·诺特尔在创造园林时所运用的法则……"严整的平面带来建筑和景观的和谐均衡是勒·诺特尔规划的关键要素,以及古典景观设计的主导方向。

[5] 勒·诺特尔被认为创造了一系列园林设计的法则,并被他的学生亚历山大·勒·布隆在1739年记录成书《造园的理论与实践》,书中提出了园林设计的法则和"语法"。勒·诺特尔式园林的法则成为古典设计的准则。这些准则如下:

[6] "花园必须同土地构成相互协调,即根据或山地或滨水或像美洲西南那样的沙漠条件,因地制宜;它的规划必须适合气候,根据冷热、潮湿或干旱,或综合考虑,因气候制宜;它必须与房屋形成一个整体,与建筑的比例、尺度和风格相协调……花园的长宽比应该是3:1,一个景致应该同另一个景致相平衡,比如树木和植坛。花园应该是多样的,花坛的设计不能重复。必须要有雕塑,水渠应布置在较低的位置,树木要种植在两侧,在建筑后面相互交织。而且,种植应当根据二十年以后的面貌进行布置……"这些法则来自勒·诺特尔独特的,却非常有秩序有组织的园林设计。这一系列法则后来被用于解释花园设计为何会出现问题,或被请教如何创造理想花园景观。

[7] 勒·诺特尔在他的景观设计中,全身心地遵守每一条"完美"自然法则,体现古典美的理念并创造完美的形象。每一处细节都在比例上完美和谐,所有事物都有宏大的比例。景观是一个人类主导自然并创造理想景致的视觉效果。在勒·诺特尔的景观中,建筑强化景观的美,尽管是人为的并受控制的,但是反之,景观也强化了建筑的美。尽管景观非常重要,但是"像凉亭、台阶、栏杆和雕塑这些建筑式小品,被认为是园林中最重要的装饰,植物被处理成附属要素"。

[8] 灌木、花卉和树木"被修剪成几何形状,花被装点成线条或色块……水池的轮廓是一开始用尺规设计在图纸上画出的形状。"在所有勒·诺特尔的规划中,他都运用并关注雕塑和建筑式设施来创造他的自然"古

· 57 ·

典美"。勒·诺特尔的数学和几何式设计来自他学院式训练,以及对法兰西学院强调线条是艺术职业的关键要素的兴趣。他将线条运用到他规划的各个方面,并且极其依赖数学比例和线条来创造他想要的无限空间的效果。现在我们将看一下勒·诺特尔如何将他的标准平面运用到上述的城堡中。

[9] 安德烈·勒·诺特尔的第一个重要作品,沃·勒·维贡特府邸庄园,是与路易·勒·沃和查尔斯·勒·布伦合作完成的,体现了他作为古典园林设计师的杰出能力。沃·勒·维贡特府邸庄园的地理景观平坦,有着广阔绿地。为了克服地势的平淡,勒·诺特尔选择运用视错觉来创造理想的景观。

[10] 平坦的场地被设计成几何式景观;灌木被修剪并模式化种植,来体现"从粗野自然中拯救的景观"。花园真实的几何布局显示出勒·诺特尔对和谐、清晰和平衡的追求。他努力平衡花园和宏伟的城堡来达到二者的协调。勒·诺特尔运用视错觉来使巨大的花园看起来比实际更小、更吸引人。他通过"从一个花园到下一个花园的入口和出口……"的设置,来鼓励游览者探索并离开主道路。勒·诺特尔园林设计的一个特点是层层台地。三个台地创造的视错觉使得城堡在尺度上没有缩小,即使观看者到城堡的距离在变大,宏伟的建筑依然是视觉的主焦点。在沃·勒·维贡特府邸庄园运用视错觉不仅增加了园林和城堡的美感,还创造了实际用途,"看起来水平的台地实际上为了排水而逐渐倾斜"。

[11] 在沃·勒·维贡特府邸庄园里,勒·诺特尔在最后的高潮处布置了洞府和水渠。洞府是城堡的平衡物,可以从庄园任何地方看到,是一路探索后令人心动的地方。洞府的大型壁龛填充着巨大的倾斜的古典雕塑。水渠和瀑布从城堡看不到,但是整个庄园都可以听到,对步行者来说这是一个鼓励继续探索的设置,直到他们到达尾声。从洞府上面的台地,以及从城堡的台地,都可以俯瞰全园,并欣赏设计的秩序和和谐。精细雕琢的植床、水池和大道显示出勒·诺特尔创造的控制自然和几何式设计的理念。庄园的每一部分都很完美,并与其他部分完全和谐。没有一个元素使另一个逊色,反而是相互辉映、相得益彰。

[12] 凡尔赛宫苑是勒·诺特尔的伟大杰作,他没有边界的限制,创造了最为宏大、17世纪以及其后几十年被推崇的规则式园林。在凡尔赛宫,勒·诺特尔又一次运用他的标准平面创造完美的理想花园,一条中轴线以宫殿作为焦点,轴线两侧为放射状道路。凡尔赛宫苑的视错觉极其明显,以至于从宫殿前主台地几乎看不到花园,花园由多层平台组成,宫殿高于花园之上。散步者每走下一层平台,都有一个新的华美的小花园。"……散步者这时才意识到,从建筑里远眺,看似平坦的花园,实际上是由层层下降的平台组成,一旦看到这些平台,台阶、水池和雕塑的这些装饰就开始展现到眼前。"主林荫道自身也是视错觉景观,当爬上林荫道,观赏者会被壮观的宫殿景观所吸引,尽管看起来主林荫道直接连接到了宫殿正面,但是要想接近宫殿,实际上还是要走过许多小花园。事实上林荫道在水池、台地和花坛附近是断开的。凡尔赛宫苑的花坛是人工自然的宏大杰作。灌木被雕琢成几何设计,花坛的每一面都刻画出理想的古典美。

[13] 花园由人工控制和主宰,自然界最好的部分被加强,直至达到理想自然。花坛中布置水池和喷泉来创造更加引人注目的极致的自然美。凡尔赛宫苑是对混沌自然最好的、让人印象深刻的控制和驯化。每一处景观都精心控制和修剪,树木被剪成朝向宫殿,花坛被造型以强调复杂的建筑立面,喷泉和水池被用来连接所有元素,来创造统一的整体和宏伟的复合体。

[14] 凡尔赛宫苑是个不小的任务,花了勒·诺特尔33年的时间和几百个工人几千小时的工作才得以建成。将原先狩猎小屋的粗野景观转变成古典美的理想境地简直就是一个奇迹。原来的土地被描述为:"不健康的沼泽地,没有景色可言,也几乎没有水……"这个不适宜的土地被精心布置成一个壮观的古典美的范例;这是一个人工战胜自然,人类最终获胜的地方。勒·诺特尔坚持他创造完美"古典景观"的法则,将花园放在最适合景观的位置,最好地捕获大自然所提供的所有美景。他借向西倾斜的地势,创造了"为潮湿寒冷的冬日获取阳光的台地;为夏日提供凉爽的林荫树和水花飞溅的喷泉;为消遣提供的户外舞厅和音乐厅,并将引人注目的台阶装点成盛装游行的背景。"在整个花园里,勒·诺特尔都结合了古典建筑、雕塑、花瓶和基座,上面描绘着太阳神的荣耀和希腊罗马的统治者和众神的伟大,例如阿波罗大喷泉。这个喷泉正是路易十四的象征,以及他与众神和古代荣耀的关联。

[15] 勒·诺特尔古典的设计和创造园林的愿望是和谐和超越自然的秩序,没有比在他宏伟的沃·勒·维贡特府邸庄园和凡尔赛宫苑将其展现得更好的了。在这些园林里,勒·诺特尔都运用了一套标准平面,并且得意

于将一个不宜居的严酷环境转变为壮丽且"无瑕的景观"。他的园林可以看作地球上虚拟的天堂,一切都是完美的,有秩序的,受控制的,并相互和谐。他的设计完全没有不合适的地方,是理想的古典美的缩影。

Reading Material B
English Landscape Gardens[1]
Robert Viau
英国风景园林

[1]　Indeed, from the end of the seventeenth century to the beginning of the nineteenth century, that is to say, during the period often referred to as the "long" eighteenth century (1660-1840), the aesthetic of garden design shifted gradually from one that stressed restraint, control, limit, and order to one that emphasized freedom and openness. From the geometrical severity of Versaille and Hampton Court[2] (Fig. 1) in the late seventeenth century to the well regulated naturalness of Blenheim[3], Castle Howard[4], and Stowe[5], by the middle of the eighteenth century designed gardens grew almost to resemble open landscape or raw nature.

Fig. 1　The plan of Hampton Court

[2]　At the opening of the eighteenth century, the dominant force in landscape design was André Le Nôtre, chief garden designer for Louis XIV[6] at Versaille. The most popular garden designs of the seventeenth and early eighteenth century were the French, Italian, and Dutch formal gardens executed to exhibit bilateral symmetry, and no one surpassed Le Notre in his realization of this rigid style.

[3]　In this garden style, the part of the garden closest to the palace or house was handled architecturally, like another room-extension of the house proper. The garden consisted of a perfectly regular series of geometrical compartments formed by closely clipped shrubs and trees and straight gravel walks, stone paths, terraces, and steps. Often the compartments were parterres de broderie[7] (plots resembling embroidery) carpeted with low evergreens (often box), flowers (actually rare until the nineteenth century), colored earth, brick dust, coal dust, white and yellow sand, etc. In the largest gardens, rigid geometry was imposed as far as the eye could see. Garden walks extended and radiated in geometrical patterns, along with canals and avenues of trees. Fountains, statues, mazes, and small woods and groves were all arranged symmetrically with reference to one central axis extending from the

exact center of the house.

[4]　The masterpiece of this style of gardening was Versaille. It became the model for princely gardens throughout Europe, and this includes the garden laid out for William Ⅲ in front of Sir Christopher Wren's[8] new east front of Hampton Court Palace.

[5]　At the end of the seventeenth century, the English inheritor of the Le Nôtre tradition was Henry Wise[9] (1653-1738), one of the principal gardeners of Hampton Court Palace. He also worked for James Brydges, the Duke of Chandos[10], at the elaborate and expensive gardens at Cannon, Middlesex[11], which Pope's [12] contemporaries believed to be Timon's Villa[13] in the "Epistle to Burlington[14]". The style of these gardens is "autocratic": palatial grandeur radiates outward from the patriarchal seat, its rigid order dominating nature and bending it to man's will.

[6]　The Great expense of maintaining Hampton Court's extensive gardens eventually led Queen Anne[15] to order Wise to reduce the cost of upkeep by two thirds. Thus in 1704 the box parterres de broderie were replaced by open lawn, in a step towards the freer landscape style that would dominate much of the rest of the century.

Transitions from Formal to Landscape Gardens

[7]　In reaction to the rigid formality of the French and Italian gardens of the late seventeenth century, a new style began to emerge which was much freer. Advocates of what eventually became the irregular landscape garden opposed symmetry, ostentation, and what they regarded as the tyranny of the French style, which they in turn associated with the tyranny of French government. Thus the growing freedom of English garden design gradually became associated with the freedom of English government. Garden aesthetics took on political meaning, sometimes, as in the case of Stowe, overt political meaning.

Charles Bridgeman & Stowe

[8]　Charles Bridgeman[16] (1690-1738) succeeded Wise as the Royal Gardener. His most famous achievement in landscape design is the famous garden at Stowe under Bridgeman's direction since 1713. This masterpiece of landscape design was added to later by Kent[17] and Capability Brown.

[9]　Bridgeman stands midway between Le Notre and Capability Brown in garden style. In the 1720's Kent took up landscape gardening in what is called the painterly manner. His most notable painterly garden is Rousham in Oxfordshire[18]. Bridgeman prepared the main lines of the garden in the 1720s, preparing the way for Kent's work in the 1730s. The painterly manner attempted to evoke something of the theatrical qualities of the landscapes of Poussin[19] and Claude[20].

[10]　Stowe (Fig. 2) is a landscape garden with political meaning. On the one hand, it celebrates the solid classical foundations of eighteenth century society, as embodied in the Neo-Palladian building[21] and the numerous Neo-Palladian garden monuments and follies. On the other hand, in its free and open treatment of garden space, Stowe also embodies the freedom which eighteenth century theorists associated with ancient British principles.

[11]　But the political meaning of Stowe is sharper and more specific still: it represents opposition politics through allegorical monuments. A large valley called the Elysian Fields[22] lies between two

ridges. On one ridge sits the Temple of Ancient Virtue[23], designed by Kent in 1734, which exhibits life-size Statues of Homer, Lycurgus, Socrates, and Epaminondas. Facing it but from lower ground stands the Shrine of British Worthies[24] (Fig. 3), also by Kent, exhibiting busts of sixteen national heroes, including modern figures like Shakespeare, Locke, Newton, and Pope as well as men of old like King Alfred. The Shrine of British Worthies literally looks up towards the Temple of Ancient Virtue in a powerful demonstration of reverence for classical ideals. For a while there was a third building nearby, the Temple of Modern Virtue[25], a ruin that allegedly satirized Sir Robert Walpole[26], the Whig[27] Minister of State whom Cobham[28], Pope, Swift[29], and many other Tory[30] writers loved to hate. The Temple of Liberty[31] is in the Gothic style[32], by Gibbs[33] (1741) associated by architects and landscape designers with ancient British ideals.

Fig. 2 Temple of Ancient Virtue in Stowe

Fig. 3 Temple of British Worthies in Stowe

Alexander Pope & His Garden in Twickenham

[12]　Pope has been called the presiding genius of the gardening revolution in the 1720s-30s. His own garden in Twickenham[34] as well as the gardens of weathly friends with whom Pope consulted testify to his remarkable influence (Fig. 4).

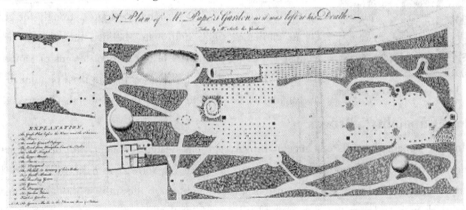
Fig. 4 Alexander Pope's garden at Twickenham

[13]　Little survives of Pope's garden. We do know from drawings and descriptions that it perfectly embodied the landscaping principles espoused in "Epistle to Burlington". To Pope, landscape gardening was an act of the imagination expressing his inner "romantic" impulses.

[14]　Bridgeman had introduced a garden design based on a relatively formal straight central axis

with flanking areas treated irregularly, so that symmetry and balance are combined with variety. Pope adapted this principle and applied it to his small garden plot across the London road from his villa in Twickenham. The bounds of the garden were concealed by dense thickets to create an enclosed irregular garden containing monuments with both ancient and modern associations. At the eastern end of the garden stood the Shell Temple, a Rococo[35] pleasure dome; at the western and darker end of the garden stood an obelisk commemorating the death of Pope's mother. From the garden a passage ran beneath the London road and into a Grotto[36] located in Pope's basement. At the garden end the Grotto looked out over an open lawn towards the Thames[37] and open country. When the doors of the Grotto were closed, it became a camera obscura reflecting thousands of images from the sparkling shells and bits of mirror in the Grotto walls, a truly remarkable and "poetic" folly of the fancy.

[15] Perhaps Pope's most remarkable indirect influence was at Stowe, Lord Cobham's 400 acre garden worked on by sixty years of landscape gardeners, architects, and sculptors: Bridgeman, Vanbrugh, Kent, Brown, and many more.

Chinese and Japanese Influences

[16] Eighteenth century garden ornaments and follies generally were either Classical or Gothic, but gradually throughout the century oriental styles began to be incorporated into landscape design, as they were into rood decoration. In the 1740s Chinese House at Shugborough[38] and the House of Confucius at Kew[39] were built. In the 1750s many pagodas, pavilions, and kiosks were built, along with Chinese style bridges such as the one across the Thames at Hampton Court (Fig. 5). By the 1750s French descriptions of the Imperial Gardens at Peking had been published in English. Architect Sir William Chambers[40] visited Canton[41], China, as a young man and in the 50s published *Designs of Chinese Buildings, Furniture, Dresses*[42], etc. (1757), followed by *Dissertation On Oriental Gardening*[43] (1772). Chambers argues strongly for great variety in garden design, and many believe that this is a reaction against the rising popularity of the garden designs of Lancelot "Capability" Brown (1716-1783), by far the most popular and prolific designer of the second half of the eighteenth century.

Fig. 5 The Pagoda at Kew Gardens

Lancelot "Capability" Brown

[17] In the "capable" hands of Lancelot Brown, gardens design lost nearly all of its formality and appearance of artifice. At Blenheim, he eliminated the great Le Notre style parterres laid out by Henry Wise and replaced it with an open expanse of lawn brought up to the walls of the house, near which he planted dark trees to frame the view of the landscape from the house. For some contemporizes such as Chamber, Brown's gardens "differ very little from common fields, so closely is common nature copied in them."

[18] Brown created this effect of the appearance of unrestrained nature by planting a vast stretch of lawn punctuated by small clusters of trees or single trees irregularly placed in wavy belts. The land

dips away from the house towards a winding lake and rise beyond to a distant woodland, completing the "landscape" (Fig. 6).

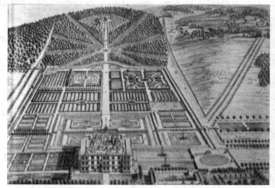

Fig. 6　Longleat, Wiltshire (1757) Before Capability Brown (left) and After Capability Brown (right)

[19]　The last stage in the development of eighteenth century gardens is the result of the powerful influence of Edmund Burke's[44] Philosophical Enquiry into *the Origin of Our Ideas of the Sublime and the Beautiful* (1757). This text profoundly influenced of the emergence of gothic literature, gothic revival architecture, and landscape design modeled on Burke's notion of the sublime or terrible in nature.

[NOTES]

[1]　本文节选自 *Eighteenth Century Garden History* 一文。

[2]　Hampton Court 汉普顿宫：全称为 Hampton Court Palace，是英国文艺复兴时期最著名的大型规则式园林，有"英国的凡尔赛宫"之称。它位于英国伦敦泰晤士河畔，建成于 1516 年，最初，它是红衣主教沃尔塞（Cardinal Wolsey）的府邸，沃尔塞去世以后归国王亨利八世（Henry Ⅷ）所有。

[3]　Blenheim 布伦海姆宫：位于英格兰牛津郡伍德斯托克（Woodstock, Oxfordshire），建于 1705 年至 1722 年之间，其苑园由亨利·怀斯（Henri Wise）设计，仍然采用勒·诺特尔式样。1764 年布朗承接改造任务，他采用以草地铺到门前的手法，重新塑造地形，将草坪一直延伸到巴洛克式宫殿前，以蜿蜒的流线形湖泊、接近自然的驳岸形成风景式苑林。

[4]　Castle Howard 霍华德庄园：位于英格兰的北约克郡（North Yorkshire），是英国最负盛名的私家府邸之一，其主体建筑建于 1699 年至 1712 年间，由著名的巴洛克建筑师约翰·温布鲁爵士（Sir John Vanbrugh）设计。庄园面积有 2 000 多公顷，地形高低起伏，园中点缀着许多珍贵的建筑物。霍华德庄园表明了 17 世纪末规则式园林开始衰落、向风景式园林演变的发展迹象。

[5]　Stowe 斯陀园：位于英格兰的白金汉郡（Buckinghamshire），1714 年由查尔斯·布里奇曼（Charles Bridgeman）设计。

[6]　Louis ⅩⅣ 路易十四（1638—1715）：法国波旁王朝国王，在位期间为了满足自己的穷奢极欲，路易十四修建了宏伟的凡尔赛宫苑（Versaille）。

[7]　parterres de broderie 刺绣花坛：指法国古典园林中常用的绣毯式花坛。

[8]　Sir Christopher Wren 克里斯托佛·雷恩爵士（1632—1723）：英国天文学家、建筑师、几何学家，其最主要的贡献是在 1666 年 9 月英国伦敦大火后，主持设计了 87 座重建教堂中的 55 座，他还设计了皇家肯辛顿宫、汉普顿宫、格林威治天文台等。

[9]　Henry Wise 亨利·怀斯：英国造园家、园林设计师，曾被任命为乔治一世和安妮王后（George Ⅰ & Queen Anne）的皇家园林师。1701 年，他主持重建了威廉三世国王汉普顿宫苑的国王花园（Privy Garden），还参与过

Castle Howard、Blenheim Palace 和 Chatsworth 等庄园的设计。

[10] James Brydges, the Duke of Chandos 詹姆斯·布里吉斯，钱多斯公爵(1673—1744)。

[11] Cannon, Middlesex 米德萨克斯郡的卡努村：米德萨克斯郡位于伦敦南部，英格兰的历史城镇。

[12] Pope 蒲柏(1688—1744)：全名为亚历山大·蒲柏(Alexander Pope)，被誉为 18 世纪英国最伟大的诗人。他曾经发表了不少有关建筑和园林设计的文章，为推动英国风景式园林的探索与形成做出了一定的贡献。

[13] Timon's Villa 帝门庄园：它是蒲伯在讽刺诗中杜撰出来的一处庄园名字。蒲伯对规则式园林极不赞赏，故在诗句中对该庄园批评道："把树木修剪成人物雕像，而雕像陈列又密如树木。"(Trees Cut to Statues, Statues thick as trees, 引自 Timon's Villa：Pope's Composite Picture, by James R. Aubrey)

[14] Epistle to Burlington 致伯灵顿的信：这是蒲伯在 1731 年写给好友伯灵顿的信。

[15] Queen Anne 安妮女王(1665—1714)：大不列颠王国女王，斯图亚特王朝末代国王。

[16] Charles Bridgeman 查尔斯·布里奇曼(1690—1738)：布里奇曼是斯陀园(Stowe)的设计师，他采用了 17 世纪 80 年代样式，既未完全摆脱法式园林的影响，但又抛弃了对称原则的束缚。1715 年后，花园规模迅速扩大，点缀着建筑和庙宇。他率先运用非整形对称的种植方式及线条柔和的园路，不再使用植物雕塑。布里奇曼还在斯陀园巨大的园地周围布置一道隐垣(Ha-Ha)，使视线得以延伸到园外的风景之中，他因此成为规则式向风景式造园发展的开拓者。

[17] Kent 肯特(1686—1748)：全名威廉·肯特(William Kent)，英国的早期风景造园家、画家，肯特将"自然嫌弃直线(Nature abhors a straight line)"这句话作为口号，打破规则的式样，对直线的园路和行道树、喷泉和绿篱等一概排斥，只保留带有不规则形状的池岸和弯曲的河流。他真实地复制自然，在肯辛顿园(Kensington Garden)中甚至种上枯树；他还补充设计了布里奇曼的斯陀园。

[18] Rousham in Oxfordshire 牛津郡的罗莎姆庄园：庄园的府邸是英国文艺复兴运动后期雅克布式风格(Jacobean style)。1719 年布里奇曼受雇对府邸的宅园进行改造，他尝试运用有别于传统的、更为自然的方式来营造景物，改造工作一直到 1737 年才结束。改造后的罗莎姆庄园，也一举成为广受赞誉的名园。

[19] Poussin 普桑(1594—1665)：全名尼古拉斯·普桑(Nicolas Poussin)，法国 17 世纪巴洛克时期的著名画家，他创作了许多风景画，这些风景画给英国风景式园林提供了启迪。《阿尔卡迪的牧人》(Les Bergers d'Arcadie)是其代表作。

[20] Claude 克劳德(1600—1682)：全名克劳德·劳伦(Claude Lorrain)，法国著名画家，尤以风景画见长。与普桑一样，他的作品对英国风景式园林的形成与发展起到了重要的影响作用。

[21] Neo-Palladian building 新帕拉迪奥式建筑：安德烈亚·帕拉迪奥(Andrea Palladio 1508—1580)，是意大利文艺复兴时期著名的建筑师。其作品充分挖掘和创造性地运用了古希腊与古罗马的建筑样式，设计风格以华丽、高雅、隽秀见长，被誉为帕拉迪奥式。1570 年他出版了《建筑四书》(I quattro libri dell'architettura)一书，被认为是西方建筑史上最有影响力的人。

[22] Elysian Fields 极乐园：Elysium(亦称 Elysian Fields)是希腊神话中死后居住的乐土和极乐世界，极乐园是斯陀园中的一个景区，可分为三个部分：最北面的部分有一条被称为 Alder 河的窄湖，由一水坝将其余第二部分分开；第二景区以名为 Worthies(或称为 Styx)河的带状湖区为主，由另一条水坝使其最后一部分相区别；第三景区以一个八角状湖池为主。

[23] Temple of Ancient Virtue 古代道德之庙：庙内设有荷马(Homer)、吕库古(Lycurgus，古希腊的一位智者、法律家)、苏格拉底(Socrates)和埃帕米浓达(Epaminondas，古希腊底比斯的将军、政治家)四人的塑像。

[24] Shrine of British Worthies 英国名人纪念龛：龛内共有 16 尊历史名人的雕像，其中包括莎士比亚(Shakespeare)、洛克(Locke，英国著名哲学家)、牛顿(Newton)、蒲柏(Pope)和阿尔弗雷德国王(King Alfred，英格兰盎格鲁-撒克逊时期的国王)等。

[25] Temple of Modern Virtue 新道德之庙。

[26] Sir Robert Walpole 罗伯特·沃波尔伯爵(1676—1745)：全称为罗伯特·沃波尔一世奥福德伯爵，英国辉格党政治家，他被认为是英国历史上第一位首相。

[27] Whig 辉格党:英国历史上的一个资产阶级政党,活跃于17世纪后期至19世纪中期。1679年,因反对具有天主教背景的约克公爵詹姆斯(即后来的詹姆斯二世)继承王位,被政敌称为"辉格"。其名称可能是"Whiggamores",意为"好斗的苏格兰长老会派教徒"一词的缩写。

[28] Cobham 科伯海姆子爵(1675—1749):英国18世纪著名政治家。蒲柏对他的宅园大为赞赏,1733年曾在诗歌 Epistle to Burlington 中称其为奇观。

[29] Swift 斯威夫特(1667—1745):全名为乔纳森·斯威夫特(Jonathan Swift),英国爱尔兰作家,讽刺文学大师。

[30] Tory 托利党:17、18世纪英国资产阶级政党,与辉格党的立场相对立。

[31] Temple of Liberty 自由之庙。

[32] Gothic style 哥特式风格:该词出现于文艺复兴时期,当时人们认为古希腊和古罗马的艺术是正统艺术,而哥特式风格则被认为是"野蛮民族"式的,故贬称为"哥特式"。事实上,哥特式建筑与哥特人并无关系。

[33] Gibbs 吉伯斯(1682—1754):全名为詹姆士·吉伯斯(James Gibbs),英国18世纪著名建筑师。

[34] Twickenham 特维臣海姆:位于伦敦西面的一个小城镇。

[35] Rococo 洛可可风格:源于18世纪法国的一种艺术风格,最初是为了反对宫廷的繁文缛节艺术而兴起的,后被新古典主义所取代。该词是由法文 Rocaille 和意大利文 Barocco 合并而来。Rocaille 是以岩石和蚌壳装饰为其特色的风格,而 Barocco 即巴洛克(Baroque)。

[36] Grotto 洞府或岩洞:这里主要是指为避暑或休闲娱乐而兴建的人工洞穴,它常是意大利花园的一个特征。

[37] Thames 泰晤士河:即 River Thames,位于英格兰东南,是英格兰最长的河流。

[38] Chinese House at Shugborough 舒巴勒庄园的中国阁:位于斯塔福德郡(Staffordshire)舒巴勒庄园内的一中式小建筑物,保持至今,附近尚保留有一中式小桥。

[39] House of Confucius at Kew 邱园的孔子亭:邱园(Kew garden)位于伦敦泰晤士河南岸,始建于1759年,由英国皇家建筑师钱伯斯(William Chambers)设计,他首次运用"中国式"手法,并在其中建造了一些中式的建筑,以中国式塔最为知名。现在的邱园已成为英国皇家植物园(The Royal Botanic Gardens)。

[40] Sir William Chambers 威廉·钱伯斯勋爵(1723—1796):苏格兰建筑师、造园家。1740—1749年,钱伯斯曾多次前往中国旅行,研究中国建筑和中国园林艺术。

[41] Canton 广东:英语中对广东(Guangdong)的旧称。

[42] Designs of Chinese Buildings, Furniture, Dresses《中国建筑、家具、服装设计》:欧洲第一部介绍中国建筑的专著。

[43] Dissertation On Oriental Gardening《论东方园林》:这是钱伯斯1772年出版的以介绍中国园林为主的书,对中国园林在英国和欧洲的流行起到了极大的推动作用。

[44] Edmund Burke 埃德蒙·伯克(1729—1797):爱尔兰政治家、作家和哲学家,曾在英国下议院担任了数年辉格党的议员。

[GLOSSARY]

formal garden	规则式园林	maze	迷园	painterly manner	绘画式手法
Gothic style	哥特式风格	landscape gardening	风景造园	Rococo	洛可可风格

[NEW WORDS]

allegedly	adv.	依其陈述,据说,据称	pagoda	n.	宝塔,塔式庙宇
allygorical	adj.	寓言的,寓意的,讽喻的	palatial	adj.	宫殿(似)的,宏伟的,壮丽的
artifice	n.	技巧,手段,诡计,欺骗	parterre	n.	花坛,花圃
autocratic	adj.	独断专横的,独裁的,专制的	patriarchal	adj.	家长的,威严的,宗法封建性的
bilateral	adj.	两边的,在两边的	plot	n.	小块土地
bust	n.	半身雕像,胸部	preside	vi.	主持,主管,掌握,管辖
camera obscura	n.	暗箱	prolific	adj.	多产的,多育的,充裕的,丰富的
carpet	vt.	用地毯铺地	punctuate	vt.	不时打断,在……中加标点符号
cluster	n.	(植物的)簇,丛	raw	adj.	天然的,未加工过的
commemorate	vt.	纪念,庆祝	resemble	vt.	与……相似,类似,像
compartment	n.	区划,划分,分隔,隔间	restraint	n.	抑制,限制,约束
contemporary	n.	同时代的人	reverence	n.	尊敬,崇敬,敬仰,敬礼
dip	vi.	向下扩展,下沉,沉没	ridge	n.	脊,山脊,分水岭
dissertation	n.	专题论文,学位论文	rigid	adj.	严格的,死板的,不变的
dominant	adj.	占优势的,统治的,支配的	rood	n.	基督受难像,十字架
eliminate	vt.	除去,除掉,消灭掉,略去,忽略	ruin	n.	废墟,破败,坍塌,毁坏,毁灭
Elysian	adj.	乐土的,像天堂的,幸福的	satirize	vt.	讽刺,讥讽
embroidery	n.	刺绣,刺绣品	severity	n.	严格,严厉,严重,严峻
enquiry	n.	询问,查询	shrine	n.	圣地,圣坛,神龛,神殿
epistle	n.	书信,书信体诗文	sparkling	adj.	闪耀的,闪闪发光的,发火花的
espouse	vt.	采纳,支持	stretch	n.	伸展,延伸,延续
evoke	vt.	产生,引起,唤起,刺激,激发	sublime	adj.	高尚的,崇高的,令人崇敬的
execute	vt.	实施,执行	surpass	vt.	胜过,超过,胜于
folly	n.	装饰性建筑	symmetry	n.	对称(性),匀称,整齐
gravel	n.	砂砾,砾石,铺石子	testify	vt.	说明,表明,作证,证明,表白
grotto	n.	岩洞,洞穴;洞室,石窟	theatrical	adj.	戏剧的,剧场的,戏剧性的
impose	vt.	强迫,强加	thicket	n.	灌木丛,丛林
inheritor	n.	继承人,后继者	tyranny	n.	暴虐,暴行,暴政,专制统治
literally	adj.	确实的,真正的,逐字的	unrestrained	adj.	不受抑制的,无拘束的
obelisk	n.	方尖石塔,短剑号,疑问记号	upkeep	n.	保养,维持,维修,保养费
ostentation	n.	炫耀,卖弄	wavy	adj.	波浪形的,波纹状的
overt	adj.	不隐瞒的,公开的	worthy	n.	杰出人物,知名人士

[参考译文]

英国风景园林
罗伯特·维奥博士

[1] 确实是,从17世纪末到19世纪初,也就是说,在经常被提及的"漫长"的18世纪(1660—1840),园林设计美学逐渐地从约束、控制、限定和讲究秩序转变为强调自由和开放。从17世纪后期凡尔赛宫和汉普顿宫严谨的几何布局,到布伦海姆宫、霍华德庄园和斯陀园布局清晰的自然风格,18世纪中期时,园林的设计几乎都发展成为对开放式景观或原始自然的模仿。

[2] 在18世纪初期,主导景观设计的人物是勒·诺特尔,即为路易十四设计凡尔赛宫苑的主设计师。17世纪和18世纪早期最受欢迎的园林设计是法国、意大利及荷兰的规则式花园,布置成两侧对称的布局。但在规则式布局的实践中,没有人能超越勒·诺特尔。

[3] 在这种风格的园林中,靠近宫殿或府邸部分的花园以建筑式手法处理,宛如建筑物的另一个延伸空间。花园由一系列完美而规则的几何空间组成,这些空间由精心修剪过的灌木和乔木、笔直的砂石步道、石板小径、台地及台阶划分。最常见的是刺绣花坛(如同刺绣图案一般的场地),它们由低矮的常绿植物(通常为黄杨)、花卉(实际上在19世纪之前还不常见)、染色的土壤、红砖粉末、煤灰粉末、白色和黄色沙子等铺就而成。在超大型的园林中,视线所及之处都遍布着这种严谨的几何构图。园路以几何图案顺着水渠和林荫道延伸和放射。喷泉、雕像、迷园、林地和树丛都以建筑正中心延伸出的中轴线为基准,对称地布置在轴线两侧。

[4] 这种造园风格的杰作是凡尔赛宫苑。它成为整个欧洲皇室园林的样板,这其中就包括为国王威廉三世所设计的园林,位于克里斯托弗·雷恩爵士的汉普顿宫新东门的前方。

[5] 在17世纪末期,勒·诺特尔风格在英国的追随者是亨利·怀斯(1653—1738),汉普顿宫的首席造园师之一。他也为钱多斯公爵詹姆斯·布里吉斯位于米德萨克斯郡卡努的精美豪华的花园工作,而该花园被与蒲柏同时期的人们认为是《致伯灵顿的信》中的帝门庄园。这类花园的风格是"专制型":宏伟壮丽从象征至高权力的座椅向外辐射,严格的秩序主宰着自然,使其屈服于人的意志之下。

[6] 维护汉普顿宏大苑园的巨额支出,最终导致安妮女王责令怀斯削减其三分之二的费用。这样在1704年,原有的黄杨植坛被开阔的草坪所取代,这是迈向更加自由的园林风格的一步,而这种自由式风格也随后主导了这个世纪剩余的大部分时间。

从规则式园林向风景式园林的转变

[7] 作为对17世纪后期严谨规则的法国和意大利式园林的回应,一种更加自由的新型园林出现了。那些最终成为这种不规则园林风格的倡导者们反对对称和炫耀,他们将法式风格与法国政府的暴政相联系,认为其过于专制。这样,在英式园林设计中不断滋长的自由形式逐渐与英国政权所体现出的自由联系在一起。园林美学具有了政治含义,有时就如斯陀园一例,具有公然的政治意味。

查尔斯·布里奇曼与斯陀园

[8] 查尔斯·布里奇曼接替怀斯成为皇家造园师。他在风景园林设计上最卓著的成就是,从1713年开始完全在其自身指引下完成的位于斯陀园的著名公园。这一风景园林的设计杰作,后来又融入了肯特和"万能"布朗的贡献。

[9] 在园林风格上,布里奇曼介于勒·诺特尔与"万能"布朗之间。在18世纪20年代肯特引领了被称为绘画式的风景造园。他最出名的绘画式园林是位于牛津郡的罗莎姆庄园。在18世纪20年代布里奇曼确定了该花园的主线,并为肯特在18世纪30年代的工作铺平了道路。绘画式手法试图营造出普桑和克劳德风景画中剧场般的效果。

[10] 斯陀园是一处具有政治寓意的风景园林。一方面,它赞美了18世纪社会坚实的阶级基础,这可以从新帕拉迪奥式建筑和大量新帕拉迪奥式园林的纪念物和装饰性建筑中体现出来。另一方面,对园林空间自由和开放的处理手法上,斯陀园也体现出了18世纪理论家根据古代英格兰原则所倡导的自由。

[11]　但斯陀园的政治色彩更加鲜明且更有针对性：它通过隐喻的纪念物表达出相反的政治诉求。一处被称为"极乐园"的开阔低洼地，位于两个山脊之间。在山脊的一端坐落着由肯特1734年设计的"古代道德之庙"，里面陈列着真人般大小的荷马、吕库古、苏格拉底和埃帕米浓达的雕像。从稍低处对着它的也是由肯特设计的"英国名人纪念龛"，其中陈列了16位民族英雄的半身像，包括现代伟人如莎士比亚、洛克、牛顿和蒲柏及古代伟人阿尔弗雷德国王。"英国名人纪念龛"确实是朝"古代道德之庙"翘首，强烈地表现出了对古典思想的崇拜。然后其附近还有第三处建筑物，即"新道德之庙"，这处废墟式建筑据说是讽刺辉格党的政府首相罗伯特·沃波尔爵士，他也是科伯海姆、蒲柏、斯威夫特和其他托利党作家深恶痛绝的对象。"自由之庙"是哥特式风格，它由充满古代英格兰理想的建筑师、园林设计师吉伯斯(1741)设计。

亚历山大·蒲柏和他的特维臣海姆花园

[12]　蒲柏被称为18世纪20—30年代造园变革的主持天才。他自己位于特维臣海姆的花园以及他为那些富有的朋友们提供咨询建议的花园都印证了其非凡的影响力。

[13]　蒲柏的花园保留下来的很少。我们从绘画和文字描述中得知，他的花园完美地体现了来自"致伯灵顿的信"中的造景原则。对蒲柏而言，风景园林的营造是一种对其内心"罗曼蒂克式"冲动的想象力表达行为。

[14]　布里奇曼所开创的园林设计手法基于相对规则的笔直中轴线，配以两侧区域的非规则式布局，由此对称与均衡同各种变化结合了起来。蒲柏改良了这一设计原则并将其应用于从他位于特维臣海姆的别墅、横穿伦敦路的自己的小花园场地上。花园的边界隐藏在茂密树丛中，从而围合出封闭的不规则式花园，里面既有古代纪念物也有现代纪念物。在花园的东部尽头伫立着一个带有洛可可式圆顶的"贝壳庙"，在西部及花园更阴暗的尽端矗立着一座方尖碑，用以纪念他逝去的母亲。花园里伦敦路的地下有一条通道，从这里可以到达位于蒲柏别墅地下室的洞府。从花园尽头的洞府向外望去，朝向泰晤士河的开阔草坪和乡间景色尽收眼底。当洞府的门关闭时，它就变成了暗箱，洞府墙壁上闪闪发光的贝壳和小块镜子呈现出无数景象，这是真正无与伦比的"富有诗意的"梦幻建筑。

[15]　也许，受到蒲柏最显著的间接影响的就是斯陀园，即科伯海姆勋爵400英亩的园林，像布里奇曼、温布鲁、肯特、布朗等许多风景造园师、建筑师、雕塑师都曾经在那里工作长达60年。

中国与日本的影响

[16]　总体上18世纪的园林装饰要么是古典式要么是哥特式，但贯穿这一世纪，东方风格逐渐开始融入到景观设计之中，例如它们曾被运用在十字架装饰中。18世纪40年代，舒巴勒庄园里建了中国阁，邱园中建了孔子庙。18世纪50年代，英国建了许多宝塔、凉亭、小屋，以及中式桥，例如汉普顿宫苑中横跨泰晤士河的桥。到了18世纪50年代，由法国人撰写的描述北京皇家宫苑的书以英文出版。建筑师威廉·钱伯斯勋爵年轻时曾到访过中国广东，并在18世纪50年代出版了《中国建筑、家具、服装设计》等书(1757)，随后又发表了《论东方园林》(1772)。钱伯斯强烈主张园林设计中的多样化，这也使许多人认为这是他对日益受到欢迎的兰斯洛特·"万能"布朗的园林设计所表现出的反对，而布朗正是当时18世纪后半叶最受欢迎且最高产的设计师。

兰斯洛特·"万能"布朗

[17]　在兰斯洛特·布朗"万能"的手中，园林设计中摒弃了几乎所有的规则形式和人工痕迹。在布伦海姆庄园，他清除了由亨利·怀斯所设计的、华丽的勒·诺特尔风格花坛，取而代之的是开阔绵延的草坪，草坪一直延伸至建筑墙脚；建筑附近栽种了深色树木，形成从建筑向外眺望的景框。对于同时代的像钱伯斯那样的人而言，布朗的花园"与一般的原野没有什么太大差距，几乎就是对普通自然的复制"。

[18]　布朗通过种植广阔的草坪，并在其间不时地点缀小丛树林，或在起伏的绿带上不规则地孤植树木，营造出仿佛未被设计过的自然景观效果。地面从建筑向蜿蜒的湖面下降，又在远处树林之外升起，完全是一幅"风景画"。

[19]　18世纪园林发展的最后阶段，是在埃德蒙·伯克1757年所著的《论崇高与美概念起源的哲学探究》巨大影响力下产生的。这部著作深刻地影响到了哥特式文学的兴起、哥特式复兴建筑和由伯克关于崇高或令人

生畏的自然的见解所模式化的景观设计。

[思考]
(1) 古埃及园林有哪几种主要形式？园林中常采用的植物有哪些？
(2) 法国勒·诺特尔式园林有哪些典型的造园要素？
(3) 英国风景式园林哪些特点？
(4) "万能"布朗对英国风景式园林的发展有哪些影响？

UNIT 4 ISLAMIC GARDEN

TEXT

Islamic Garden
伊斯兰园林

[1]　The first Muslims came from the Arabian Desert around sixth century AD. As far as is known, there is no history of garden as we understand them in Arabia at this period. It was not until Islam conquered other countries with civilizations of their own, in particular, Persia, that the Islamic garden can be said to have been born. Here, Islam absorbed the already well-established tradition of hunting-parks and royal pleasure gardens and invested them with a new spiritual vision.

[2]　The typical Islamic garden is a representation of heaven on earth. It developed on the Persian plain, which was an arid desert. The environment was harsh—a paradise garden was its opposite. The Islamic notion of paradise included water, shade, flowers and fruit trees. It was an enclosed garden, shutting out the harshness of the surrounding landscape. Water is regarded as the life of garden and the soul of gardening.

[3]　The basic form of an Islamic garden is based upon intersecting canals forming four quadrants. This was based on an ancient cosmological idea that the universe was made of four quarters divided by two great rivers (See the Biblical description of Eden). The creation of water scenery and the gardening art spread to North Africa and all over of Spain with the expedition of the Islamic Army. By the 13th century it was spread to Northern India and Kashmir.

[4]　Throughout the early history of Islamic gardens, there are two developing periods, Moorish gardens[1] and Mughal gardens[2].

[5]　In 711 C. E. the first Muslims from North Africa (Moors) crossed the Straight of Gibraltar[3] to Spain and the Caliph of al-Andalus[4] was established in 750 C. E. Granada[5] was to remain in Moorish hands until the Moors were expelled by the Christian in 1492 when gardens developed Islamic traditions constantly, absorbed some Roman features and formed their own features. During the period the Arabs made sure to emphasize a strong Arab culture in al-Andalus and gardens and architecture were a personification of this attempt. This meant that gardens in Spain would have embodied some Spanish-Christian, Roman, and even possibly Jewish ideas that had been transformed into an Islamic layout, specifically the type from Persia. Two most significant Moorish gardens are the Alhambra[6] and the Generalife[7] built during 13th and 14th centuries, both outside Granada, in al-Andalus.

[6]　The Alhambra (Fig. 1) is a palace and fortress complex of the Moorish rulers of Granada in southern Spain, occupying a hilly terrace on the southeastern border of the city of Granada. It was completed towards the end of Muslim rule in Spain by Yusuf Ⅰ (1333-1353) and Muhammed Ⅴ, Sultan of Granada (1353-1391). The Alhambra mixes natural elements with man-made ones, and is a testament to the skill of Muslim craftsmen of that time. The Patio de los Arrayanes[8] (Fig. 2) and the Patio de los Leones[9] (Fig. 3) are its two main courts. In the Patio de los Arrayanes there are a shallow and flat rectangular reflecting pond. On both sides of the pond with two lines of the hedge Myrtaceae, which is the origin of the court's name. The Patio de los Leones is a classic Arab-style garden. It is divided into four sections by two channels. In the centre of the court is the Fountain of Lions, a alabaster basin supported by the figures of twelve lions in white marble, not designed with sculptural accuracy, but as symbols of strength and courage. Water spilled from the mouth of lions that flows through these two channels together in the Circuit Court of the four corridors.

Fig. 1　Alhambra　　　　Fig. 2　Patio de los Arrayanes　　　　Fig. 3　Patio de los Leones

[7]　The Generalife (Fig. 4) was the summer palace and country estate of the Nasrid sultans[10] of Granada (Fig. 3). The palace and gardens were built during the reign of Muhammad Ⅲ (1302-1309) and redecorated shortly after by Abu I-Walid Isma'il (1313-1324). The Patio de la Acequia[11] (Water-Garden Courtyard) is the central garden and the highest point of all gardens in Generalife. A narrow and almost 50 meters long pond is the garden's focal point. There are four flowerbeds along both sides of the pond. The four beds are divided by a path, which crosses the pond at about the center of the garden. This path created a cross axis from a small mosque which no longer exists; it also allowed worshippers easy access to the pond for their ablutions. The Patio de la Acequia is thought to best preserve the style of the medieval garden in Al-Andalus.

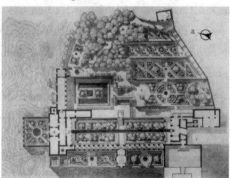

Fig. 4　Generalife garden and Patio de la Acequia

[8] There are notable differences from earlier Islamic gardens. These Spanish gardens were not built in the desert, therefore there was less inclination to shut out the external environment.

[9] The next great phase of garden-building began after the 16th century when Babur set up what was called the 'Mughal' Empire. The Mughal Emperors—Babur, Akbar, Jahangir and Jahan—were all extremely keen on creating gardens. The Islamic architectural styles were also expressed in their gardens. Two of the most famous works—Shalimar Gardens[12] in Lahore and the Taj Mahal[13], Agra—were completed by Mughal emperor Shah Jahan in the middle of the 17th century.

[10] The Shalimar Gardens (Fig. 5) are laid out in the form of an oblong parallelogram, surrounded by a high brick wall, which is famous for its intricate fretwork. It had a very good water supply and drainage system. In the garden there are three descending terraces from south to north, which are elevated 4-5 meters one by one. To irrigate the Gardens, a Royal canal was brought from Rajpot (present day Madhpur in India). The canal intersected the Gardens and discharged into a large marble basin in the middle terrace. From this basin and the canal, rise 410 fountains.

Fig. 5 Shalimar Gardens

Fig. 6 Taj Mahal

Fig. 7 The Fin Garden

[11] Taj Mahal (Fig. 6) is a mausoleum, built by Shah Jahan in memory of his wife, Mumtaz Mahal. It is actually an integrated complex of a white domed marble mausoleum, garden and outlying buildings. The Taj Mahal is considered the finest example of Mughal architecture, a style that combines elements from Persian, Ottoman, Indian, and Islamic architectural styles.

[12] At the same time as the Mughal Emperors were building their many gardens, there was a fresh golden age of garden-building in Iran too. A powerful ruler, Shah Abbas (1587-1629) transformed Isfahan[14] into a famously beautiful city where the centre was largely composed of gardens. Its great central square is little changed today. The Fin Garden[15] (Fig. 7) at Kashan was also built by him.

[13] Fine Islamic gardens continued to be built during the 18th and 19th centuries and existing gardens were modified. But the most creative periods for garden design are those mentioned.

[14] Across the Islamic world these gardens show a great variation of style, reflecting practical and environmental factors, as well as indigenous cultural ones; such as topography, availability of water and purpose or type of garden. These range from the vast open gardens to the smaller, inward-looking courtyard gardens of traditional cities such as Fes and Damascus. Like Islamic art and architecture, however richly varied these garden styles maybe, they nevertheless retain the same principles and are expressions of the Islamic spirit.

[NOTES]

[1] Moorish gardens 摩尔园林:摩尔人是中世纪伊比利亚半岛(今西班牙和葡萄牙)、马格里布和西非的穆斯林居民,即阿拉伯人,历史上摩尔人主要指在欧洲的伊斯兰征服者,他们创造的摩尔园林最具代表性的即阿尔罕布拉宫(Alhambra)和格内拉里弗宫(Generalife)。

[2] Mughal gardens 莫卧儿园林:莫卧儿王朝(1526—1857),又名蒙兀儿王朝,是由蒙古人在印度建立的强大帝国,信奉伊斯兰教。莫卧尔园林是印度伊斯兰园林发展的鼎盛时期,最具代表性作品为泰姬陵(Taj Mahal)。

[3] Straight of Gibraltar 直布罗陀海峡。

[4] Caliph of al-Andalus 安达卢西亚的哈里发王朝:安达卢西亚(Andalusia)是西班牙最南边的一个地区,也是西班牙面积最大的一个自治区,其名称来源于阿拉伯语的 Al-Andalus。在公元8世纪到公元16世纪之间,安达卢西亚是古代阿拉伯帝国的一个行省,是古代后倭马亚王朝的主要领土。其首府科尔多瓦正是后倭马亚王朝的首都,同时也是古代伊斯兰文明、基督教文明和犹太教文明相互交融的中心。在那里,伊斯兰文明得到了发扬广大,同时,在穆斯林、基督教和犹太教徒的共同努力下,科尔多瓦成了当时西方世界最为繁荣的文明中心。

[5] Granada 格拉纳达:是西班牙安达卢西亚自治区内格拉纳达省的省会,位于内华达山脉北麓,风景如画,建筑多姿多彩,尤其是具有伊斯兰风格的阿拉伯式建筑,阿尔罕布拉宫(Alhambra)便位于此地。

[6] Alhambra 阿尔罕布拉宫:是13—14世纪摩尔人在西班牙建立的格拉纳达王国的王宫,是伊斯兰世界中保存得比较完整也比较典型的一所宫殿。这座由大小6个庭院和7个厅堂组合而成的宫苑位于地势险峻的山上,宫内园林以庭园为主,采用罗马宅园四合院庭的形式,其中最精彩的是狮子院、桃金娘庭院和石榴院。"阿尔罕布拉"阿拉伯语意为"红堡",因此阿尔罕布拉宫也被称作红堡。

[7] Generalife 格内拉里弗宫:建于1319年,作为苏丹穆罕默德三世避暑的离宫,位于阿尔罕布拉宫东边的山丘上,具有典型的伊斯兰园林的风格,尤以少胜多的理水技巧见长。格内拉里弗宫花园包括水渠庭院(Patio de la Acequia)以及苏丹花园(Jardín de la Sultana)。

[8] Patio de los Arrayanes 桃金娘庭院。

[9] Patio de los Leones 狮子院。

[10] the Nasrid sultans 奈斯尔苏丹:奈斯尔王朝是中世纪安达卢西亚阿拉伯人在格拉纳达建立的最后一个伊斯兰王朝(1232—1492)。苏丹即伊斯兰王朝的君主。

[11] Patio de la Acequia 水渠中庭:格内拉里弗宫内最有代表性的一个庭院。

[12] Shalimar Gardens 夏利玛园:位于巴基斯坦拉合尔(Lahore),建于1637年。

[13] Taj Mahal 泰姬陵:位于印度阿格拉(Agra),建于1637年,是印度伊斯兰园林最著名的代表作品,也是印度知名度最高的古迹之一,世界文化遗产。

[14] Isfahan 伊斯法罕:伊斯法罕是伊朗的第三大城市,早在玛代王国(前728—前550)就已存在,17世纪在萨非王朝阿巴斯一世的统治下发展至顶峰,一度成为首都。现存著名的景观及建筑有四十柱宫,伊玛姆广场及伊玛姆清真寺。

[15] Fin Garden 菲茵花园:位于伊朗卡珊(Kashan),建于1590年,是伊朗现存最古老的园林。

[NEW WORDS]

ablution	n.	净身礼,洗手礼,沐浴	corridor	n.	走廊,通道
alabaster	n.	条纹大理石,雪花石膏,汉白玉	cosmological	adj.	宇宙的
arid	adj.	干旱的,干燥的,不毛的,贫瘠的	descend	vi.	下降,下斜,下倾,下去
Biblical	adj.	《圣经》的,《圣经》中的	discharge	vi.	放出,流出,排出
Christian	n.	基督徒,信徒	dome	n.	圆屋顶

continued

drainage	n.	排水,放水,排水系统	Moorish	adj.	摩尔人的,摩尔人风格的
Eden	n.	(圣经)伊甸园,乐园	Moors	n.	摩尔人(非洲西北部伊斯兰教民族)
elevate	vt.	举起,提高			
estate	n.	地产,房产	mosque	n.	清真寺,伊斯兰教寺院
expedition	n.	远征,征战	Muslim	n.	穆斯林
expel	vt.	驱逐,逐出	Myrtaceae		桃金娘科植物
fortress	n.	堡垒,要塞	notion	n.	观念,看法,奇想,意图
fretwork	n.	回纹饰,回纹装饰,交错饰纹	oblong	adj.	长方形的,长椭圆形的
harsh	adj.	严厉的,严酷的,刺眼的,粗糙的	Ottoman	adj.	土耳其帝国的,土耳其人的
hedge	n.	绿篱	parallelogram	n.	平行四边形
inclination	n.	倾向,趋向	patio	n.	天井,院子,庭院
indigenous	adj.	天生的,土生的,固有的	personification	n.	化身,象征,人格化,拟人化
intersect	vi.	交叉,相交	quadrant	n.	象限,圆或其圆周的四分之一
intricate	adj.	错综复杂的	rectangular	adj.	长方形的
invest	vt.	授予,投入,投资	significant	adj.	重要的,重大的,有意义的,显著的
irrigate	vt.	灌溉	terrace	n.	台地,平台
Islam	n.	伊斯兰教,伊斯兰教徒	testament	n.	证明,见证
marble	n.	大理石	worshipper	n.	崇拜者,礼拜者
mausoleum	n.	陵墓			

[参考译文]

伊斯兰园林

[1] 第一批穆斯林来自约公元6世纪的阿拉伯沙漠。就目前所知,在此期间阿拉伯没有园林史。直到伊斯兰民族征服其他有着各自文明的国家,特别是波斯,那时才可以说伊斯兰园林得以诞生。这样,伊斯兰民族吸收了早已建立完善的猎园和皇室游乐园传统并赋予了它们新的精神理念。

[2] 典型的伊斯兰园林是人间天堂的代表。它发展于有着干旱沙漠的波斯平原,虽然外界环境恶劣,天堂园却与之相反。伊斯兰信徒向往的天国包括水、绿荫、鲜花和果树。它是一个与周围严酷景观相隔离的封闭式花园。水被认为是花园的生命和造园的灵魂。

[3] 伊斯兰园林的基本形式是基于相交水渠所分出的四部分。这源于一个古老的宇宙观,即宇宙是由两条大河分成的四个部分组成(见《圣经》中描述的伊甸园)。其所创的水景和造园艺术随着伊斯兰军的远征传播到北非和西班牙各地。到13世纪它传播到了印度北部和克什米尔。

[4] 纵观伊斯兰园林的早期历史,有两个发展时期,即摩尔园林和莫卧儿园林。

[5] 公元711年来自北非的穆斯林(摩尔人)越过直布罗陀海峡到达西班牙,公元750年安达卢西亚的哈里发王朝建立。直到1492年摩尔人被基督徒驱逐出去,格拉纳达一直被摩尔人占有,而园林的发展不断延续着伊斯兰的传统,在吸收了一些罗马特色的基础上形成了自己的特征。在此期间,阿拉伯人试图在安达卢西亚发扬强大的阿拉伯文化,园林和建筑被赋予了这种使命。这意味着这些位于西班牙,原本应该体现一些西班牙基督徒、罗马人甚至可能是犹太人思想的园林,却变成了伊斯兰布局,特别是来自波斯的园林风格。两个最重要

的摩尔园林是建于13和14世纪安达卢西亚的阿尔罕布拉宫和格内拉里弗宫,它们都位于格拉纳达城外。

[6] 阿尔罕布拉宫是一个摩尔人统治者在西班牙南部格拉纳达所建的宫殿和城堡综合体,位于格拉纳达市东南部边界一个丘陵台地上。它大约于尤素夫一世(1333—1353)和格拉纳达的苏丹穆罕默德五世(1353—1391)在西班牙的穆斯林统治末年竣工。阿尔罕布拉宫融自然元素与人工元素为一体,是那个时期穆斯林工匠技艺的充分证明。桃金娘庭院和狮子院是其两个主要庭院。在桃金娘庭院有一个浅而平的矩形反射池,池的两边各植有两排桃金娘绿篱,而庭院也因此得名。狮子院是一个经典的阿拉伯风格园林。它被两个水渠分为四个部分,庭院中心是狮子喷泉,它是一个由12只白色大理石狮子托起的石质水钵,这些石狮的设计并不追求雕塑的精确性,而是作为力量和勇气的象征。水从石狮的口中泻出,流经这两条水渠及围合中庭的四个走廊。

[7] 格内拉里弗宫是格拉纳达的奈斯尔苏丹的夏宫和庄园别墅,这个宫殿和花园是在穆罕默德三世统治时期(1302—1309)修建并由阿布尔·瓦立德伊斯梅尔(1313—1324)在不久后重新装修。水渠中庭是中心花园,亦是格内拉里弗宫内所有花园的最高点。一条狭窄的近50米长的水池是花园的焦点。沿水池两侧有四个花床。这四个花床由一条位于花园中心横跨水池的道路分割而成。这条道路从一个已不复存在的小清真寺起形成了一条横轴,它使得礼拜者方便到达水池做洗礼。水渠中庭被认为是安达卢西亚保存最完好的中世纪风格园林。

[8] 这些西班牙园林与早期的伊斯兰园林有着明显的差异,它们不是建在沙漠中,因此很少倾向于与外部环境隔离。

[9] 园林建设的另一个伟大阶段是在16世纪巴布尔建立了所谓的"莫卧儿"帝国后开始的。莫卧儿帝王巴布尔、阿克巴、查罕杰和沙贾汗都非常热衷于造园。伊斯兰建筑风格也体现在他们的园林中。两个最有名的作品是在17世纪中叶由莫卧儿帝王沙贾汗所建造的位于拉合尔的夏利玛园和位于阿格拉的泰姬陵。

[10] 夏利玛园是由高砖墙所围合而成的长方形,这道砖墙因错综复杂的格子细工而著称。它有一个非常好的给排水系统。在公园里有三个自南向北下降的平台,一个比一个逐次高出4至5米。为了灌溉花园,一条皇家运河从瑞普(现在为印度的马多普尔河)引出。运河与花园交汇并流入中间平台上一个大的大理石盛水池中。在这个盛水池和运河中,分布着410个喷泉。

[11] 泰姬陵是沙贾汗为纪念他的妻子蒙泰姬·玛哈尔而建的一个陵墓。它实际上是一个由白色穹顶大理石陵墓、园林和外围建筑物所组成的综合体。泰姬陵被认为是莫卧儿建筑最好的典范,综合了波斯、土耳其、印度的元素和伊斯兰的建筑风格。

[12] 在莫卧儿帝王们营建众多园林的同时,伊朗也迎来了一个全新的园林建设的黄金时代。一个强大的统治者尚·阿巴斯(1587—1629)把伊斯法罕变成了一个中心由花园组成的著名的美丽城市。它巨大的中心广场至今几乎没有什么变化。而在卡珊的菲茵花园也是由阿巴斯所建。

[13] 精美的伊斯兰园林在18和19世纪不断地建成,现存的园林也有一些改变。但是,园林设计最有创造力的时期就是以上所提及的那些时期。

[14] 纵观伊斯兰世界的这些园林,其风格变化多种多样,反映出实践和环境方面的因素,以及本土文化的影响,例如地形、水的运用以及花园的用途或类型。从大型的开放式园林到传统城市如菲斯和大马士革的较小的、内向型的庭园都会有所不同。像伊斯兰的艺术和建筑一样,不管这些园林风格怎样千变万化,它们仍然保留着相同的原则并表达着伊斯兰精神。

Reading Material A

Islamic Gardens of Spain
西班牙的伊斯兰园林

[1] The leaders of the Muslim army that invaded Spain in 711 C. E. were inspired by the hope of

plunder as much as by religious enthusiasm. After defeating the Visigoths, the western Goths who had in the fourth century C. E. claimed the parts of the Roman Empire now occupied by France and Spain, the Arabs established themselves within the remains of their enemies' settlements and the extensive ruins of former Roman colonies. Then, in 750, after a palace revolution, a surviving member of the Umayyad caliphate, the first dynasty of Arab caliphs (661-750), fled Damascus to found the independent emirate of al-Andalus in the southern Spanish region that is known today as Andalusia. Geographically distant, therefore, from Islamic culture in the Middle East, the Arabs in Spain formed their own brilliant and erudite society. Here in Spain, Muslim, Christian, and Jewish communities coexisted under Arab rule. The Arabs had brought with them Persian artistic concepts along with Arabic science, and they married these with Greek philosophy. In this way, Córdoba became in the Middle Ages a center of learning, renowned for such philosophers as Ibn Rushd, known in the West as Averroës[1] (1126-1198), a rationalist who translated Aristotle and anticipated Thomoas Aquinas[2] (1226-1274).

[2] While southern Spain has the same hot climate as many other parts of the Islamic geographical sphere, its topography is for the most part more rugged and its land well watered and more fertile. Moorish engineers, rebuilding and extending the remains of Roman aqueducts all around them, created intricate irrigation systems by introducing the *noria*, a bucket scoop waterwheel that raised water into elevated canals. This greatly increased the amount of arable land and facilitated the construction of numerous gardens. Agriculture prospered, and famous treatises on the subject were written, such as the one by Abu Zakariya, a thirteenth-century agronomist and botanist in Seville. Farmers introduced sugarcane and brought into cultivation rice, cotton, flax, silkworms, and many kinds of fruit. Both the valley of the Guadalquivir and the plain of Granada became densely settled zones with a rich tapestry of agriculture and gardening, their air fragrant with the scent of orange, lemon, and citron trees.

[3] Topography and contact with the abundant remains of Iberian Rome served to modify the conquerors' Syrian architectural heritage with its previously assimilated Roman and Persian forms. But Spanish Muslim architecture never aspired to the monumental three-dimensionality of architecture in the West. Instead, its aim was atmospheric, and its highest achievement lay in the interweaving of indoor and outdoor spaces in sensuous and sophisticated ways. Thus, in a typical Moorish house and garden, the patio and the structural spaces surrounding it interpenetrate one another in a casual yet studied congress of architecture and nature. Ceramic tiles, with their brightly glazed, reflective surfaces, produced an effect of cool airiness.

[4] The spatial interpenetration of indoors and outdoors was not confined to fortress palaces or homes such as are found today in the Albaicín district of Granada, where the durable plan of the Roman peristyle was appropriated and reinterpreted as the carmen, the inward-focused house and garden of Arabic Spain. The same planning principle can be seen in the design of the Great Mosque at Córdoba[3] (Fig. 1), begun in 785-786. The mosque (now a cathedral), has a hypostyle interior in which the columns supporting a vast number of horseshoe-shaped double arches are evenly placed throughout like rows of trees in an orchard. Originally the wall on this side of the mosque opened via

a facade of arches to a 3-acre courtyard planted in avenues of orange trees, which were aligned with the columns of the mosque's interior. An underground cistern supplied the fountains and irrigation channels in this court, called the Patio de los Naranjos, or Court of the Oranges. Unfortunately, the underground cistern has long since been converted into an ossuary, and Christian chapels have filled in the arches that once opened onto this patio.

[5]　The Alcázar[4] (Fig. 2) in Seville, which dates from about a hundred years after the Christian conquest of that city in 1248, exemplifies the extensive Islamic influence upon subsequent Spanish architecture. Using the considerable resources of Islamic craftsmanship, the king of Castile, Pedro the Cruel (ruled 1334-1369), built his fortified palace on the ruins of the former Islamic citadel. The Alcázar gardens consist of a series of enclosed patios. Though altered in subsequent centuries, the Moorish origins of these gardens are evident in their high enclosing walls, geometrical layout, raised paths, numerous fountains, brightly glazed tiles, and plantings of cypress, palm, orange, and lemon trees.

Fig. 1　Great Mosque at Córdoba

Fig. 2　Alcázar

[6]　Granada was the last Arabic city recaptured by Christian monarchs, falling to Ferdinand and Isabel in 1492. Its remaining Arabic gardens—the Alhambra and the Generalife—which date from the thirteenth and fourteenth centuries, are among the oldest gardens in the world and enjoy legendary status.

[NOTES]

[1]　Averroës 阿威罗伊：生于西班牙科尔多瓦，著名的阿拉伯哲学家，他因曾对亚里士多德的著作写了一长篇有见解的评论而闻名。

[2]　Thomas Aquinas 托马斯·阿奎纳：意大利神学家，欧洲中世纪经院哲学代表人物。他是自然神学最早的提倡者之一，也是托马斯哲学学派的创立者，成为天主教长期以来研究哲学的重要根据。他所撰写的著作《神学大全》是基督教巨著之一。

[3]　Great Mosque at Córdoba 科尔多瓦清真寺：位于西班牙南部古城科尔多瓦市内，具有摩尔建筑和西班牙建筑的混合风格，是西班牙伊斯兰教最大的神圣建筑之一。公元前786年前后，白衣大食王国国王阿卜杜勒·拉赫曼一世欲使科尔多瓦成为与东方匹敌的宗教中心，在罗马神庙和西哥特式教堂的遗址上修建了这座清真寺。

[4]　Alcázar 阿尔卡扎宫：位于西班牙塞维利亚市中心，最早是座建于公元1世纪的阿拉伯城堡，1364年扩建成西班牙国王的行宫，由于阿尔卡扎宫随着朝代更迭和建筑风格的变化而不断修补扩建，因而混合着哥特式教堂风格、伊斯兰教风格和巴洛克风格等，是塞维利亚最为精美的建筑之一。

[NEW WORDS]

agronomist	n.	农学家,农艺学家	Goth	n.	哥特人
airiness	n.	通风,空气流通	Guadalquivir	n.	瓜达尔基维尔河
Albaicín	n.	阿尔巴辛(西班牙格拉纳达一个区)	hypostyle	adj.	多柱式建筑(的)
Andalusia	n.	安达卢西亚	Iberian	adj.	伊比利亚的,伊比利亚人的
aqueduct	n.	沟渠,导水管	interpenetrate	v.	互相渗透,贯穿,扩散
arable	adj.	可耕的,适于耕种的	monarch	n.	君主,帝王
assimilate	v.	吸收,消化,同化	noria	n.	戽水车
botanist	n.	植物学家	ossuary	n.	藏骨罐;藏骨堂;骨灰瓮
Caliph	n.	哈里发(伊斯兰领袖的称号)	peristyle	n.	柱廊式,列柱走廊,列柱围廊式
caliphate	n.	伊斯兰王权	plunder	n.	抢劫,偷盗
Castile	n.	卡斯提尔(古代西班牙北部一王国)	rationalist	n.	唯理论者,理性主义者
cathedral	n.	大教堂	recapture	vt.	夺回,取回,收复
ceramic	adj.	陶器的,陶瓷的	rugged	adj.	高低不平的,崎岖的
chapel	n.	小礼拜堂,礼拜	ruin	n.	毁坏,破败,废墟,遗迹
cistern	n.	水塔,蓄水池	scoop	n.	铲子
citadel	n.	城堡,堡垒	sensuous	adj.	感觉上的,给人美感的
citron	n.	香橼,圆佛手柑	Seville	n.	塞维利亚(西班牙古都)
Cordoba	n.	科尔多瓦(西班牙城市)	sugarcane	n.	甘蔗,糖蔗
cypress	n.	柏树,柏树属植物	Syrian	adj.	叙利亚的,叙利亚人
Damascus	n.	大马士革(叙利亚首都)	tapestry	n.	织锦,挂毯
emirate	n.	阿拉伯酋长之职位或阶级,酋长国	tile	n.	瓷砖,瓦片
erudite	adj.	博学的	treatise	n.	专题著作,专题论文,专著
facade	n.	(房屋的)正面,外观,表面	Umayyad	n.	倭马亚王朝
flax	n.	亚麻	Visigoth	n.	西哥特人
fortified	adj.	加强的,增强防御的	waterwheel	n.	水车,吊水机,水轮

[参考译文]

西班牙的伊斯兰园林

[1]　公元711年穆斯林军队的领袖在宗教热情和劫掠期冀的激发下入侵西班牙。西哥特人在4世纪曾宣称部分罗马帝国现在由法国和西班牙占有,在战胜了西哥特人之后,阿拉伯人在西哥特人的遗址上和前罗马殖民地的大片废墟中建立了自己的家园。然后,在750年的一场宫廷政变之后,阿拉伯哈里发王朝(661—750年)一个幸存的倭马亚哈里发成员逃离大马士革,在西班牙南部地区建立了独立的安达鲁斯酋长国,即现在的安达卢西亚。因此,与中东伊斯兰文化的地理隔离,使阿拉伯人在西班牙形成了他们自己辉煌博大的社会文化。西班牙穆斯林、基督教和犹太教社会在阿拉伯统治下共存。阿拉伯人带来了波斯艺术及阿拉伯科学,并将其与希腊哲学结合。通过这种方式,科尔多瓦成为中世纪的知识中心,有着闻名世界的哲学家,如唯理论者伊本·鲁世德,西方人称之为阿威罗伊,他翻译了亚里士多德的作品,并介绍了托马斯·阿奎那。

[2]　尽管南部西班牙有着与伊斯兰世界大部分地区同样炎热的气候,但大部分的地形却更崎岖,土地水分更充足,土壤更肥沃。摩尔工程师重建并延长了罗马遗留下来的沟渠,创建了错综复杂的灌溉系统,通过戽水车,一种斗勺水车,将水提升至抬高的运河。这种方法大大增加了可耕作土地,并促进了大量的园林建设。农业繁荣发展,促使了农业论著的出现,例如13世纪塞维利亚农学家及植物学家阿布·扎卡里亚的著述。农民们引

进了甘蔗,开垦种植水稻、棉花、亚麻、蚕和各类水果。瓜达尔基维尔山谷和格拉纳达平原成为人口密度很高的聚居地,农田和花园犹如织锦一般,空气中弥漫着橘子、柠檬和香橼树的芳香。

[3] 相似的地形以及与伊比利亚罗马建筑遗存的诸多关联,使叙利亚被征服以后的建筑遗产被改造并同化,并吸收以前的罗马和波斯形式。但是西班牙伊斯兰建筑从未追求西方建筑的宏大的立体三维性。相反,它的目标是创造气氛,建筑的最高成就在于室内和室外空间在感官上的相互交织和设计上的精细程度。因此,在一个典型的摩尔式建筑和园林里,中庭及围合中庭的结构空间相互渗透,看似随意,却精心考虑了建筑和自然关系。有着反光表面、明亮的釉面砖,产生了凉爽的通风效果。

[4] 室内外的空间渗透不只局限于现在发现的格拉纳达阿尔拜辛地区的城堡宫殿或住宅,在这个地区,耐久的罗马柱廊园设计被采用,并重新诠释为卡门,即阿拉伯西班牙的内向式房屋和园林风格。在科尔多瓦清真寺(始建于785—786年)的设计中可以看到同样的规划原则。清真寺(现为大教堂)是多柱式内部结构,柱子支撑大量马蹄形双拱,均匀排列,犹如果园的树列。最初,清真寺的这面墙通过立面上的拱门向外开口,朝向一个3英亩的庭院,院内布置着橘子树林荫道,橘子树与清真寺内部的柱子相平齐。一个地下水箱为这个被称为橘子中庭的喷泉和灌溉渠提供水源。不幸的是,地下水箱长期被用于藏骨堂,基督教堂封上了这个曾经面向中庭的拱门。

[5] 位于塞维利亚的阿尔卡扎宫,建于基督徒1248年征服这个城市一百年后,证明了伊斯兰文化对之后西班牙建筑的长远影响。佩德罗一世,卡斯提尔王在以前伊斯兰城堡的废墟上建造他的城堡宫殿时,运用了大量的伊斯兰工艺。阿尔卡扎花园由一系列围合的中庭构成。尽管在以后几个世纪有所改变,这些花园的摩尔起源在他们围合的高墙,几何式布局,抬高的小径,众多的喷泉,明亮的釉面砖和柏树、棕榈、橘子和柠檬树得到证明。

[6] 格拉纳达是最后一个被基督教君主收复的阿拉伯城市,1492年被费迪南和伊萨贝拉女王收回。其中仍留存的阿拉伯花园——阿尔罕布拉宫和格内拉里弗宫——可以追溯到13和14世纪,是世界上最古老的花园之一并享有传奇般的地位。

Reading Material B

Taj Mahal—Mughal Garden
泰姬陵——莫卧儿园林

[1] The Taj Mahal (Fig. 1-Fig. 3) is a mausoleum located in Agra, India, built by Mughal Emperor Shah Jahan in memory of his favorite wife, Mumtaz Mahal. It is an integrated complex of structures composed of a white domed marble mausoleum, a Mughal garden and mosque.

The tomb

[2] The focus of the Taj Mahal is the white marble tomb, which stands on a square plinth consisting of a symmetrical building with an iwan[1], an arch-shaped doorway, topped by a large dome. Like most Mughal tombs, basic elements are Persian in origin. The base structure is a large, multi-chambered structure. The base is essentially a cube with chamfered edges and is roughly 55 meters on each side. On

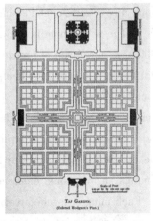

Fig. 1 Ground layout of the Taj Mahal

the long sides, a massive pishtaq[2], or vaulted archway, frames the iwan with a similar arch-shaped balcony.

[3] On either side of the main arch, additional pishtaqs are stacked above and below. This motif of stacked pishtaqs is replicated on chamfered corner areas as well. The design is completely symmetrical on all sides of the building. Four minarets, one at each corner of the plinth, facing the chamfered corners, frame the tomb. The main chamber houses the false sarcophagi of Mumtaz Mahal and Shah Jahan; their actual graves are at a lower level.

Fig. 2　The mausoleum of the Taj Mahal

Fig. 3　Walkways besides reflecting pool

[4] The marble dome that surmounts the tomb is its most spectacular feature. Its height is about the same size as the base of the building, about 35 meters, and is accentuated as it sits on a cylindrical "drum" of about 7 meters high. Because of its shape, the dome is often called an onion dome (also called a guava dome). The top is decorated with a lotus design, which serves to accentuate its height as well. The shape of the dome is emphasized by four smaller domed chattris (kiosks) placed at its corners. The chattri domes replicate the onion shape of the main dome. Their columned bases open through the roof of the tomb and provide light to the interior. Tall decorative guldastas (spires) extend from edges of base walls, and provide visual emphasis to the height of the dome. The lotus motif is repeated on both the chattris and guldastas. The dome and chattris are topped by a gilded finial, which mixes traditional Persian and Hindu decorative elements.

[5] The main dome is crowned by a gilded spire or finial. The finial, made of gold until the early 1800s, is now made of bronze. The finial provides a clear example of integration of traditional Persian and Hindu decorative elements. The finial is topped by a moon, a typical Islamic motif, whose horns point heavenward. Because of its placement on the main spire, the horns of moon and finial point combine to create a trident shape, reminiscent of traditional Hindu symbols of Shiva.

[6] At the corners of the plinth stand minarets, the four large towers each more than 40 meters tall. The minarets display the Taj Mahal's penchant for symmetry. These towers are designed as working minarets, a traditional element of mosques as a place for a muezzin to call the Islamic faithful to prayer. Each minaret is effectively divided into three equal parts by two working balconies that ring the tower. At the top of the tower is a final balcony surmounted by a chattri that mirrors the design of those on the tomb. The minaret chattris share the same finishing touches, a lotus design topped by a gilded finial. Each of the minarets was constructed slightly outside of the plinth, so that in the event

of collapse, a typical occurrence with many such tall constructions of the period, the material from the towers would tend to fall away from the tomb.

The garden

[7] The complex is set around a large 300-meter square charbagh. The garden uses raised pathways that divide each of the four quarters of the garden into 16 sunken parterres or flowerbeds. A raised marble water tank at the center of the garden, halfway between the tomb and gateway, with a reflecting pool on North-South axis reflects the image of the Taj Mahal. Elsewhere, the garden is laid out with avenues of trees and fountains. The raised marble water tank is called al Hawd al-Kawthar, in reference to "Tank of Abundance" promised to Muhammad. The charbagh garden, a design inspired by Persian gardens, was introduced to India by the first Mughal emperor Babur. It symbolizes four flowing rivers of Paradise and reflects the gardens of Paradise derived from the Persian paridaeza, meaning "walled garden". In mystic Islamic texts of Mughal period, paradise is described as an ideal garden of abundance with four rivers flowing from a central spring or mountain, separating the garden into north, west, south and east.

[8] Most Mughal charbaghs are rectangular with a tomb or pavilion in the center. The Taj Mahal garden is unusual in that the main element, the tomb, instead is located at the end of the garden. With the discovery of Mahtab Bagh or "Moonlight Garden" on the other side of the Yamuna[3], Archaeological Survey of India interprets that the Yamuna itself was incorporated into the garden's design and was meant to be seen as one of the rivers of Paradise. The similarity in layout of the garden and its architectural features such as fountains, brick and marble walkways, and geometric brick-lined flowerbeds with Shalimar's suggest that the garden may have been designed by the same engineer, Ali Mardan Khan[4]. Early accounts of the garden describe its profusion of vegetation, including roses, daffodils, and fruit trees in abundance.

Outlying buildings

[9] The Taj Mahal complex is bounded by crenellated red sandstone walls on three sides with river-facing side open. Outside these walls are several additional mausoleums, including those of Shah Jahan's other wives, and a larger tomb for Mumtaz's favorite servant. These structures, composed primarily of red sandstone, are typical of the smaller Mughal tombs of the era. The garden-facing inner sides of the wall are fronted by columned arcades, a feature typical of Hindu temples later incorporated into Mughal mosques. The wall is interspersed with domed kiosks (chattris), and small buildings that may have been viewing areas or watch towers like the Music House, which is now used as a museum.

[10] The main gateway (darwaza) is a monumental structure built primarily of marble and is reminiscent of Mughal architecture of earlier emperors. Its archways mirror the shape of tomb's archways, and its pishtaq arches incorporate calligraphy that decorates the tomb. It utilizes bas-relief and pietra dura (inlaid) decorations with floral motifs. The vaulted ceilings and walls have elaborated geometric designs, like those found in the other sandstone buildings of the complex.

[11] The Taj Mahal is considered the finest example of Mughal architecture, a style that combines

elements from Persian, Ottoman, Indian, and Islamic architectural styles. In 1983, the Taj Mahal became a UNESCO World Heritage[5] Site and was cited as "the jewel of Muslim art in India and one of the universally admired masterpieces of the world's heritage."

[NOTES]

[1] iwan 伊万:是指圆顶的大厅或场所,三面墙,另外一面完全开放。伊万是波斯萨珊王朝建筑的一个标志,后来才被引入到伊斯兰建筑中。在塞尔柱帝国时期,伊万的引入达到顶峰,同时也成为伊斯兰建筑中一个基本元素。典型的伊万对着中间的庭院,并被广泛用于公共建筑和住宅。

[2] pishtaq 圆拱门:指"伊万"前的入口门面。

[3] Yamuna 亚穆纳河:又名朱木纳河,是印度北部主要河流之一、恒河的支流。全长1 370千米。起源于北阿坎德邦,在安拉哈巴德注入恒河。

[4] Ali Mardan Khan 阿里·马丹汗:泰姬陵花园的设计者。

[5] UNESCO World Heritage 联合国教科文组织世界遗产:全称为联合国教育、科学及文化组织(United Nations Educational, Scientific and Cultural Organization)。

[NEW WORDS]

accentuate	vt.	使突出,强调	kiosk	n.	小亭
arcade	n.	连拱廊,拱形走道,拱形建筑物	mausoleum	n.	陵墓
archaeological	adj.	考古学的	minaret	n.	清真寺的尖塔,宣礼塔,光塔
archway	n.	拱门,拱道,拱廊	motif	n.	图形,基本图案;主题,主旨
balcony	n.	阳台	muezzin	n.	报告祈祷时刻的人
bas-relief	n.	浅浮雕,浅浮雕品	parterre	n.	花坛,花圃
bronze	n.	青铜,青铜色,赤褐色	penchant	n.	爱好,嗜好
calligraphy	n.	书法,笔迹,美术字(体)	pietra dura		佛罗伦萨马赛克饰面
chambered	adj.	有房间的,隔成房间的	plinth	n.	基座,基角
chamfered	adj.	倒角的,倒棱的,斜切的	profusion	n.	大量,丰富,充沛
charbagh	n.	(波斯语)波斯式样的庭院	reminiscent	adj.	提醒的,暗示的,像……的
crenellated	adj.	有垛口的,有雉堞的	replicate	v.	复制
daffodil	n.	黄水仙	sarcophagus	n.	石棺(复 sarcophagi)
elaborate	vt.	精心制作	Shiva	n.	印度教毁灭之神湿婆
finial	n.	顶尖,尖顶饰,顶端饰,屋脊饰	spire	n.	(教堂的)塔尖,尖顶
finishing	adj.	表面加工,表面修饰	sunken	adj.	下沉的,凹陷的,下陷的
gateway	n.	出入口,门	surmount	vt.	居于……之上,在……顶上;克服,战胜
gilded	adj.	镀金的,涂上金色的,装饰的	symmetrical	adj.	对称的,均衡的
heavenward	adj.	朝向天空的,朝向天国的	symmetry	n.	对称(性),匀称,整齐
Hindu	adj.	印度教的,印度人的	tank	n.	(印度、巴基斯坦等处的)大水池
inlaid	adj.	镶嵌的	trident	adj.	三叉的
intersperse	vt.	点缀,散布	vaulted	adj.	拱状的,圆顶的

[难句]

1. This meant that gardens in Spain would have embodied some Spanish-Christian, Roman, and even possibly Jewish ideas that had been transformed into an Islamic layout, specifically the type from Persia. 这意味着这些位于西班牙,原本应该体现一些西班牙基督徒、罗马人甚至可能是犹太人思想的园林,却变成了伊斯兰布局,特别是来自波斯的园林风格。[虚拟语气](would have…ideas 是虚拟语气,意为"原本应该体现……")

2. The Arabs had brought with them Persian artistic concepts along with Arabic science, and they married these with Greek philosophy. 阿拉伯人带来了波斯艺术以及阿拉伯科学,它们与希腊哲学相互联姻。[直译法](married 直译成联姻,更能反映原文的修辞与意境)

[参考译文]

泰姬陵——莫卧儿园林

[1] 泰姬陵是位于印度阿格拉城内的陵墓,是莫卧儿王朝沙贾汗为了纪念他的爱妻蒙泰姬·玛哈尔而建。它是由白色大理石穹顶陵墓、莫卧儿式花园和清真寺组成的复合体。

陵墓

[2] 泰姬陵的中心是白色大理石墓穴,伫立于一个方形基座之上,基座是由一个带有伊万结构的对称建筑组成,伊万是一个圆顶的拱形门廊。与大多数莫卧儿陵墓一样,基本元素源自波斯。建筑的基本结构是大型多室结构。基底基本上是一个有着倒角边缘的立方体,每边大致55米。在长边上,有一个巨大的圆拱门,以一个类似拱形的楼厅将伊万框住。

[3] 在主拱门的两侧,还有额外的圆拱门上下堆叠。这个堆叠圆拱门的图案,在基座倒角面上也重复出现。建筑每一面的设计都完全对称。基座的四个角各伫立着一个尖塔,它们朝向倒角面,将陵墓围合。主要的大厅放着蒙泰姬·玛哈尔和沙贾汗的假石棺,他们真正的墓穴位于下面一层。

[4] 墓穴之上的大理石穹顶是陵墓最壮观的部分。它的高度与建筑基座基本相等,约35米高,坐落于7米高圆柱形"鼓"之上,以加强穹顶高度。因其外观,穹顶常常被称为洋葱顶或番石榴顶。穹顶的顶部装饰有莲花设计,也是用来加强高度的。穹顶的形状由位于四角的四个小穹顶亭强调。小亭的穹顶也重复了主穹顶的洋葱形。小亭的圆柱形基础从顶部就开放,为建筑室内提供光线。高高的有装饰的尖顶从墙缘开始伸展,在视觉上强调了穹顶的高度。莲花图案在小亭和尖顶都得以重复。穹顶和小亭顶部有镀金尖顶饰,混合了波斯和印度的传统装饰元素。

[5] 主穹顶由镀金尖顶或尖顶饰装饰。尖顶饰直到19世纪早期都是由黄金制成,现在改用青铜。尖顶饰提供了一个波斯和印度传统装饰结合的明显例子。尖顶饰的顶部装饰有角点朝向天空的月亮,一个典型的伊斯兰图案。由于布置在主尖顶,月亮的尖角和尖顶饰的顶点一起形成了一个三叉戟的形状,暗示了印度教湿婆的象征。

[6] 在基座拐角矗立着40多米高的四个大型尖塔。尖塔体现了泰姬陵对对称布局的偏好。这些塔被设计成具有实际用途的清真寺传统要素,是伊斯兰信徒虔诚祷告的地方。每一个尖塔都由两个包围塔的阳台分成三等份。塔上部是一个覆盖小亭的尖顶饰阳台,与墓穴顶部的设计一样。尖塔小亭有着同样的饰面,莲花形顶部有着镀金尖顶饰。每一个尖顶塔建造得都稍超出基座,这样在崩塌的情况下,这一时期高大建筑会遇到的典型事件,塔的建筑材料会落向远离墓穴的方向。

花园

[7] 墓园设在大约300米长的方形园林中。抬高的小径将花园四个部分切分成16个下沉花坛。一个抬高的大理石水池位于花园中心,墓穴到入口的中间位置,南北轴线上的倒影池倒映着泰姬陵的身影。花园其他地方布置着林荫道和喷泉。抬高的大理石蓄水池代表着穆罕默德允诺的来自天堂的河水。这个源自波斯的园林,是由第一个莫卧儿皇帝巴布尔引入的。它象征着四条流经天堂的河流,反映来自波斯"围墙花园"的天堂园理

念。在莫卧儿时期神秘的伊斯兰文化中,天堂园被描述为理想花园,四条河流从中部泉眼或山脉流出,将花园分成东西南北四个部分。

[8] 大部分的莫卧儿花园是矩形的,中心是墓穴或凉亭。泰姬陵花园却不一般,里面的主要元素墓穴,却位于花园的尽头,随着亚穆纳河另一侧"月光花园"的发现,印度考古调查解释了亚穆纳河自身就被纳入花园的设计,并被看作天堂园的河流。夏利玛园布局和建筑小品,如喷泉、砖块和大理石步道、几何形砖块收边的花床,与泰姬陵非常相似,说明它们可能出自同一个工程师阿里·马丹汗之手。早期对于花园的的描述记载了丰富的植物,包括大量的蔷薇花、黄水仙和果树。

外围建筑

[9] 泰姬陵墓园的三边都被有垛口的红色砂岩墙环绕,面向河流的一侧是开放的。这些墙外是几个其他陵墓,包括沙贾汗其他的妻子,一个较大的是蒙泰姬最喜爱的仆人的陵墓。这些建筑物主要是由红色砂岩筑成,是这个时代典型的小型莫卧儿陵墓。花园朝着的内墙墙面,陈列着柱廊,这个印度寺庙的典型特征,后来被纳入莫卧儿清真寺的设计中。围墙还散点着穹顶凉亭和小型建筑,可以作为观景区域或者像音乐厅的瞭望塔,现在这里被用作博物馆。

[10] 墓园主入口是一座纪念式建筑,主要由大理石筑成,使人想起早期帝王的莫卧儿建筑。它的拱廊与墓穴拱廊的形状一样,圆拱门上刻有书法来装饰墓穴,采用花卉图案的浅浮雕和镶嵌装饰。拱形的天花板和墙体有着精美的几何形设计,正如在其他砂岩建筑中发现的一样。

[11] 泰姬陵被认为是莫卧儿建筑最优秀的典范,综合了波斯、土耳其、印度和伊斯兰建筑风格。在1983年,泰姬陵成为联合国教科文组织世界遗产,被描述为"穆斯林艺术在印度和全世界遗产的瑰宝"。

[思考]

(1)波斯天堂园形成的背景是什么?
(2)伊斯兰园林的基本要素有哪些?其基本形式是什么?
(3)伊斯兰园林如何反映环境因素并解决实际问题?

PART III

Development of Landscape Architecture in Modern Time
第三部分 近现代风景园林发展

UNIT 5 LANDSCAPE PLANNING AND PROTECTION

TEXT

Landscape planning[1]
Julius Gy. Fabos[2]
景观规划

[1] The entire American populations depends daily on our national landscape for work, for food and for recreation. More than three million acres of land are convert from rural to urban use in the United States each year. We build about one and a half million homes and various commercial, industrial, institutional and recreational facilities annually. We dispose millions of tons of waste into the air, water and soil and crisscross the country with roads, railroads, transmission lines and pipelines. With all these activities, we as a nation deeply affect the quality and value of the environment we occupy.

[2] The rich bounties of this country as well as its varied and ample landscape can provide a sound and beautiful environment, but only if we extensively apply landscape planning principles. Landscape planning takes place in many ways. Currently, the most widely used application occurs within the realm of project site planning. Increasingly, more developers around the country are employing the services of landscape architects in such endeavors. The results are more attractive, more environmentally suitable and make more economic sense than those without such planning expertise.

[3] Frederick Law Olmsted (Fig. 1), the father of American landscape architecture, was absorbed in all aspect of landscape planning. He was most instrumental initiating the successful national park movement. He planned the first model community, preserving riverfront for public open space in Riverside[3], Ⅲ. (1868-1870). He also linked three communities in Boston by transforming the Muddy River[4] into a spectacular linear park in 1880s. His pupils Charles Eliot and Eliot's nephew, Charles Eliot Ⅱ, expanded Olmsted's vision into a statewide open space plan for Massachusetts. This visionary concept is still being implemented.

[4] After World War Ⅱ, Olmstedian tradition was continued at every level. Ian McHarg[5] (Fig. 2) emerged as the most important voice of landscape planning during the 1960s. His seminal book, *Design with Nature* (1969), established contemporary principles of landscape planning ranging from shaping development in response to nature to preserving critical landscape resource. Another leader of landscape planning at this time was Philip H. Lewis, Jr.[6], who initiated a creative statewide recrea-

UNIT 5　LANDSCAPE PLANNING AND PROTECTION

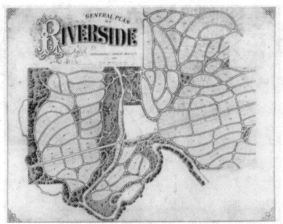

Fig. 1　Frederick Law Olmsted, and his planning of Riverside, Ⅲ. (1868—1870)

tion plan in Wisconsin that established environmental corridors throughout the state. The environmental movement[7] provided landscape planners with many opportunities during the turbulent decade of the 1960s. They were called on to make critical land use decisions and assess the visual and cultural features of large regions. They developed sophisticated procedures to assess diverse landscape qualities. And they helped determine the amount and type of development for urbanizing regions.

Fig. 2　Ian McHarg and his *Design with Nature*(1969)

[5]　Since the 1970s landscape planners have worked increasingly with scientists to obtain relevant findings about the impact of humans on the landscape. Thus, planning has become even more complex. Landscape planning research groups have sprung up to develop procedures capable of synthesizing scientific knowledge to provide a foundation for more intelligent land use decision. Current landscape planning is responding to the ever-increasing amount of new scientific information and turned to computers for help. Jack Dangermond[8] (Fig. 3), through his firm, ESRI, the current leader in this trend toward computerization, has developed the most advanced geographic information systems to deal with the myriad spatial data now essential for determining land use policies at all levels. Other leading landscape planners, including Laura Muessig and Allen Robinette, are using these procedures in forecasting and managing land use changes in the state of Minnesota. Current research has also been exploring the utility of computer technology for landscape planning at the community level. We will soon see these emerging procedures used in many types of site planning.

[6]　Landscape planning has a bright and challenging future. Planning the landscape has become the most significant of all land use objectives. In the information age, landscape planning can help us use our resources more intelligently than we have ever before.

Fig. 3 Jack Dangermond and geographic information systems

[NOTES]

[1] Landscape 指风景,地形,前景,景观。从词源分析来看,land 是一块地,-scape 是"看起来"或"样子"的意思,也就是景,因此英文 landscape 的含义是指视野中的一片土地,中文常翻译为风景、地景、景观。从视觉美学意义上来讲可理解为"风景""景色";作为地理学概念可理解为"地形""地物""景观";作为生态学概念也可理解为"景观"。文中 landscape planning 涉及三方面内容,因此,兼顾三层含义,此处译为"景观规划"为宜。

[2] Julius Gy. Fabos 朱利叶斯·Gy·法布士:美国著名景观规划理论家、马萨诸塞大学景观设计与区域规划系教授,1964 年毕业于哈佛大学,美国风景园林师协会会员,1997 年获得风景园林师协会设计奖章,景观评价与规划 METLAND 系统的主要创始人,著有《绿道:国际性运动的开始》(*Greenways*: *The Beginning of an International Movement*, 1996)。以法布士教授命名的 Fabos 景观规划及绿道国际研讨会是三年一届的景观界盛会,汇聚了景观规划界具有影响力的专家学者,以及从地方到国际的相关政策和绿道设计优秀案例。

[3] Riverside 里弗塞得社区:也译作河滨庄园,是奥姆斯特德与沃克斯自 1868 年花费两年时间规划的芝加哥德斯普兰斯河河畔的一个社区,面积 1 600 英亩。主要规划思想是保留足够的休闲空间,保证所有的居民都能到达优美的河畔美景,为当时的社区规划提供了很好的蓝本。

[4] Muddy River 莫迪河:波士顿公园系统("翡翠项链")中一部分,奥姆斯特德将一系列溪流和池塘联通而形成的壮观的带状绿地。

[5] Ian McHarg 伊恩·麦克哈格(1920—2001):英国著名园林设计师、规划师和教育家,其著作《设计结合自然》(*Design with Nature*)是有关生态规划和分析的具有里程碑意义的专著。

[6] Philip H. Lewis Jr. 菲利普·H.刘易斯:他在威斯康星州进行了全州休闲规划,并建立遍及全州的环境保护廊道,还提出环境廊道的概念。

[7] environmental movement 环境保护运动:20 世纪 60—70 年代在美国发生了一次以生态观为主旨的规模空前的群众性的环保运动,是美国历史上自然和资源保护运动的发展和继续,它的直接起因是人们对日趋严重的环境污染的不满和恐惧,同时它也有着更为广阔的社会背景,环境保护运动对美国的社会经济以及人们的环境意识和环保实践等诸多方面都产生了很大的影响。

[8] Jack Dangermond 杰克·丹杰蒙:对 GIS 方法论的发展、GIS 软件市场开发、GIS 技术研究和相关分析方法的创建都有着重大影响,于 1969 年创建了美国环境系统研究所(Environmental Systems Research Institute),ESRI 是全球地理信息系统软件开发与应用领域最大的公司,从创建之初就引领着世界 GIS 技术的潮流,1981 年推出首个商业版本的 ARC/INFO 软件之后,又开发了 ArcView、ArcIMS、ArcSDE、MapObjects、ArcPad 等软件。

[GLOSSARY]

environmental movement	环境保护运动	landscape planning	景观规划
geographic information system(GIS)	地理信息系统	model community	示范社区

UNIT 5 LANDSCAPE PLANNING AND PROTECTION

[NEW WORDS]

absorbed	adj.	全神贯注的，一心一意的	linear	adj.	线的，直线的
assess	vt.	评估,评定	myriad	adj.	无数的,种种的
bounty	n.	慷慨,宽大,施舍	procedure	n.	程序,手续
community	n.	社区,团体,社会	recreation	n.	消遣,娱乐
computerization	n.	计算机化	relevant	adj.	有关的,相应的
contemporary	adj.	当代的,同时代的	rural	adj.	乡下的,田园的
crisscross	vt.	画十字形于	seminal	adj.	种子的,(喻)启发性的
critical	adj.	危急的,临界的	significant	adj.	有意义的,重大的,重要的
decade	n.	十年,十	sophisticated	adj.	先进的,复杂的
dispose	v.	处理,除去,处置	spatial	adj.	空间的
diverse	adj.	不同的,变化多的	spectacular	adj.	引人入胜的，壮观的
endeavor	n.	努力,尽力	transform	vt.	改造,转换,改变
expertise	n.	专家的意见,专门技术	transmission	n.	播送,发射,传输
implement	vt.	贯彻,实现	turbulent	adj.	狂暴的,吵闹的
in response to		响应,适应	urbanize	vt.	使城市化
instrumental	adj.	有帮助的	visionary	adj.	有远见的,富于想象力的

[参考译文]

景观规划

朱利叶斯·Gy.法布士

[1] 所有美国人每天都要依赖我们的景观进行工作、获取食物并开展娱乐活动。在美国,每年有超过三百万英亩的农业用地转化为城市用地。每年我们建造一百五十多万所住宅以及各种商业、工业、企事业和娱乐设施。我们不仅将成百万吨的废弃物排放到了空气、水体和土壤中,同时也将公路、铁路、传输线和管道纵横分布于整个国家。我们所有的这些行为,深深地影响了我们所处的这些环境的质量和价值。

[2] 只有广泛地运用景观规划原则,才能使富饶的国土和丰富多样的景观创造出健康优美的环境。景观规划有多种方式。目前,运用最广泛的是工程项目中的场地规划。全国越来越多的土地开发者开始请风景园林师进行这方面的工作。这种趋势的结果是,经过景观规划的土地开发,比起那些没有经过专业规划的开发,更具吸引力、更符合环境需求,也更加符合经济原则。

[3] 弗雷德里克·劳·奥姆斯特德,美国风景园林之父,致力于景观规划的各个方面。在成功发起的国家公园运动中,他起着最重要的推动作用。在伊利诺斯州里弗塞得,他规划了第一个示范社区,其中保留了滨河区域作为公共开放空间(1868—1870)。19 世纪 80 年代,他还将莫迪河改造成为一个壮观的线形公园,以此连接波士顿的三个社区。他的学生查尔斯·艾略特和艾略特的侄子小查尔斯·艾略特将奥姆斯特德的理念延伸到马萨诸塞州全州范围的开放空间规划中。这个富有远见的理念至今仍被实施。

[4] 二战之后,奥姆斯特德的传统思想在各个层面的景观规划中都被继续沿用。20 世纪 60 年代,伊恩·麦克哈格的观点占据了最为重要的位置。其具有深远影响的书——《设计结合自然》(1969),奠定了从结合自然的形态设计到保护关键景观资源的当代规划原则。在这个时期,另一个景观规划的领袖人物是菲利普·H.小

刘易斯,他在威斯康星州发起了一场全州范围的休闲娱乐区规划,这个创造性的规划建立了贯穿该州的环境廊道。在纷乱的20世纪60年代,环境保护运动为景观规划师提供了很多工作机会。他们的工作包括作出关键的土地利用决策,以及对大区域的视觉和文化特征做出评估。他们创建了复杂的流程来评估不同的景观特质,还帮助决策区域城市化过程中所涉及的开发的数量和类型。

[5] 从20世纪70年代起,景观规划师逐渐开始与科学家合作,从而获取有关人类对景观影响的研究成果。因此,规划变得更加复杂。众多景观规划研究团体涌现出来以研究开发能够结合科学知识的程序,从而为更明智的土地利用决策提供基础。当前的景观规划正是基于这些不断增长的科学新信息来进行的,并且求助于计算机获得帮助。杰克·丹杰蒙,当前计算机化趋势的领袖人物,通过他的公司ESRI,开发了最先进的地理信息系统来处理繁多的空间数据,这些数据对于所有层面的土地利用决策都是不可缺少的信息。包括劳拉·麦西格和艾伦·罗宾奈特在内的其他景观规划领袖人物,正是运用这些程序来预测和管理明尼苏达州土地利用情况的变化。当前的研究还包括探索计算机技术运用于社区层面的景观规划。我们将很快看到这些出现的新兴程序被运用于各种类型的场地规化。

[6] 景观规划拥有一个光明且充满挑战的未来。规划景观已变成所有土地利用目标中最重要的内容。在信息时代,景观规划能够帮助我们比以往任何时候都更明智地利用我们的资源。

Reading Material A

Landscape Scenery

Wayne G. Tlusty

自然风景

[1] Landscape scenery has become a highly valued resource. For well over a century, major changes in attitudes toward scenic landscape have generally paralleled the development of landscape architecture as a profession. Recognizing the innate qualities of pastoral beauty, Andrew Jackson Downing[1] (Fig. 1) influenced American sensitivity to landscape appreciation. His landscapes illustrated how gardens could be linked visually to the countryside. Downing was also an early advocate for bringing natural beauty to urban areas through public parks. Frederick Law Olmsted and Calvert Vaux designed most of Central Park as idealized natural scenery, emulating the best examples of meadows, forests, hills lakes and streams. Their approach to creating parks was widely followed throughout America and fostered a new awareness of landscape scenery.

Fig. 1 Andrew Jackson Downing and his plan of Llewellyn Park

UNIT 5　LANDSCAPE PLANNING AND PROTECTION

[2]　During the mid-19th century, writers, painters and naturalists began discovering and focusing on spectacular American landscapes in the Hudson River Valley[2] (Fig. 2) and the West. Their work in the West led to federal recognition of the need to preserve natural scenery for all citizens. In 1865 Frederick Law Olmsted developed a management philosophy for Yosemite[3], addressing both preservation and public use in the world's first scenic landscape reserved for public enjoyment. Olmsted's philosophy for the preserve served as a model for our present national park system.

[3]　Broadening the concept from public to private lands, landscape architect Frank Waugh[4] sought recognition also for scenic beauty in the countryside, advocating that all citizens had an inalienable right to enjoy fine rural scenery. In his 1910

Fig. 2　Hudson River Valley

book he wrote about the need to maintain pastoral, utilitarian scenery and the inherent beauty of agricultural landscapes. Waugh maintained that "beauty does not interfere with utility, nor utility with beauty—the two are sisters."

[4]　The first applied wilderness policy[5] evolved from efforts to protect Trappers Lake in Colorado. As the landscape architect for the U. S. Forest Service[6], Arthur H. Carhart was directed in 1919 to provide recommendations for seasonal homes along the undeveloped shoreline. His proposal that the area's scenic beauty should remain pristine and permanently free of development for all generations was adopted.

[5]　Scenic preservation received a boost in 1954, when the U. S. Supreme Court (in Berman v. Parker) recognized that aesthetics alone was a sufficient reason to regulate land development. The environmental movement of the 1960s brought about more public awareness of landscape scenery in America. The National Conference on Natural Beauty in 1965 placed landscape scenery on the nation's environmental agenda. During this period many landscape architects made significant contributions to protecting and enhancing the aesthetic quality of America's landscape. The Wisconsin studies by Philip Lewis (Fig. 3) in the early 1960s established a process to identify environmental corridors[7] that included areas of notable scenery (Fig. 4). A 1964 study for Sea Ranch[8] (Fig. 4) in California by Lawrence Halprin was a model for protecting open space and scenery while allowing planned development. Ian McHarg used ecological determinants[9] to structure land use planning and protect visual character.

[6]　In 1969 the National Environmental Policy Act[10] stressed the "continuing responsibility" of the federal government to ensure aesthetically and culturally pleasing surroundings. It required the development of methods to measure how landscape contributes to the quality of life. As one example, Forest Service landscape architects in 1974 developed a visual resource management process, which was applied to 190 million acres of public land. Other approaches were designed to address visual resource needs for roads, shorelines, agricultural areas, utility corridors and development in the urban

fringe. Since 1964 the Interior Department has designated a number of natural areas deemed to be of national significance as Registered Natural Landmarks[11].

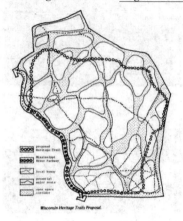

Fig. 3 Wisconsin's Heritage Trails Proposal (Lewis 1964)

Fig. 4 Lawrence Halprin and his Sea Ranch, CA (1965—1972)

[7] Today regard for landscape scenery has evolved into a broad recognition of the landscape as a vital national resource. In many areas, well-established policies now guide economic development to preserve scenic character, mitigate adverse intrusions and establish desired visual qualities. Landscape architects have provided critical leadership in protecting landscape scenery.

[NOTES]

[1] Andrew Jackson Downing 安德鲁·杰克逊·唐宁(1815—1852):园艺家、建筑师、美国第一位伟大的风景园林家,著有《园林的理论与实践概要》,该书是美国风景园林发展史中进行美学意义探索的第一次真正的尝试,他的著作在美国和欧洲都有着普遍的影响力。他还一直致力于编辑以"田园艺术和田园风格"为主题的杂志《园艺家》。唐宁在19世纪中叶为美国创造了景观园林艺术,成为简洁、自然、永恒的自然主义风格流派的伟大代表,对美国社会生活的许多方面都带来了深远的影响。

[2] Hudson River Valley 哈德逊河谷:位于纽约北部的,自然风景优美,被誉为"美国的莱茵河",沿河谷分布众多传统或现代豪华庄园以及著名的西点军校。the West 美国西部。19世纪中期,作家、画家和自然学家对哈德逊河谷和西部壮丽风景的积极探索和研究,使国家认识到保护自然风景的必要性,对国家公园的产生起到推动作用。

[3] Yosemite 约塞米蒂地区:位于美国加州旧金山内华达山中心,风景极为壮丽。1865年奥姆斯特德为该地区建立了一个致力于促进保护和公共使用的管理体系,约塞米蒂地区因此成为世界上第一个为公众娱乐而开发的风景游览区,这个管理体系成为现在国家公园系统的原型。约塞米蒂国家公园建于1890年,是世界上最早的国家公园之一。

[4] Frank Waugh 弗兰克·沃(1869—1943):倡导乡土化设计,寻求对乡村自然美景的认知和农业景观的保护,他和考利斯(Henry Chandler Cowles, 1869—1939)在全美公路网建设中,运用乡土植物群落展现地方景观特色,因为造价低廉并有助于保护生态环境的延续,有效解决了公路两侧的美化和护坡问题。

[5] wilderness policy 荒野保护政策:美国荒野保护是在国家公园和森林保护的基础上逐渐形成的,1964年荒野法(Wilderness Act)建立了国家荒野保护系统 National Wilderness Preservation System (NWPS),以保护未受人类文明入侵的自然荒野。荒野保护政策从保护科罗拉多州特的拉帕湖发展而来,由阿瑟·H.坎哈特提出,保留科罗拉多州特的拉帕湖的自然面貌,并且世代永久地免于开发。

UNIT 5 LANDSCAPE PLANNING AND PROTECTION

[6] U. S. Forest Service 美国林业局。
[7] environmental corridors 环境廊道：环境廊道的概念是由 20 世纪 60 年代菲利普·H. 刘易斯(Philip H. Lewis Jr.)在对威斯康星州规划中提出，环境廊道可归属于生态基础设施理论体系。
[8] Sea Ranch 海滨农场住宅区：位于旧金山北部，1967 年由劳伦斯·哈普林事务所设计，是劳伦斯·哈普林(Lawrence Halprin, 1916—2009)率先引入生态主义理念的设计作品，哈普林认为住宅区不仅要提供住宅和居住场地，而且居民应该能够在其中享受到野外粗犷的风景和自然的地形，并确保土地条件不受破坏，野生资源能够获得保护。
[9] ecological determinants 生态决策：由伊恩·麦克哈格(Ian McHarg, 1920—2001)提出的生态规划方法。
[10] National Environmental Policy Act 美国《国家环境政策法》：制定于 20 世纪 60 年代，作为国际上最早出现的环境基本法，是针对 20 世纪 50—60 年代美国频频爆发的重大环境污染事件而制定的法律对策。
[11] Registered Natural Landmarks 注册自然地标：由内政部指定的对国家意义重大的自然区域。

[GLOSSARY]

ecological determinants	生态决策	landscape scenery	自然风景	scenic landscape	自然景观
environmental corridor	环境廊道	national park system	国家公园系统	scenic preservation	风景保护

[NEW WORDS]

address	vt.	从事,忙于	meadow	n.	草地,牧场
adverse	adj.	不利的,敌对的,相反的	mitigate	v.	减轻
aesthetics	n.	美学,美术理论,审美学	parallel	vi.	平行,相应
agenda	n.	议程	pastoral	n.	牧歌,田园诗
awareness	n.	认识,意识	permanently	adv.	永存地,不变地
boost	v.	推进	preservation	n.	保护,保存
corridor	n.	走廊	preserve	vt.	保护,保存
deem	v.	认为,相信	pristine	adj.	质朴的
determinant	adj.	决定性的	recognition	n.	认知,识别,赞誉
emulate	n.	仿效	registered	adj.	已注册的,已登记的
foster	vt.	养育,培养,鼓励,促进	regulate	vt.	管制,控制
fringe	n.	边缘	sensitivity	n.	敏感,灵敏(度)
illustrate	vt.	阐明,举例说明,图解	shoreline	n.	海岸线
inalienable	adj.	(指权利等)不能剥夺的	significance	n.	意义,重要性
innate	adj.	先天的,天生的	supreme	adj.	最高的,极度的,极大的
interfere	vi.	干涉,干预,妨碍	undeveloped	adj.	不发达的,未开发的
interior	adj.	内部的,内的	utilitarian	adj.	实用的,功利的
intrusive	adj.	打扰的,插入的	valued	adj.	有价值的,宝贵的
landmark	n.	地标,里程碑,划时代的事	wilderness	n.	荒野,茫茫一片

· 93 ·

【难句】

1. His proposal that the area's scenic beauty should remain pristine and permanently free of development for all generations—was adopted. 他提议将这个地区的自然美景保留其质朴的面貌,并且世代永久地免于开发,这个提议被采纳了。

2. Scenic preservation received a boost in 1954, when the U.S. Supreme Court (in Berman v. Parker) recognized that aesthetics alone was a sufficient reason to regulate land development. 1954 年,美国最高法院认为仅美学价值这一点就足以证明控制自然风景区土地开发的重要性,风景保护因此得到了推进。

3. Ian McHarg used ecological determinants to structure land use planning and protect visual character. 伊恩·麦克哈格运用生态决策方法来进行土地利用规划及视觉特征的保护。

[参考译文]

自然风景
韦恩·G.特鲁斯蒂

[1] 自然景观已成为非常宝贵的资源。一个多世纪以来,人们对于自然风景态度的转变主要伴随着风景园林职业发展而并行发展。安德鲁·杰克逊·唐宁在认识到田园美景的天然特质后,以其观点影响了美国风景欣赏的敏感度。他的景观概念阐述了花园如何从视觉上与周边乡村景观建立联系。唐宁也是最早倡导通过建立公共公园将自然美景带到城市中来的人之一。弗里德里希·劳·奥姆斯特德和卡尔·沃克斯将中央公园的大部分区域设计成为理想的自然风景,将自然界中最理想的草地、森林、山脉、湖泊和溪流景观作为仿效的范例。他们创建公园的方法在美国受到广泛运用,并且使人们对自然景观形成一个新的认识。

[2] 在 19 世纪中期,作家、画家和自然学家开始探索和关注美国哈德逊河谷和西部的壮丽风景。他们在西部的研究使国家认识到为国民保护自然风景的必要性。1865 年,弗里德里希·劳·奥姆斯特德为约塞米蒂地区建立了一个致力于促进保护和公共使用的管理体系,约塞米蒂地区成为世界上第一个为公众娱乐而开发的风景游览区。奥姆斯特德的旨在保护的管理体系成为我们现在国家公园系统的原型。

[3] 景观设计师弗兰克·沃将这个理念由公共用地拓展到私人用地,寻求对乡村自然美景的认知,倡导所有民众都享有享受美好田园风光不可剥夺的权利。在他 1910 年出版的书中写到保护田园乡村、利用田园风景和农业景观内在美感的必要性。沃还提到"自然美景不会影响人类的使用,人类的使用也不会影响自然美景,他们是相辅相成的一对"。

[4] 最早采用的荒野保护政策就是从保护科罗拉多州特的拉帕湖发展而来。作为美国森林服务总局的景观设计师,亚瑟·H.卡哈特在 1919 年受命为沿未开发海岸线的季节性景观提出规划建议。他提议将这个地区的自然美景保留其质朴的面貌,并且世代永久地免于开发,这个提议被采纳了。

[5] 1954 年,美国最高法院认为仅美学价值这一点就足以证明控制自然风景区土地开发的重要性,风景保护因此得到了推进。20 世纪 60 年代的环境运动促使更多的公众了解美国的自然风景。1965 年的国家自然美景会议中,将自然风景列入了国家环境议程。在这个时期,许多景观设计师对保护和加强美国自然风景的美学质量作出了重要贡献。20 世纪 60 年代早期,菲利普·刘易斯对威斯康星州的研究建立了鉴定包含著名风景区的环境走廊的方法。1964 年劳伦斯·哈普林对加利福尼亚海滨牧场的研究就是一个保护开放空间和自然风景,同时允许规划开发的典范。伊恩·麦克哈格运用生态决策方法来进行土地利用规划及视觉特征保护。

[6] 1969 年国家环境保护政策法案强调,政府在确保令人愉悦的自然环境的美学和文化价值方面应负有"持续的责任"。这就要求建立衡量景观如何改善生活质量的方法。1974 年森林服务总局的景观设计师建立的视觉资源管理方法就是一个例子,这方法曾用于一亿九千万英亩的公共土地管理中。其他还有城市边缘区的道路、海岸线、农业区、可用廊道和开发区的视觉资源管理方法。自从 1964 年开始,内政部指定了许多对国家意义重大的自然区域作为注册自然地标。

[7] 今天,对于自然风景的关注,已经发展到将其作为国家重要资源的一个更广阔的认识。现在在许多地

区,完善的政策引导经济朝着保护风景特征、减轻负面干扰、建立理想视觉质量的方向发展。景观设计师在保护自然风景方面起到了关键的领导作用。

Reading Material B

National Parks

Raymond L. Freeman

国家公园

[1]　The world's first national park was established in 1872 when Congress designated more than two million acres in Wyoming as Yellowstone National Park[1] (Fig. 1). By the 1980s, more than 300 diverse parks had been incorporated into America's world-renowned national park system[2]. The concept of such reserves, pioneered in this country, was a significant contribution to world civilization, and other nations eventually followed this inspiring model.

Fig. 1　Yellowstone national park

[2]　The history of our national parks and the profession of landscape architecture have long been intertwined. Today, about 200 landscape architects have a vital role in providing stewardship for many of our nation's most cherished natural and cultural resources through the National Park Service[3], a part of the U.S. Department of the Interiors[4].

Fig. 2　Yosemite Valley　　　　　　　Fig. 3　Mariposa Big Tree

[3]　In 1864 President Abraham Lincoln signed legislation setting aside the magnificent Yosemite

Valley(Fig. 2) and *Mariposa Big Tree*[5] (Fig. 3). Groves are to be held by the state of California for "public use, resort and recreation inalienable for all time". Frederick Law Olmsted was appointed a commissioner for these reservations and supervised the preparation of an influential report for their administration. In addition to his skillful plan for managing this park, he advocated a policy of establishing national parks across the nation and laid the foundation for our current national park system.

[4] To fill the need for a separate division to oversee these parks, the secretary of the interior[6] in 1910 recommended creating a Bureau of National Parks and Resorts[7] specifically to employ landscape architects for their expertise in planning park development. Three years later the new position of general superintendent of the national parks was created. Mark Daniels, a practicing landscape architect from California, filled this job for two years. His most significant accomplishment was bringing sensitive design[8] into park administration and planning.

[5] It was not until 1916, however, that the National Park Service was formally established. That year the annual meeting of the American Society of Landscape Architects had passed a resolution supporting the National Park Service bill. The ASLA also addressed park issues requiring landscape architectural expertise, including the delineation of boundaries in consonance with topography and landscape units and the development of comprehensive plans for managing natural and developed areas.

[6] Stephen Mather, the first National Park Service director, took ASLA recommendations to heart. He and subsequent directors, relying heavily on landscape architects in guiding the development of the national park system, together with an array of consultants including Frederick Law Olmsted, Jr., James Pray, Warren H. Manning, Harold Caparn and James Greenleaf (all former ASLA presidents), continued the accelerated work of establishing boundaries, campgrounds, buildings, roadways, bridges and other park facilities. During this period, a National Park Service landscape architect, Daniel P. Hull, developed a distinctive nonintrusive, rustic parks building design, sometimes referred to as "parkitecture"[9]. Built in the early 1920s, the Ranger Club House[10], designed by staff landscape architect Charles P. Punchard, Jr., for the Yosemite Nation Park in California, and Hull's Administration Building[11], now a museum in the Sequoia National Park (Fig. 4) in California, became important symbols of this look for park buildings.

Fig. 4 museum in the Sequoia National Park

[7] By 1922 the landscape architecture division of the National Park Service, originally located at Yosemite, was moved to San Francisco with Thomas Vint as its influential director. Under Vint the task of creating master plans for each park began in earnest, often with staff landscape architects taking the lead.

[8] During the Depression, many landscape architects were employed temporarily to carry out Civilian Conservation Corps[12] programs managed by landscape architect Conrad L. Wirth, later to become Park Service director. Programs during this period shifted from a focus on natural sites to histor-

ic resources and parkways such as the Blue Ridge Parkway in Virginia and North Carolina and battlefields and historical parks such as Gettysburg in Pennsylvania (Fig. 5).

Fig. 5　Gettysburg in Pennsylvania

[9]　Years of low funding and lack of interest left the national park system in a state of disrepair after Word War Ⅱ. Conrad Wirth proposed a major conservation program that would implement a nationwide effort to assess, reorganize and restore all parts of the park system. Known as *Mission 66*[13] and initiated in 1956 as a *10-year national park renewal*, it led to significant federal support, which in turn raised the standards of all national parks. A planning staff, including three landscape architects headed by William Carnes, developed the program. Campgrounds were restored and sometimes even relocated to more appropriate sites (a campsite in Yellowstone National park was discovered to have been located on the path of migrating grizzly bears), new roads and trails were built, visitor centers for the first time were established and management policies were instituted. By 1966 the park system had evolved fully into a nationally recognized network of parks.

[10]　In this time of explosive demands for use of the parks, the job of protecting fragile natural, historic, cultural and scenic resources has become increasingly complex for the National Park Service. There is probably no other governmental agency in which landscape architects have had more influence in this process.

[NOTES]

[1]　Yellowstone National Park 黄石国家公园:建于1872年,世界上第一个国家公园,位于美国西部北落基山和中落基山之间的熔岩高原上,占地约9 000平方千米,地跨怀俄明(Wyoming)、蒙大拿(Montana)和爱达荷(Idaho)三州,拥有峡谷、瀑布、温泉及间歇喷泉等优美的自然景观,以及各类野生动植物资源。

[2]　national park system 美国国家公园系统:包括20个类别,国家公园(National Park)、国家历史公园(National Historical Park)、国家休闲娱乐区(National Recreation Area)、国家历史地(National Historic Site)、国际历史地(International Historic Site)、国家纪念地(National Monument)、国家纪念碑(National Memorial)、国家战场(National Battlefield)、国家战场公园(National Battlefield Park)、国家战场遗址(National Battlefield Site)、国家军事公园(National Military Park)、国家海滨(National Seashore)、国家湖滨(National Lakeshore)、国家河流(National River)、国家公园道(National Parkway)、国家风景小径(National Scenic Trail)、国家荒野风景河流及河道(National wild and Scenic River and Riverway)、国家保护地(National Preserve)、国家保留地(National Reserve)及其他(Other Designations)。

[3] National Park Service 美国国家公园管理局:建于1916年,隶属于美国内政部,主要负责美国境内的国家公园、国家历史遗迹、历史公园等自然及历史保护遗产。

[4] U. S. Department of the Interiors 美国内政部。

[5] Mariposa Big Tree 美利坡撒大树林:或译作马里波萨大树林,临近约塞米蒂河谷(Yosemite Valley),并与其一起在1864年保留作为保护区,以巨大的红杉林为特色。

[6] secretary of the interior 内政部长。

[7] Bureau of National Parks and Resorts 国家公园和度假区管理局。

[8] sensitive design 敏感性设计:马克·丹尼尔(Mark Daniels)在对国家公园的管理和规划中提出的保护性规划方法。

[9] parkitecture 公园式建筑:由丹尼尔·P. 赫尔(Daniel P. Hull)建立的富有特色、非入侵式的公园建筑设计形式。

[10] Ranger Club House 护林员会所:位于约塞米蒂国家公园内。

[11] Hull's Administration Building 赫尔行政管理建筑:现在的加州红杉国家公园(Sequoia National Park)中的博物馆。

[12] Civilian Conservation Corps 公民保护团:是美国在1933年至1942年间,对18至25岁单身失业男性推行的就业方案。公民保护团(CCC)是罗斯福"新政"改革项目之一,通过该项目,800多个国家公园、州公园和森林得到保护和提升。

[13] *Mission 66* 66计划:二战后国家公园普遍处于荒废失修状态,1956年康拉德·沃斯提出了10年国家公园恢复计划(10-year national park renewal),该计划使国家公园获得大量联邦政府的资助,从而提高了所有国家公园的质量水平。

[GLOSSARY]

historical park	历史公园	natural and cultural resources	自然和人文资源	reserve	保护区或保留地
national park	国家公园	parkway	公园道	sensitive design	敏感性设计

[NEW WORDS]

accomplishment	n.	成就,完成	conservation	n.	保存,保护
annual	adj.	一年一次的,每年的	consonance	n.	一致,调和
appoint	vt.	任命,委任	consultant	n.	顾问
battlefield	n.	战场,沙场	delineation	n.	描绘
bill	n.	议案,法案	designate	vt.	指定,任命
campground	n.	野营地,露营场所	disrepair	n.	失修,塌毁,荒废
campsite	n.	营地	distinctive	adj.	与众不同的,有特色的
cherish	vt.	爱护,珍爱	fragile	adj.	易碎的,脆的
commissioner	n.	委员,专员	grizzly	adj.	略灰色的,呈灰色的
comprehensive	adj.	全面的,综合的,广泛的	grove	n.	树丛,树林
Congress	n.	美国国会	incorporate	vi.	合并,混合

continued

institute	vt.	制定,创立	rustic	adj.	乡村的
intertwine	v.	缠结,纠缠	secretary	n.	秘书,书记
legislation	n.	立法,法规	sequoia	n.	红杉
magnificent	adj.	壮丽的,宏伟的	set aside	v.	留出
migrate	vi.	迁徙,移居,移植	sign	vt.	签署
oversee	v.	监管,视察,俯瞰	stewardship	n.	职位,工作
relocate	v.	重新部署	subsequent	adj.	后来的,并发的
renewal	n.	恢复,更新	superintendent	n.	主管,负责人
renowned	adj.	有名的,有声誉的	supervise	v.	监督,管理,指导
resolution	n.	坚定,决定	topography	n.	地形学
restore	vt.	恢复,修复,重建	vital	adj.	至关重要的,重大的

【难句】

1. By the 1980s, more than 300 diverse parks had been <u>incorporated</u> into America's world-renowned national park system. 到20世纪80年代,已有300多个不同类型的国家公园被<u>纳入</u>世界闻名的美国国家公园系统。

2. Groves are to be held by the state of California for "public use, resort and recreation inalienable for all time." 树林归属加利福尼亚州管理,以供"公众使用,任何时候都可以在此度假和休闲。"

3. The ASLA also addressed park issues requiring landscape architectural expertise, including the delineation of boundaries in consonance with topography and landscape units and the development of comprehensive plans for managing natural and developed areas. 美国风景园林师协会还表明国家公园中一些问题的解决需要风景园林师的专业知识,包括根据地形和景观单体来设定边界、制订管理自然区域和开发区域的综合规划计划。

[参考译文]

国家公园

雷蒙德·L.弗里曼

[1] 1872年美国国会将怀俄明州200万英亩的土地划定为黄石国家公园,建成了世界上第一个国家公园。到20世纪80年代,已有300多个不同类型的国家公园被纳入世界闻名的美国国家公园系统。由美国所倡导的保护区概念,对世界文明做出了重大贡献,并且其他国家最终也采用了这个鼓舞人心的模式。

[2] 国家公园的历史长期以来都和风景园林职业相互交织,不可分开。目前,大约有200位风景园林师通过内政部国家公园管理局,在为我们国家最受喜爱的自然和人文资源的管理中发挥着巨大的作用。

[3] 1864年亚伯拉罕·林肯总统签署了保留壮丽的约塞米蒂谷和美利坡撒大树林作为保护区的法案。树林归属加利福尼亚州管理,供"公众使用,任何时候都可以在此度假和休闲"。奥姆斯特德被任命为这些保护区的管理专员,并为制定具有影响力的保护区行政管理报告提供指导。除了为公园管理做出精心规划外,他还主张在全国建立国家公园的政策,这为我们当前的国家公园系统奠定了基础。

[4] 为了满足创建单独的分支机构以监管这些公园的需要,内政部部长在1910年提议建立国家公园和度假区管理局,专门聘请风景园林师为公园的规划发展提供专业的服务。三年后,国家公园中建立了总管这样一个新的职位。加州从业风景园林师马克·丹尼尔,在这个职位上工作了两年。他最重要的成就就是将敏感性设

计带到公园的管理和规划中。

[5] 然而直到1916年,国家公园管理局正式成立。这一年,美国风景园林师协会年会通过了一项支持国家公园管理局法案的决议。美国风景园林师协会还表明国家公园中一些问题的解决需要风景园林师的专业知识,包括根据地形和景观单体来设定边界、制订自然区域和开发区域的综合管理规划。

[6] 斯蒂芬·玛斯尔,第一个国家公园管理局主管,极重视美国风景园林师协会的建议。他及其继任者,都重点依靠风景园林师来指导国家公园系统的建设,同时在小奥姆斯特德、詹姆斯·普瑞、沃伦·H.曼宁、哈罗德·卡彭和詹姆斯·格林利夫(他们均担任过ASLA的会长)等大批顾问的帮助下,不断加速创建边界、营地、建筑、道路、桥梁及其他公园设施的工作。在这个时期,一个国家公园管理局的风景园林师,丹尼尔·P.赫尔,建立了一个富有特色、非入侵式的乡村公园建筑设计,有时也被称作"公园式建筑"。由风景园林师小查尔斯·P.潘查德设计,建于20世纪20年代早期,加州约塞米蒂国家公园的护林员会所,以及赫尔行政管理建筑,即现在的加州红杉国家公园博物馆,都成为这类公园建筑的代表。

[7] 到1922年,原先位于约塞米蒂的国家公园管理局风景园林师分局,迁至旧金山,托马斯·文特成为其具有影响力的主管。在文特的管理下,对每一个公园制定总体规划的任务认真地展开了,这些工作往往由风景园林师牵头进行。

[8] 在经济大萧条期间,许多风景园林师临时受雇执行由康拉德·L.沃斯管理的公民保护团项目,后来沃斯成为公园管理局主管。这个时期项目的焦点由关注自然场所转移到关注历史资源以及公园道上来,例如像弗吉尼亚州和北卡罗来纳州蓝色山脊这样的公园道,以及像宾夕法尼亚州盖茨堡这样的战场和历史公园。

[9] 二战之后的资金不足和对景观缺乏兴趣的年代中,国家公园系统处于荒废状态。康拉德·沃斯提出了一个对全国范围的公园系统进行评估、改造和恢复的重要保护项目。1956年发起的以66计划著称的10年国家公园恢复计划,使其获得大量联邦政府的资助,从而提高了所有国家公园的质量水平。以威廉·卡恩尼斯为首,包括三名风景园林师在内的规划团队执行了这个计划。营地被修复,有时甚至重新布置到更适合的场所(在黄石国家公园就曾发现一个野营地位于灰熊迁徙的路径上);创建新的道路和小径;首次建立游客中心,并且制定管理政策。到1966年,公园系统已经全面发展到全国性的公园网络。

[10] 在当前人们对于公园使用需求剧烈增长的时期,保护脆弱的自然、历史、文化和风景资源的工作,对于国家公园局来说已经变得越来越复杂。可能没有其他的政府机构能够像国家公园局一样,在那里风景园林师对这个过程拥有更大的影响力。

[思考]

(1)景观规划的意义和目标是什么?

(2)20世纪60年代的环境保护运动和自然风景保护运动对景观规划理念的发展起到了什么作用?

(3)简述美国国家公园从产生、发展到体系化的过程。

(4)美国的景观规划理念的发展对中国的风景园林产生了什么影响?

UNIT 6 URBAN SPACE

TEXT

Metropolitan Open Space[1]
Roger B. Martin
都市开放空间

[1] As with many aspects of landscape architecture, precedents for metropolitan park systems[2] can be traced to Europe. In early 19th century Germany, for example Prince Puckler-Muskau[3], a benevolent aristocrat, converted his ancestral lands in Silesia[4] into an extensive parklike system[5]. His desire was to surround the town of Muskau "in such a way that it would become merely part of the park." (Fig. 1). As American cities expanded, several early landscape architects, inspired by the work of Puckler-Muskau and others, envisioned vast open space systems extending through and around the nation's growing urban regions.

Fig. 1 Muskauer Park

[2] Linear parks, or parkways[6], connecting significant parkland within a metropolitan area were pioneered by Frederick Law Olmsted, Calvert Vaux and H. W. S. Cleveland[7]. The latter in 1869 recommended a 14-mile "grand avenue," similar to the tree-lined boulevards of Europe, connecting Chicago's parks. An early project by Olmsted and Vaux-Brooklyn's Eastern Parkway[8], (1870) and another by Olmsted and his firm, Arborway, part of Boston's urban park system[9] (1878-1895)-incorporated this feature. To accommodate American needs, however, the two landscape architects provided a generous strip of land for carriages, equestrian trails and pedestrian walks.

[3] As early as 1872 H. W. S. Cleveland advocated a bold metropolitan park system for the growing

cities of Minneapolis and St. Paul. Included would be the nearby Mississippi River bluffs (Fig. 2) and land encompassing the region's picturesque lakes, hills and valleys, plus additional small inner-city parks. He also proposed radial avenues and treelined boulevards plunging deep into the central city.

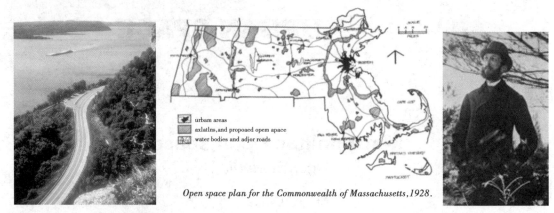

Fig. 2　Mississippi River bluffs　Fig. 3　Charles Eliot and his metropolitan system of open spaces for greater Boston

[4]　In 1890 Charles Eliot[10] suggested a metropolitan system of open spaces for greater Boston (Fig. 3). The following year, Massachusetts established the Trustees of Public Reservations[11] to acquire and hold land for public use. As this organization's secretary and later its consulting landscape architect, Eliot worked to establish a Metropolitan Park Commission and formulated an elaborate park plan. His park system included ocean frontage, islands in the inner bay, tidal estuaries, wild forest areas and numerous small squares, playgrounds and parks; boulevards and parkways were added in 1894 as connecting links. While unique to Boston's geography, the plan established the framework for other metropolitan open space networks in America.

[5]　A system of parkways with sizable areas for recreation was achieved on a regional scale in Westchester County, N.Y., during the early 20th century. This second-generation metropolitan park system used a parkland corridor to link major park facilities and provide a continuous recreational experience. The financial success of the automobile parkway along the Bronx River resulted in legislation establishing the Westchester County Park Commission in 1922. As a result of his accomplishments on the Bronx River Parkway[12] during the second decade of this century, Gilmore D. Clarke was appointed landscape architect for the county, and he skillfully coordinated planning for thousands of acres of parks and parkways (Fig. 4). The linear park system connected large recreational facilities incorporating golf, swimming, amusements, hiking and other activities.

[6]　The evolution of metropolitan park systems during the mid-20th century saw continuing emphasis on providing active recreation in natural setting. Hundreds of miles of hiking trails within large suburban park reserves[13] throughout the United States provided opportunities for exploring nature in close proximity to home and work. Northern California's East Bay Regional Park District[14] illustrates further refinement of the metropolitan open space system concept. During the 1960s, under the direction of landscape architect William Penn Mott (later director of the National Park Service), the system grew to more than 70 000 acres of diversified parkland with more than 1 000 miles of trails. In-

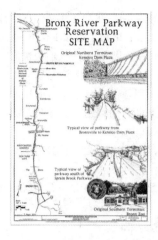

Fig. 4　Bronx River Parkway

cluded were lands that protected unique components such as geological formations and vegetation of interest to science, as well as natural and cultural features, and provided a wide range of recreational opportunities for the 1500-square-mile area's growing population.

[7]　In the future, landscape architects will continue to bring their vision, creativity and understanding of the landscape to the metropolitan open space needs of America's growing urban population. Current metropolitan park systems in rapidly expanding urban areas are exploring new and creative ways to find and permanently protect regional open space. In the Chicago area, vital regional facilities will be provided well into the 21st century through the Open Lands Project[15], a nonprofit agency working with the Lake County Forest Preserve District[16], under the direction of landscape architect Jerry Soesbe. More than 12 000 acres of open land have been preserved through private funding since the project was initiated in 1963. More of this type of protection will be needed to ensure a balance in our urban settings between development and open space.

[NOTES]

[1]　metropolitan open space 都市开放空间：兴起于19世纪，一直延续至今，其相关基础理论不断发展演化，由早期的公园道、公园系统开始，然后直到开放空间系统和绿色通道，例如奥姆斯特德(F. L. Olmsted)的波士顿"翡翠项链"和艾略特(Charles Eliot)的波士顿都市开放空间系统，以及法布士(Julius Gy. Fabos)的新英格兰地区绿色通道规划，这些规划案例反映了景观空间结构规划的思想发展历程。

[2]　metropolitan park system 都市公园系统。

[3]　Prince Puckler-Muskau 平克勒·穆斯考亲王(1785—1871)：英国的浪漫主义造园风格的推崇者，耗尽毕生精力和钱财建造了极富浪漫色彩的穆斯考林苑，即现在的穆斯考公园(Muskauer Park)。此外，他还设计了著名的勃兰尼茨风景公园(Park Branitz)，并在1835年出版了颇具影响力的《造园指南》(Hints on Landscape Gardening)一书。

[4]　Silesia 西里西亚(中欧一地区)。

[5]　parklike system 类似公园的系统：即穆斯考公园(Muskauer Park)，是位于波兰和德国两国边界横跨尼斯河的一处风景公园，面积约560公顷，是1815年到1844年由平克勒·穆斯考亲王建立，公园采用"用植物作画"的概念，运用当地植物来增强现有景观的内在品质，并与周围的乡村景观完全相融，先进的设计理念成为风景园林的先驱，并影响了欧洲和美洲的风景园林发展。

[6]　Linear parks, or parkways 带状公园或公园道:最早由奥姆斯特德提出的公园道,主要是指两侧树木郁郁葱葱的线性通道。这些通道连接着各个公园和周边的社区,宽度也不大,仅能够容纳马车道和步行道,道路应具有优美的曲线、宽敞的空间、避免出现尖锐的折角这种理念,形成适于人们游憩、思考,且令人们愉快而宁静的环境。从 19 世纪 60 年代开始,奥姆斯特德就开始尝试用公园道或其他线形方式来连接城市公园,或者将公园延伸到附近的社区中,从而增加附近居民进入公园的机会,如在芝加哥的里弗塞得庄园(Riverside)、波士顿公园系统"翡翠项链"中莫迪河(Muddy river)段,以及和沃克斯共同设计的 Buffalo 公园道和芝加哥开放空间系统。

[7]　H. W. S. Cleveland 克里夫兰(1814—1900):美国近代风景园林师之一,与奥姆斯特德共同致力于美国早期的景观规划与土地规划。

[8]　Brooklyn's Eastern Parkway 布鲁克林东部公园道:穿越纽约布鲁克林的景观大道,1866 年由奥姆斯特德与沃克斯所构想。

[9]　Boston's urban park system 波士顿城市公园系统:奥姆斯特德规划的"翡翠项链"。

[10]　Charles Eliot 查尔斯·艾略特(1834—1826):艾略特是奥姆斯特德的学生,参与了奥姆斯特德在波士顿的主要项目。他将奥姆斯特德的思想进一步完善和发展,并运用到波士顿大都会,创建了大波士顿都市开放空间系统(metropolitan system of open spaces for greater Boston)。该规划将 3 条主要的河流(包括 Charles 河)和 6 个大的城市郊区的开放空间连接到一起,为波士顿地区增加了 250 平方英里的开放空间,艾略特因此被誉为"波士顿开放空间系统之父"。正当处于事业顶峰的时候,艾略特突然故去,艾略特的侄子 Eliot II 接替艾略特的工作,将这一任务一直持续到 20 世纪 30 年代。艾略特最杰出的贡献体现在两个方面,一是对自然景观的保护;二是他提出了著名的"先调查后规划"理论,该理论将整个风景园林学从经验导向系统和科学,该方法一直影响到 20 世纪 60 年代以后的刘易斯和麦克哈格的生态规划。

[11]　Trustees of Public Reservations 公共保护协会。

[12]　Bronx River Parkway 布朗克斯河公园道。

[13]　suburban park reserves 郊野公园保护区。

[14]　East Bay Regional Park District 东部海湾区域公园协会。

[15]　Open Lands Project 芝加哥公共土地计划:开始于 1963 年,以"寻求休闲娱乐资源的保护和发展"为主旨,众多的森林、河流等开放空间通过该计划得到保护。

[16]　Lake County Forest Preserve District 雷克郡森林保护区。

[GLOSSARY]

linear park	带状公园	metropolitan park system	都市公园系统	parkway	公园道
metropolitan open space	都市开放空间	open space system	开放空间系统	suburban park	郊野公园

[NEW WORDS]

amusement	n.	娱乐,消遣	benevolent	adj.	慈善的
ancestral	adj.	祖先的,祖传的	bluff	n.	断崖,绝壁
aristocrat	n.	贵族	bold	adj.	大胆的
automobile	n.	汽车	boulevard	n.	林荫大道,干道,大街
avenue	n.	林荫路,道路,大街	coordinate	vt.	调整,协调
bay	n.	海湾	elaborate	adj.	详细阐述的,精细的

continued

envision	vt.	想象，预想	proximity	n.	接近，亲近
estuary	n.	河口，江口	refinement	n.	精确，明确表达
evolution	n.	进展，发展，演变，进化	sizable	adj.	相当大的，大的
formulate	vt.	阐明	strip	n.	条，带，狭长的一块土地等
hiking	n.	徒步旅行	tidal	adj.	潮汐的，定时涨落的
illustrate	vt.	举例说明，阐明，作图解	trace to	v.	上溯，回溯
metropolitan	adj.	大城市的，大都会的，首府的	trail	n.	小径，路线，踪迹，痕迹
nonprofit	adj.	非赢利的，不以赢利为目的的	vast	adj.	巨大的，辽阔的
precedent	n.	先例	vegetation	n.	[植]植被，(总称)植物、草木

[参考译文]

都市开放空间
罗杰·B. 马丁

[1] 与风景园林在许多方面的发展一样，都市公园系统的先例也可以追溯到欧洲。例如，在19世纪早期的德国，一位慈善贵族平克勒·穆斯考亲王就曾将其祖先留下的西里西亚的土地改造成了一个广阔的类似公园的系统。他的愿望是"以成为公园组成部分的方式"将穆斯考镇包围。随着美国城市的扩张，几位早期的风景园林师受到平克勒·穆斯考及其他一些人做法的启发，构想了一个巨大的开放空间系统贯穿并环绕这个国家不断增长的城市区域。

[2] 弗里德里克·劳·奥姆斯特德，卡尔弗特·沃克斯和H.W.S.克里夫兰倡导，在都市区域，以带状公园或公园道连接主要公园用地。1869年克里夫兰建议修建一条14英里长的类似于欧洲林荫大道的"大林荫道"来连接芝加哥的各个公园。奥姆斯特德和沃克斯的早期项目布鲁克林东部公园道(1870)，以及奥姆斯特德和其公司项目艾伯路(波士顿城市公园系统的一部分)也结合了这个特征。然而，为了满足美国人的使用需求，这两位风景园林师设计了宽阔的带状用地用于车道、骑马道和步行道。

[3] 早在1872年H.W.S.克里夫兰为发展中城市明尼阿波利斯和圣保罗市倡导了一个大胆的都市公园系统。其中包括附近的密西西比河断崖和围绕在这个区域风景如画的湖泊、山脉和谷地周围的土地，以及其他小型的城内公园。他还提议以放射状的大街和林荫道插入城市中心。

[4] 1890年，查尔斯·艾略特建议为大波士顿规划一个都市开放空间系统。此后，马萨诸塞州建立了公共保护协会以获取和保留公共土地。作为协会秘书长以及协会后来的风景园林顾问，艾略特致力于建立一个都市公园委员会并完成详尽的公园规划。他的公园系统包括海滨、内海湾的岛屿、潮汐口、原始森林区以及众多的小广场、运动场和公园，在1894年，还增加了林荫道和公园道作为连接通道。虽然波士顿的地理条件独特，该规划仍然为美国其他的都市开放空间体系建立了框架。

[5] 20世纪早期，在纽约州威斯特郡，规划了一个区域性的包含大面积休闲娱乐区的公园道系统。这个第二代都市公园系统运用公园廊道连接主要的公园设施，同时提供了一个连续的休闲体验过程。布朗克斯河沿岸的机动车公园道在财政上的成功使得1922年立法建立威斯特郡公园委员会。本世纪20年代完成的布朗克斯河公园道规划使得基尔默尔·D.克拉克被任命为该郡的景观规划师，他成功地协调规划了几千英亩公园和公园道。这个带状公园系统连接了包括高尔夫、游泳、娱乐、徒步旅行和其他活动在内的大型娱乐设施。

[6] 从20世纪中期都市公园系统的演变可以看出其重心不断倾向于在自然环境中提供具有活力的娱乐场所。遍布美国郊野公园保护区中几百英里的徒步旅行路线为家庭和工作提供了近距离探索自然的机会。北加

利福尼亚东部海湾区域公园协会对都市开放空间系统的概念做了更精确的阐述。20 世纪 60 年代,在威廉·佩恩·莫特(后来的国家公园局主管)的指导下,这个系统发展为包括 1 000 多英里小径的 7 万多英亩不同类型的公园用地。该系统包括受到保护的独特景观,如具有科研价值的地质构造和植被,以及自然和文化景点,并为那些在 1 500 平方英里内不断增长的人口提供了广泛的娱乐机会。

[7]　未来,风景园林师将继续运用他们的想象力、创造力以及对景观的理解,来满足美国不断增长的城市人口对都市开放空间的需求。当前,城市区域内快速扩大的都市公园系统正在探索新的、创造性的方式来寻求和永久地保护区域开放空间。在芝加哥地区,重要的区域设施将由风景园林师杰瑞·索斯比领导的致力于雷克郡森林保护区工作的非赢利机构所参与的公共土地计划提供,直至 21 世纪。自从该项目于 1963 年启动后,有 12 000 英亩的开放空间通过私人基金得到保护。为了能够确保城市开发与开放空间之间的平衡,我们需要更多这样的保护。

Reading Material A

City Planning

Lance M. Neckar

城市规划

[1]　Landscape architects had a seminal role in the development of city planning in the United States. Many landscape architects would date the involvement of their profession in city planning from the turn of the century, citing the role of <u>Frederick Law Olmsted</u>, Jr., as a member of the <u>McMillan Commission</u>[1], which set about to redesign Washington D. C., in 1901(Fig. 1). Others would turn to the older Olmsted's plan for the <u>1893 World's Columbian Exposition in Chicago</u>[2] (Fig. 2). However, the urban planning role of landscape architects goes back even further. Olmsted and Vaux's design of Central Park in New York City in 1858 and the subsequent development of urban park systems in the late 19th and early 20th centuries provided significant concepts for urban growth. The mid-19th-century suburban town designs by Olmsted, <u>Jed Hotchkiss</u>[3], H. W. S. Cleveland and others constituted the first organized visions of urbanization.

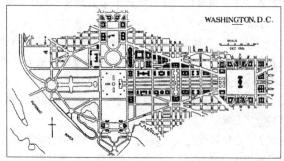 

Fig. 1　Frederick Law Olmsted, Jr. and the plan for Washington D. C., in 1901

[2]　The primary impetus for creating the discipline and profession of city planning came in the decades following the <u>McMillan Plan</u>. This was the era of the <u>City Beautiful</u>[4], a phrase coined by <u>Charles Mulford Robinson</u>, a journalist-turned-planner and author of the influential *Modern Civic Art, or the City Made Beautiful* (1903). Most projects were executed by landscape architects and archi-

tects. The ideal of these early plans was embodied in the civic center, a monumental space designed for city government buildings. Ideally, this civic center, by its location and grandeur, would symbolize the improvement of the whole city. Many landscape architects also provided schematic park system, waterfront and suburban subdivision plans as a regular component of their services.

Fig. 2　The Administration Building at the 1893 World's Columbian Exposition in Chicago

[3]　While Robinson and Daniel H. Burnham[5], were arguably the most famous planners of the day, landscape architects were the preeminent organizers of the profession of city planning. In 1909 the first national conference on city planning was held, with the younger Olmsted and John Nolen[6], making significant presentations. That year Nolen persuaded Wisconsin's legislature to adopt the first legislation authorizing cities to create planning commissions and prepare city plans. That same year, James S. Pray offered the first course in city planning at Harvard University's School of Landscape Architecture[7].

[4]　By the decade's end the grandeur of the City Beautiful had paled, and planners began examining the social, economic and functional aspects of the city. Again, landscape architects were among the first to try "scientific" processes. Frederick Law Olmsted, Jr., attempted to incorporate statistical analysis in his report to the Civic Improvement Committee of New Haven[8] in 1910. Warren H. Manning[9] (Fig. 3), taking an entirely different task, examined landscape resources such as forest cover and soil in his planning work of the same period for Billerica, Mass.

[5]　Science contributed to the growth of professionalism. It was becoming clear that no single established profession could plan a city alone. By 1915, 14 landscape architects took the lead among other professions in establishing the American City Planning Institute[10], now the American Planning Association. Harvard in 1923 offered a city planning option[11] in landscape architecture and in 1929, under Henry V. Hubbard's leadership, created the first separate School of City Planning.

Fig. 3　Warren H. Manning

[6]　In the 1920s and 1930s, the notion of interdisciplinary planning was emphasized in designing industrial and residential communities; landscape architects were cast as leaders or participants in many teams. As the discipline and profession of planning gained strength and legislative support in subsequent decades, many landscape architects turned to government employment or became consultants in urban landscape design, and the growth of several prominent firms in practice today can be

traced to this specialization.

[NOTES]

[1] McMillan Commission 麦克米伦委员会：19 世纪末，由国会议员麦克米伦（Senator James McMillan, 1938—1902）提出整顿首都华盛顿建设，恢复朗方规划，并于 1902 年组成了麦克米伦委员会，对华盛顿进行重新规划，即麦克米伦规划（McMillan Plan）。小奥姆斯特德作为麦米伦委员会成员参与了华盛顿规划，此后几十年他一直从事区域规划和城市规划工作。

[2] 1893 World's Columbian Exposition in Chicago 1893 年芝加哥世界哥伦比亚博览：此届世界博览会是为纪念哥伦布发现美洲大陆 400 周年，奥姆斯特德为此场地做了总体场地规划。

[3] Jed Hotchkiss 杰德·哈奇凯斯（1828—1899）：19 世纪中期，杰德·哈奇凯斯与奥姆斯特德、克里夫兰等人共同设计了郊区城镇，形成第一个有组织的城市化观念。

[4] City Beautiful 城市美化：出现于 1903 年，由专栏作家查尔斯·马尔福德·罗宾逊（Charles Mulford Robinson, 1869—1917）创造，作为一名非专业人士（后来学习风景园林和城市规划），他乘借 1893 年芝加哥世博会的巨大的城市形象冲击，呼吁城市的美化与形象改进，并倡导以此来解决当时美国城市的物质与社会脏乱差的问题。后来，人们便将在他倡导下的所有城市改造活动称为"城市美化运动"（City Beautiful Movement）。城市美化运强调规则、几何、古典和唯美主义，而尤其强调把这种城市的规整化和形象设计作为改善城市物质环境和提高社会秩序及道德水平的主要途径。在本世纪初的前十年中，城市美化运动不同程度地影响了美国和加拿大几乎所有的主要城市。但它在美国实际上却只风行了 16 年的时间（从 1893 年的芝加哥博览会到 1909 年的美国第一届全国城市规划会议）。尽管如此，这一阶段在城市规划和风景园林史上有着重要的意义，其影响至今尤存。从积极的方面来讲，它促进了城市设计专业和学科的发展，开始改善城市形象，也促进了景观和城市规划设计师队伍的形成。但由于"城市美化"往往被城市建设决策者的极权欲、开发者的金钱欲和规划师的表现欲所偷换，把机械的形式美作为主要的目标进行城市中心地带大型项目的改造和兴建，并试图以此来解决城市和社会问题，从而使城市美化运动迷失方向，使倡导者美好的愿望不能实现。在 1909 年的首届全美城市规划大会上，城市美化运动很快被科学的城市规划思潮所替代。

[5] Daniel H. Burnham 丹尼尔·H. 伯纳姆（1846—1912）：美国建筑师和城市规划大师，被尊为美国现代城市规划之父。

[6] John Nolen 约翰·诺伦（1869—1937）：美国近代重要的风景园林师、规划师。

[7] Harvard University's School of Landscape Architecture 哈佛大学风景园林学院。

[8] New Haven 纽黑文市。

[9] Warren H. Manning 沃伦·H. 曼宁（1860—1938）：曼宁是以生态学为基础进行风景园林规划设计的先行者之一。在自己开业之前，曼宁一直为奥姆斯特德工作，在 1893 年芝加哥世界博览会之后的 40 年中，曼宁的实践有效拓展了奥姆斯特德所创立的风景园林行业的范围。曼宁是以自然资源和社区参与为基础的规划设计先驱，也是提倡应用乡土植物的重要代表，主张以乡土植物为主要素材设计野生花园。他的思想被广泛应用于今天的公园设计、社区开发、新城规划、区域规划等土地利用规划的各个方面，并具有深远的影响。1912 年曼宁首次利用透射板进行地图叠加作为分析手段，获得综合信息，为马萨诸塞州比勒里卡（Billerica）做了一个开发与保护规划。当时，美国正在绘制可以供大众使用的国家资源地图，曼宁收集了数百张关于美国土壤、河流、森林和其他景观要素的地图，将其叠在透射板上，基于这样的方法，他做了一个大胆的最具独创性的全美景观规划，这个规划包括了今天一个完整的景观规划所需要的所有内容。他还规划了未来的城镇体系、国家公园系统和休憩娱乐区系统，还规划了今天所使用的主要高速公路系统和长途旅行步道系统。

[10] American City Planning Institute 美国城市规划学会：建立于 1915 年，即现在的美国规划协会（American Planning Association）。

[11] city planning option 城市规划方向。

[GLOSSARY]

| city planning / urban planning | 城市规划 | City Beautiful | 城市美化 |

[NEW WORDS]

arguably	adv.	可论证地,可以认为地	monumental	adj.	纪念碑式的,不朽的,非常的
authorize	v.	批准	participant	n.	参与者,共享者
cite	vt.	引用,引证	preeminent	adj.	卓越的
discipline	n.	学科,纪律	primary	adj.	主要的,第一位的,根源的
embody	vt.	体现,代表,具体表达	prominent	adj.	卓越的,显著的,突出的
grandeur	n.	庄严,伟大	schematic	adj.	示意性的
impetus	n.	推动力,促进	statistical	adj.	统计的,统计学的
interdisciplinary	adj.	多领域的,各学科间的	subdivision	n.	细分,一部
journalist	n.	新闻记者	symbolize	vt.	象征

[难句]

Many landscape architects would <u>date</u> the involvement of their profession in city planning from the turn of the century, <u>citing</u> the role of Frederick Law Olmsted, Jr., as a member of the McMillan Commission, which set about to redesign Washington D. C., in 1901. <u>Others</u> would turn to the older Olmsted's plan for the 1893 World's Columbian Exposition in Chicago. 从新世纪开始,许多景观设计师参与到城市规划中来,例如在1901年,弗里德里克·劳·小奥姆斯特德就作为麦克米伦委员会的成员,以城市规划的角色,对华盛顿进行重新设计。又如,1893年老奥姆斯特德为芝加哥世界哥伦比亚博览会做了总体规划。

[参考译文]

城市规划
兰斯·M. 耐卡

[1] 风景园林师在美国城市规划发展过程中起着开启性的作用。许多风景园林师从新世纪之初就开始参与到城市规划中,例如1901年弗里德里克·劳·小奥姆斯特德作为麦克米伦委员会的成员,对华盛顿进行重新设计。又如,老奥姆斯特德为1893年芝加哥世界哥伦比亚博览会做的总体规划。然而,风景园林师的城市规划角色可以追溯到更早以前。1858年奥姆斯特德和沃克斯设计的纽约中央公园以及此后在19世纪末20世纪初城市公园系统的建立,提供了重要的城市发展理念。19世纪中叶,奥姆斯特德、杰德·哈奇凯斯、H. W. S. 克里夫兰以及其他规划师的郊区城镇设计形成了第一个系统的城市化观念。

[2] 建立城市规划学科和职业的主要推动力来自麦克米伦规划之后的几十年。这是城市美化的时期,由查尔斯·摩尔福德·鲁宾逊创造的时代。鲁宾逊从新闻记者转向规划师,创作了极具影响力的书——《现代城市艺术:或城市创造美》(1903)。这个时期大多数城市规划项目都是由风景园林师和建筑师完成的。这些早期规划理念在城市中心区得到体现,即为城市政府建筑设计一个丰碑式的空间。城市中心区以其特殊的位置和庄严宏大的气势,理想地体现着整个城市的进步。许多风景园林师还进行公园系统、滨水区和城郊区域的总体规划,并将其作为他们常规的工作内容。

[3] 鲁宾逊和丹尼尔·H.伯恩汉姆可以说是当时最著名的规划师,此时,风景园林师已经在城市规划职业中充当着卓越的组织者角色。1909年,首届全国城市规划会议召开,小奥姆斯特德和约翰·诺伦做了重要的发言。那一年,诺伦说服威斯康星州的立法机构采纳第一个批准城市创建规划委员会和筹备各项城市规划的立法。同年,詹姆斯·普瑞在哈佛大学风景园林学院开设了第一门城市规划课程。

[4] 到本世纪末,宏大的城市美化运动逐渐衰退,规划师开始审视城市的社会、经济和功能方面的问题。风景园林师再次参与到最早的"科学规划"尝试过程。1910年,在纽黑文市城市改造委员会的报告中,小弗里德里克·劳·奥姆斯特德曾尝试结合统计分析的方法。同一时期,在为马萨诸塞州布莱瑞卡的规划中,沃伦·H.曼宁,进行了完全不同的工作,他对诸如森林覆盖和土壤等景观资源进行了调查。

[5] 科学对规划专业的发展起到推进作用。越来越明显,没有一个已有的专业能够单独地规划一个城市。到了1915年,14位风景园林师牵头,与其他专业人士一起,建立了美国城市规划学会,即现在的美国规划协会。1923年,哈佛大学在风景园林专业中开设了城市规划方向,1929年,在亨利·V.哈伯德的主持下,创办了第一个独立的城市规划学院。

[6] 20世纪20年代和30年代,跨学科合作规划的观念在工业和居住社区设计中得到强化;风景园林师在许多规划团队中担当着领导者或参与者身份。在后来的十年中,随着规划学科和职业的力量变得强大以及立法的支持,许多风景园林师都转到政府部门工作或成为城市景观设计顾问,目前几个著名的公司的发展可以上溯到这个专业化的过程。

Reading Material B

Parkway and Recreational Areas

公园道路及休闲区

Parkway[1]

Harley E. Jolley

[1] "God grant me the serenity to accept the things I cannot change, courage to change the things I can and wisdom always to know the difference." This "alcoholics' prayer" easily could be that of landscape architects. They, too, have dreams of converting the ugly and devastating into the beautiful and exhilarating. Their demons are degraded environments. Their stimulants are marrying beautiful and utility to create "the way more beautiful"—the parkway[1]—for society's pleasure.

[2] The late U.S. Senator Harry F. Byrd once captured the essence of a parkway: "It is a wonder way over which the tourist will ride comfortably in his car while he is stirred by a view as exhilarating as the aviator may see from the plane." To ensure that exhilaration, a parkway is designed for pleasure, artfully simulating nature's open spaces.

[3] William Penn[2] purportedly incorporated a parkway into his beloved Philadelphia, and Olmsted and Vaux's Central Park classically demonstrated the benefits of the pleasure way. But true parkways are a 20th-century phenomenon, twin-born with the automobile. The rural crest of the Blue Ridge Highway[3] (Fig. 1) between Virginia and North Carolina gave birth to the idea in 1909. However, the suburban parkways of Westchester County[4], N.Y. (1913-1938), were the mold into which most others have been cast.

[4] Today's parkways are remarkably diverse in design and function, as exemplified by the George

Fig. 1 Blue Ridge Highway

Washington Memorial (commemorative)[5] and Colonial (historical) parkways in Virginia[6] and Natchez Trace (heritage) Parkway in Mississippi[7], approved by Congress in 1928, 1930 and 1938, respectively (the three were the products of National Park Service design teams). All, however, continue objectives stated by the Westchester planners: "To preserve for present and future generations some of the charm and natural beauty... to provide for the refreshment of the mind and body plus the well being and happiness of the people." They share, too, the dictum of historian Hiram M. Chittenden, designer of the Yellowstone National Park roads: "They must lie lightly upon the ground." Common, also, are motivations: easing a traffic headache, alleviating inner-city deterioration, rehabilitating a polluted river basin or converting an environmental wasteland into a playland. Out of such considerations were born the Bronx River Parkway of Westchester County, N. Y. and its sister, the Hutchinson River, Sawmill River and Cross County parkways[8]. Likewise, the Blue Ridge Parkway in Virginia was a Depression child, designed to rehabilitate people and land.

[5]   These two parkways—one metropolitan-suburban, the other rural—provided prototypes for most others: curvilinear alignment, limited access, elimination of grade crossings, exclusion of commercial traffic, satellite parks, scenic casements, a consummate blending of natural and cultural features, and the conversion of the entire corridor into a park. Their landscape architects—Gilmore D. Clarke for the Bronx River, begun in 1935 and following the old Blue Ridge Highway—worked with vastly different media: Clarke with the leavings of many years of city despoliation, Abbott with the jaded remnants of two centuries' land abuse. But masterpieces emerged, classic models of design, preservation and conservation.

[6]   Their legendary accomplishments inspired other parkways at the federal, state and regional levels, such as the George Washington Memorial, Garden State[9] (late 1940s) in New Jersey and Mississippi River (planned in the Mid1950s), constructed from Minnesota to Louisiana. All were conservation, preservation and recreation oriented, designed to provide a new visual, mental and spiritual experience for a vast public exasperated with tension-producing "rare gems in the necklace". Thanks to their visionary skills, millions today are able to say, joyfully, "Let's take a break and go on the parkway!"

Recreational Areas

Miriam Easton Rutz

[7] Of the endless range of recreational activities, many take place in an attractive landscape setting. Arboretums and zoological gardens of kings and nobility in Egypt, Mesopotamia and Mexico, as well as places of leisure in many other cultures, have long been described in literature. Taking a cue from their classical Roman ancestors, the Italians in the 16th centuries and the French in the 17th and 18th centuries built luxurious dwellings on extensive landscaped grounds that served as settings for recreation and amusement. In the American colonies, the gardens of southern plantations were developed for pleasure and leisure activities.

[8] As cities grew, open space for recreation became more difficult to find. Cemeteries[10], with their parklike settings, turned into popular places for picnics and strolls. Urban, national and state parks were developed to provide areas for people to spend their leisure time. However, public spaces did not offer all the desired recreational facilities, and, with the increasing prosperity enjoyed in the United States from the mid-19th century, private, sometimes nonprofit organizations began providing additional parks, zoos, gardens, resorts, clubs and camps.

[9] Traditionally, landscape architects have been involved in the design of these recreational areas, because planning is needed to accommodate people while protecting the existing environment. In zoos, for example, the creation of an animal's proper ecological setting or exhibits combining a region's plant and animal collections have become important contemporary concepts. Exposition and world's fair grounds[11] such as Balboa Park[12] (Fig. 2) (1926) in San Diego were early examples of recreational spaces that incorporated parks, cultural events and amusement areas. Many of their design characteristics can be recognized in now-popular theme parks, including a prominent view of an important structure from the entry gate, well-planned circulating patterns and consideration for visitor comfort.

Fig. 2 Balboa Park

[10] The natural environment and vistas are important in planning resorts and clubs. Clubs usually focus on one or two primary activities such as golf, tennis, swimming, boating, horseback riding or skiing, combined with a dining facility or clubhouse. Resorts usually have a combination of housing units including hotels, condominiums and second homes, along with recreational activities. Because such places are set in locations of natural beauty, the search for the proper setting is a major task. The Grand Hotel[13] (1887) near San Diego are good examples of turn-of-the-century resorts that still maintain their original purpose. Resorts created today such as the Homestead[14] near Traverse City, Mich., designed by Johnson, Johnson and Roy in the early 1970s, are carefully planned to provide much more for the visitor, along with landscaped grounds and carefully protected natural features.

[11]　　Recreation entails rejuvenation of the sprit as well as the body, so outdoor spaces for recreation must refresh the whole person. This requires using land, water, plants, views, breezes and other site amenities. But recreation often also calls for fantasy and nostalgia—places such as the turn-of-the-century <u>Main Street at Disneyland</u>[15] (1957) (Fig. 3), the ethnic farms at Old World Wisconsin (1976) and the varied geographical settings at *Seattle's* Woodland Park Zoo, redesigned in the late 1970s by Jones and Jones, also testify to the landscape architect's role in maximizing planned recreational experiences.

Fig. 3　the Main Street at Disneyland

[NOTES]

[1]　　parkway 公园道路:公园道路的规划思想最早是由奥姆斯特德在波士顿"翡翠项链"中提出的用一些连续不断的绿色空间将各个公园连接,奥姆斯特德所说的公园道路,主要是指两侧树木郁郁葱葱的线性通道。这些通道连接着各个公园和周边的社区,宽度也不大,仅能够容纳马车道和步行道。奥姆斯特德和沃克斯在晚期的作品中大量使用这种表现方式,包括 Buffalo 的公园道路和芝加哥的开放空间系统等。稍晚的时候在英国也相继独立出现了一些相关的概念,如霍华德(Ebenezer Howard)田园城市(Garden City)、绿带(Greenbelt)等思想。在霍华德的田园城市这个理想规划中,有420英尺宽的林荫大道环绕着中心城市(Howard, 1902)。由于奥姆斯特德生活的时代汽车还未大量使用,他所强调的交通方式是马车和步行;1920年以后的公园道路建设虽然继承了奥姆斯特德的思想,但主要强调汽车以及道路两旁的景观所带来的行车愉悦感。

[2]　　William Penn 威廉·佩恩(1644—1718):是北美殖民地时期的一位重要政治家、社会活动家,宾夕法尼亚殖民地的开拓者。威廉·佩恩1683年的费城规划是美国最早的城市规划之一。

[3]　　Blue Ridge Highway 蓝色山脊公路或蓝岭公路:是由美国国家公园管理局(National Park Service)管理,沿着风景优美的蓝色山脊山脉(Blue Ridge Mountains)的山路,从维吉尼亚州的仙纳度国家公园(Shenandoah National Park)开始,一直到北卡罗来纳州及田纳西州交界的大烟山国家公园(Great Smoky Mountains National Park)为止,绵延大约470英里。

[4]　　the suburban parkways of Westchester County 纽约威斯特郡的郊野公园道路。

[5]　　George(historical) Washington Memorial 乔治华盛顿纪念馆公园道路(纪念性公园道路)。

[6]　　Colonial parkways in Virginia 维吉尼亚殖民地公园道路(历史性公园道路)。

[7]　　Natchez Trace(heritage) Parkway in Mississippi 纳齐兹部族公园道路(遗产性公园道路)。

[8]　　Hutchinson River, Sawmill River and Cross County parkways 哈琴森河公园道路、索米尔河公园道路和克罗斯郡公园道路。

[9]　　Garden State 花园州公园道路。

[10]　　Cemeteries 墓园:是美国早期公共休闲的区域之一。1796年 Josiah Meigs 设计的 New Haven 墓园,改变传统坟地和教堂墓地的荒凉气氛,创造了美国第一个经过设计的墓地景观,1831年马萨诸塞州园艺协会在波士顿建造第一个"乡村"花园式墓园 Mount Auburn,设计师 Dr. Jacob Bifelow 受英国花园构筑物的启发设计了埃及式大门、歌特式小教堂和诺曼式塔楼,后来的 Laurel Hill(1836)、Green-Wood(1838)、Holly-Wood(1848)、Bellefotaine 墓园(1848),基本上都是模仿它而建。墓园因其公园般的景色,成为人们野餐和散步的休闲场地。

[11] world's fair grounds 世界博览会。
[12] Balboa Park 巴尔波亚公园:圣地亚哥最大的城市文化娱乐公园,以西班牙建筑风格和十五个主题各异的博物馆而著称,还包括一处野生动物园。
[13] Grand Hotel 格兰特酒店。
[14] Homestead 霍姆斯德特度假区。
[15] Main Street at Disneyland 迪斯尼乐园大街。

[GLOSSARY]

cemetery	墓园	recreational area	休闲区
parkway	公园道路	theme park	主题公园

[NEW WORDS]

abuse	n.	滥用	exhilarating	adj.	令人喜欢的,使人愉快的
alignment	n.	排成直线,队列	gem	n.	宝石,珍宝,精华
alleviate	vt.	减轻	grade crossing	n.	平面交叉,平交道
amenity	n.	愉快,舒适,便利设施	inner-city	n.	市中心区,内城
arboretum	n.	植物园	jaded	adj.	疲倦不堪的,厌倦的
aviator	n.	飞行员,飞行家	luxurious	adj.	奢侈的,豪华的
beloved	adj.	心爱的	masterpiece	n.	杰作,名著
cemetery	n.	墓地,公墓	Mesopotamia		美索不达米亚
condominium	n.	分契式公寓,共管公寓	mold	n.	模子,铸型
crest	n.	顶部,顶峰	motivation	n.	动机
cue	n.	暗示,提示	nostalgia	n.	怀旧之情,乡愁
curvilinear	adj.	曲线的	orient	vt.	确定方向,导向
degraded	adj.	被降级的,退化的	picnic	n.	野餐
demon	n.	魔鬼,噩梦	prototype	n.	原型
despoliation	n.	掠夺	purportedly	adv.	据称
deterioration	n.	退化,堕落	rehabilitate	v.	使恢复,使复原
devastating	adj.	破坏性的,毁灭性的	rejuvenation	n.	返老还童,恢复活力
dictum	n.	格言	remnant	n.	残余,残迹,剩余
dwelling	n.	住处	respectively	adv.	分别地,各个地
elimination	n.	排除,除去,消除	serenity	n.	平静
entail	vt.	使蒙受,使承担,使必需	stimulant	n.	刺激物
ethnic	adj.	人种的,种族的,异教徒的	stroll	n.	漫步,闲逛
exasperate	v.	激怒	vistas	n.	狭长的景色,远景,展望
exemplify	vt.	例证,作为……例子	wasteland	n.	荒地,废弃地

[难句]

1. "God grant me the serenity to accept the things I cannot change, courage to change the things I can and wisdom always to know the difference." This "alcoholics' prayer" easily could be that of landscape architects. 上帝给予我平静，使我能接受我所不能改变的事情；给予我勇气，使我能改变我所能改变的事情；并给予我智慧去分辨这两者的区别。"这所谓的"醉鬼祈祷"也是风景园林师们所想的。

2. Their landscape architects—Gilmore D. Clarke for the Bronx River, begun in 1935 and following the old Blue Ridge Highway—worked with vastly different media: Clark with the leavings of many years of city despoilation, Abbott with the jaded remnants of two centuries' land abuse. But masterpieces emerged, classic models of design, preservation and conservation. 他们的景观设计师——基尔默尔·D. 克拉克，在1935年为布朗克斯河，以及之后为老的蓝色山脊公路采用了完全不同于其他的方法：克拉克保留了城市多年来掠夺式发展而产生的残余物，艾伯特保留了两个世纪来土地滥用而遗留的残迹。但是这却产生了杰作和一流的设计模式，即保护和保留。

[参考译文]

公园道路及休闲区

公园道路

哈利·E. 贾利

[1] "上帝给予我平静，使我能接受我不能改变的事情；给予我勇气，使我能改变我所能改变的事情；并给予我智慧去分辨这两者的区别。"这所谓的"醉鬼祈祷"也是风景园林师们所想的。他们也梦想着改变丑陋破败的环境使之成为美丽愉快的新环境。他们的噩梦则是退化的环境。社会休闲需求刺激他们将美观与实用相结合，创造出更美丽的道路——公园道路。

[2] 已故的美国参议员哈瑞·F. 伯德曾经这样阐述公园道路的本质："它是能让旅行者在汽车上一边舒服地驾驶，一边能像飞行员从飞机上看地面的那种愉悦心情看风景的非同一般的道路。"为了确保这种愉悦，公园道路要设计成为休闲道路，并巧妙地模拟自然开放空间。

[3] 据称，威廉·佩恩曾在他最喜欢的费城建造了一条公园道路，奥姆斯特德与沃克斯的中央公园也诠释了这种休闲道路的益处。但是真正的公园道路出现在20世纪，伴随汽车的出现而出现。1909年弗吉尼亚和北卡罗来纳州之间乡村山顶的蓝色山脊公路（蓝岭公路）诞生了这个理念。然而，纽约威斯特郡的郊野公园道路（1931—1938）成为其他地方纷纷效仿的蓝本。

[4] 今天的公园道路在设计和功能上多种多样，例如乔治·华盛顿纪念馆（纪念性）、维吉尼亚殖民地（历史性）和纳齐兹部族（遗产性）公园道路，分别于1928年、1930年和1938年通过国会认证（这三个都是国家公园管理局设计团队的作品）。然而，所有这些公园道路所追求的目标都如威斯特郡规划师所述："为当代人和后代保存一些迷人的自然美景……使人身心畅快并且舒适健康、精神愉悦。"他们也遵循历史学家、黄石国家公园道路的设计者海勒姆·M. 奇滕登的格言："他们必须轻轻地铺展于地面上。"他们也有着共同的动机：减少令人头痛的交通问题，减轻市中心衰退，恢复被污染河流流域或者将废弃地改造成游戏场，正是出于这些考虑，才诞生了纽约州威斯特郡布朗克斯河公园道路，以及她的姐妹哈琴森河公园道路、索米尔河公园道路和克罗斯郡公园道路。同样的，弗吉尼亚蓝色山脊公园道路也曾是一处大萧条时期的产物，设计的动机就是让人们和土地修复。

[5] （布朗克斯河公园道路和蓝色山脊公园道路）这两种公园道路，一个位于城市郊区，另一个位于乡村，为其他公园道路提供了样本：弯曲的线形，有限的入口，没有平面交叉，没有商业交通，拥有卫星公园和景窗，自然和文化特性完美融合，整个道路廊道变成一个公园。他们的风景园林师——基尔默尔·D. 克拉克，1935年开始设计布朗克斯河公园道路，在老的蓝色山脊公园道路之后，采用了完全不同的方法：克拉克保留了城市多年来掠夺式开发产生的残余物，艾伯特（的蓝色山脊公园道路）保留了两个世纪来土地滥用而遗留的残迹。但是

· 115 ·

这却产生了杰作和一流的设计模式,即保护和保留。

[6] 他们传奇般的成就激发了其他国家、州和区域的公园道路建设,例如新泽西州乔治·华盛顿纪念地公园道路和花园州公园道路(20世纪40年代末)、穿越明尼苏达州和路易斯安那州的密西西比河公园道路(规划于20世纪50年代中期)。所有这些公园道路都以保留、保护和休闲娱乐为导向,为广大的处于精神紧张和焦虑的人们提供一个新的视觉、精神和心灵上的体验,创造犹如"珍珠项链"般的美好风景。感谢他们的远见卓识,使今天的百万民众能够愉快地说"我们休息下,去公园道路吧!"

休闲区

米里亚姆·阿斯顿·鲁茨

[7] 无穷的休闲活动中,大部分均发生在有吸引力的景观环境中。埃及、美索不达米亚和墨西哥国王和贵族的植物园、动物园,以及其他文明中的休闲场所,长久以来都被文学作品所描述。从他们古老的罗马人祖先那里得到提示,16世纪的意大利人,17、18世纪的法国人在广阔的风景中建造奢华的住处,作为他们休闲的地方。在美国殖民地,南方种植园也发展成为消遣娱乐的场所。

[8] 随着城市的发展,用于休闲的开放空间越来越难找到。墓地,因其拥有与公园相似的环境,成为受欢迎的野餐和散步地。新建的城市公园、国家公园以及州公园为人们度过闲暇时间提供了去处。然而,公共空间却不能提供所有人们想要的娱乐设施,随着美国自19世纪中期以来的繁荣发展,私人或非营利机构开始提供更多的公园、动物园、花园、度假区、俱乐部和野营地。

[9] 从传统来看,风景园林师早已涉及这些娱乐区的设计,因为这些区域需要能够容纳人同时又保护现有环境的规划。例如在动物园中,创造适合动物生存的生态环境或馆所,并搜集某个区域的植物或动物种类,已成为当前重要的设计理念。展览会以及世界博览会,如圣地亚哥的巴尔波亚公园(1926)是较早的休闲区结合公园、文化活动及娱乐区的案例。他们的许多设计特点在现在流行的主题公园中可以看到,包括重要而醒目的大门结构、精心规划的交通流线以及使游客舒适的设计考虑。

[10] 自然环境和视景在度假区和俱乐部的规划中很重要。俱乐部关注一到两个主要的活动,比如说高尔夫球、网球、游泳、划船、骑马或滑雪,结合餐饮设施或会所。度假区通常会有多种住宿单元,包括宾馆、共管公寓和第二寓所,也结合休闲活动。由于这样的地方处于风景优美的自然环境中,设计合适的环境是很重要的任务。19世纪初的圣地亚哥附近的格兰特酒店(1887)就是个优秀例子,它现在依然保持着原有的用途。现在新建的风景胜地,比如19世纪70年代初期由詹森·詹森和罗伊设计的密歇根州特拉弗斯城附近的霍姆斯德特,就为游客进行了精心的规划,其环境景观也被精心保护,并保持着自然的风貌。

[11] 娱乐使人的精神放松,身体恢复活力,所以供人娱乐的户外空间必须要使人恢复一新。这就需要用到土地、水体、植物、视线、呼吸以及其他令人愉快的场所要素。但是娱乐活动也常常需要幻想和怀旧,就像20世纪初迪斯尼乐园大街(1957)、威斯康星州的旧世界农场(1976)以及19世纪70年代被琼斯重新设计的西雅图森林动物园里的各种地形环境,都印证了风景园林师在娱乐体验最优化的规划中所起到的作用。

[思考]

(1)西方的都市开发空间理论对当前绿道理论发展有什么影响?
(2)美国的城市规划专业和学科发展历程是怎样的?
(3)美国的城市美化运动是谁发起的?对城市规划的发展有哪些影响?
(4)公园道路概念是由谁开创的?公园道路有哪些主要类型和案例?
(5)娱乐区作为为公众服务的开放空间,其规划应考虑哪些方面的内容?

UNIT 7　URBAN GREEN SPACE

TEXT

Urban Parks
Anne Whiston Spirn[1]

城市公园

[1]　The first parks were royal hunting grounds. The word "park" itself meant an enclosed tract of land for keeping beasts of the chase and, later, deer, cattle and sheep. The Boston Common[2] (Fig. 1), set aside as common pasture in 1630, was America's first public park. From these origins in animal husbandry and stewardship, the park became associated with the health of both people and the city itself-as if a park were "the lungs of the city".

Fig. 1　the Boston Common

[2]　Other early American "parks" were town squares. William Penn's plan for Philadelphia[3] in 1683, for example, and James Oglethorpe's plan for Savannah[4] in 1733 (Fig. 2) set aside squares that were ultimately surrounded by residences and institutions. These represented a vision of urbanity inspired by new town planning ideas in England.

Fig. 2　James Oglethorpe's plan for Savannah in 1733

[3]　All across America, for a half century from 1858 on, cities built large urban parks, designed in a pastoral style: Central Park[5] (1858, Olmsted and Vaux) in New York City (Fig. 3); Fairmount Park[6] (1865) in Philadelphia; Prospect Park[7] (1866, Olmsted and Vaux) in Brooklyn (Fig. 4),

Forest Park[8] (1876) in St. Louis; Golden Gate Park[9] (1870) in San Francisco.

Fig. 3　Central Park (1858, Olmsted and Vaux) in New York City

Fig. 4　Prospect Park (1866, Olmsted and Vaux) in Brooklyn

[4]　By the end of the 19th century, the park stood for certain democratic ideals. Frederick Law Olmsted, for example, saw the park as common ground where all cities might have access to "the best scenery of the region." This idea found its expression in Olmsted's work in Boston on what sometimes is referred to as the Emerald Necklace[10] (1878-1895) (Fig. 5). One of the nation's first and most completely developed urban park systems, it connected the heart of the city with new suburbs and outlying farmland and integrated parks and parkways with a streetcar line and storm drainage. As systems, parks shaped the growth of American cities, inspired the city planning movement in the United States and increased adjacent land values, contributing to the city's economic health.

UNIT 7　URBAN GREEN SPACE

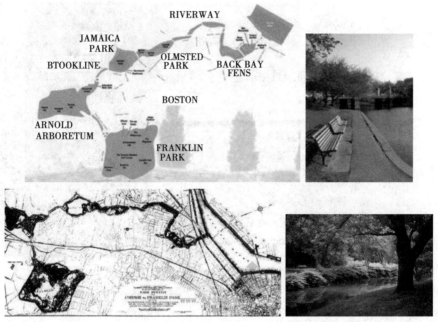

Fig. 5　Emerald Necklace (1878-1895) of Boston park systems

[5]　The large 19th century parks have been adapted again and again to meet changing needs and values. As American society has changed, so have the role and form of the urban park. Older urban parks are now composed of layers of erasures and additions. Many have been sites of expositions, whose buildings still remain, others have accommodated the addition of facilities for intensive recreation, such as stadium, skating rinks and swimming pools.

[6]　Sports and games became an increasingly important part of parks during the 20th century. Charlesbank[11] (1892, Frederick Law Olmsted) in Boston was among the first urban playgrounds. By the 1920s metropolitan parks such as New York's Jones Beach[12] (1929, Long Island Park Commission) were built for the recreational needs of enormous numbers of city residents who arrived not by streetcar, but by automobile. Since the development of the Emerald Necklace in the last century, parks have often been associated with transportation projects. The esplanade of Carl Schurz Park[13] in New York City is cantilevered out over the East River with a highway on the lower level.

[7]　We owe many urban parks to a concern for health and safety. The establishment of Fairmount Park in Philadelphia was motivated, in part, by a desire to protect the city's water supply, the Boston system's Fens and Riverway[14] by a plan to improve water quality and reduce floods. The San Antonio Riverwalk[15] (1939-1941) was also a response to a flood hazard.

[8]　Today, urban parks range from tiny "pocket" parks to entire districts. Paley Park[16] (Fig. 6) (1965-1968, Zion and Breen) in New York City, the size of a single building lot, is an oasis in midtown Manhattan. Independence National Historical Park[17] (1956) in Philadelphia includes historic buildings as well as plazas, landscaped blocks and gardens. More recently, entire portions of cities in need of economic revitalization have been designated as parks: Lowell[18], Mass., for example.

[9]　The park has been an extremely fertile idea that has continued to evolve and to spawn many other landscape types such as the parkway, the garden suburb, the playground, the amusement park

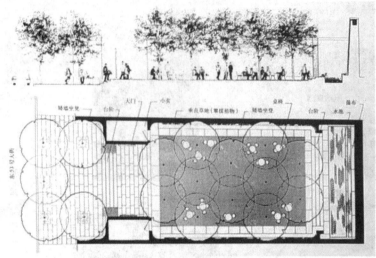

Fig. 6　Paley Park (1965-1968, Zion and Breen) in New York City

and the industrial park. Today, urban parks carry with them many associations: stewardship, health, scenery, sport, economic revitalization and the framework for human settlement. These responsibilities are a complex burden for a single type of open space to fulfill and present a challenge for plotting the present and future role of the urban park in American society.

[NOTES]

[1]　Anne Whiston Spirn 安·维斯特·斯本:美国宾夕法尼亚大学风景园林教授,著有《景观的语言》(Language of Landscape)。

[2]　Boston Common 波士顿公园:建于1634年,是美国最早的城市公园,位于波士顿市中心,最初用于放牧牛羊,还曾是殖民地民兵的军事训练场。后来为公众开放作为公共公园,现已被列为美国国家历史地标。

[3]　William Penn's plan for Philadelphia 威廉·佩恩费城规划:1683年威廉·佩恩的费城规划是美国最早的城市规划之一,他所划下的费城整齐的街道,一直到今天,格网没有任何的改动。他首先派测量师在宾夕法尼亚州的中心位置进行测量,然后把它划成整整齐齐的格子,北美大陆早期的很多规划都是测量师在测量图上完成的。由于威廉·佩恩的开拓,费城很快就变成美国早期规模最大的城市,最后成为革命的首都。费城规划反映了美国城市规划的历程,测量师开拓的规划模式,以及严格按照土地制度、法律制度和财产管理制度的出发点。

[4]　James Oglethorpe's plan for Savannah 詹姆士·奥格尔索普的沙瓦那规划(1733)。

[5]　Central Park 纽约中央公园:在1857年举办的纽约中央公园设计比赛中,奥姆斯特德(Olmsted)与沃克斯(Vaux)的"绿草地"方案成为得奖设计,并于1858年公园始建。纽约中央公园是奥姆斯特德新型城市公园思想的集中体现,也因此掀起美国公园建设的高潮,进而发展成为遍及全美的城市公园运动。

[6]　Fairmount Park 费城费尔蒙特公园(1865):是一个自然式大型公园,也是美国最大的市区公园,面积约为35平方千米,公园以自然景观为主,极少人工修饰。费尔蒙特公园还是1875年美国独立一百周年纪念博览会的会场,一直到现在里面还保存着当时许多国家参加展览的展品。

[7]　Prospect Park 布鲁克林希望公园(1866):是奥姆斯特德完成纽约中央公园工作之后,与沃克斯共同设计的位于布鲁克林中心地带的城市公园。希望公园占地2.38平方千米,是一个供人们休闲娱乐、重返自然的地方。

〔8〕 Forest Park 圣·路易斯森林公园(1876)：是一个大型综合性城市森林公园，占地530多公顷，有众多的河池、自行车道、运动场和历史遗址、纪念雕塑，包括美国最古老和最大的圣路易斯动物园(St. Louis Zoo)、室外音乐厅——城市剧院(Municipal Theater)、科学中心(Science Center)、杰佛逊纪念馆(Jefferson Memorial)、历史博物馆(History Museum)、谬尼歌剧院(Muny Opera)及艺术博物馆(Art Museum)及高尔夫球场、网球场等。

〔9〕 Golden Gate Park 旧金山金门公园(1870)：旧金山大型城市公园，占地410多公顷，从城市中心区一直绵延至太平洋海岸，面积广阔，气势磅礴。

〔10〕 Emerald Necklace 波士顿"翡翠项链"：波士顿城市公园系统(Boston urban park systems)，奥姆斯特德将波士顿主要的公园绿地以林荫道或河道绿化串联，连接了波士顿(Boston)、布鲁克兰(Brookline)和坎布里奇(Cambridge)，并与查尔斯河(Charles River)相连，长度达25千米，其中包括查尔斯河岸(Charlesbank)、巴克贝沼泽(Back Bay Fens)、河道公园(Riverway)、莱弗里特公园(Leverett Park)、牙买加池塘(Jamaica Pond)、阿诺德植物园(Arnold Arboretum)、富兰克林公园(Franklin Park)以及相连的公园道路。

〔11〕 Charlesbank 波士顿查尔斯滨河(1892)。

〔12〕 Jones Beach 纽约琼斯海滩(1929)。

〔13〕 Carl Schurz Park 纽约卡尔·舒尔茨公园。

〔14〕 Fens and Riverway 巴克贝沼泽公园及河道公园。

〔15〕 San Antonio Riverwalk 圣·安东尼奥滨河道(1939—1941)。

〔16〕 Paley Park 纽约帕雷公园(1965—1968)：20世纪50—60年代，西方发达国家，尤其是美国，建造了大量的高层建筑，对城市环境造成了巨大的破坏，城市中的绿地犹如沙漠中的绿洲，于是出现了一些见缝插针的小型城市绿地——袖珍公园(vestpocketpark)，很快受到公众的欢迎，位于纽约的帕雷公园是这类袖珍公园中的第一个，它曾被称赞为20世纪最有人情味的空间设计之一。

〔17〕 Independence National Historical Park 费城国家独立历史公园：位于费城特拉华河西岸，是美国历史上举足轻重的历史区域，《独立宣言》和美国第一部宪法都在此诞生。1948年经美国国会决定，将费城的独立大厅以及周围的所有历史性建筑辟为国家独立历史公园，公园内保存有当年进行独立革命活动的场所、文物，独立后第一任总统和美国联邦议会厅，最早的哲学馆、图书馆、教堂等。

〔18〕 Lowell 洛厄尔：马萨诸塞州一城市。

[GLOSSARY]

amusement park	游乐园，娱乐公园	storm drainage	排水系统
garden suburb	花园郊区	urban park	城市公园
industrial park	工业园，产业园	urban park system	城市公园系统
new town planning	新镇规划，新城规划	water supply	供水系统，水源

[NEW WORDS]

cantilever	vt.	使……伸出悬臂	esplanade	n.	平坦的空地，游憩场，散步路
chase	n.	追赶，追击，追逐	fertile	adj.	富饶的，多产的，丰富的
drainage	n.	排水，排水系统，排水区域	husbandry	n.	畜牧，饲养，耕种，管理
emerald	n.	翡翠，绿宝石，翠绿色	midtown	n.	市中心区
enclose	vt.	把……围起来，围绕，圈起	motivate	v.	激发

continued

oasis	n.	绿洲,宜人的地方,乐土	rink	n.	溜冰场,滑冰场
outlying	adj.	边远的,偏僻的	spawn	n.	(鱼、蛙等)大量产(卵);引发
pastoral	adj.	田园风光的,田园生活的	stewardship	n.	管理工作
pasture	n.	草原,牧场,牧地	streetcar	n.	有轨电车
plotting	n.	测绘,描图,绘图	tract	n.	广阔的地面,大片土地,地域
revitalization	n.	新生,复兴	urbanity	n.	文雅,雅致,都市风格

[参考译文]

城市公园
安·维斯特·斯本

[1] 最早的公园由皇家狩猎场转化而来。"公园"这个词本身原意是为狩猎而圈养牲畜,后来为饲养鹿、牛和羊所围合的大片土地。波士顿公园在1630年被用作公共牧场,是美国最早的公共公园。公园源于对动物的畜养和管理,而后变成了与人们和城市自身健康相关联的事物,犹如"城市的肺脏"。

[2] 美国其他早期的"公园"是城镇广场。例如,威廉·佩恩1683年为费城做的规划,以及詹姆士·奥格尔索普1733年为沙瓦那做的规划,都留出了广场,这些广场最终被居住区和公共机构所包围。它们体现了英国新镇规划思想所激发的都市化理念。

[3] 从1858年以后的半个世纪,美国所有的城市都建立了大型城市公园,并设计成田园式风格,如纽约中央公园(1858,奥姆斯特德和沃克斯)、费城费尔蒙特公园(1865)、圣·路易斯森林公园(1876)、旧金山金门公园(1870)。

[4] 在19世纪之前,公园代表某种民主思想。例如,弗雷德里克·劳·奥姆斯特德将公园看作公共区域,在那里所有的市民都能够到达"该区域最好的风景"。这个理念在奥姆斯特德为波士顿所做的规划中得到体现,这个规划有时被称为波士顿"翡翠项链"(1878—1895)。作为美国最早并且最完善的城市公园系统之一,它将城市中心与新郊区、远郊农田相连,并以一条有轨电车线路和排水系统将公园和公园道路相连。作为系统,公园塑造了美国城市的发展,激发了美国城市规划运动,提升了相邻土地的价值,促进了城市经济健康发展。

[5] 19世纪的大型公园都被不断改造来满足不断变化的需求和价值取向的转变。如同美国社会所发生的变化,城市公园的作用和形式也发生了变化。老城市公园现在是由许多删除和增加的内容层叠而成。许多原来是展览会的场所,其建筑依然保留;还有些为集中的休闲活动增设了一些设施,比如露天运动场、溜冰场和游泳池。

[6] 在20世纪,运动和比赛变成公园中越来越重要的内容。波士顿查尔斯滨河(1892,弗雷德里克·劳·奥姆斯特德),就是最早的城市运动场之一。到20世纪20年代,都市公园比如纽约琼斯海滩(1929,长岛公园委员会)是为了满足开汽车而不是乘坐电车而来的数量庞大的城市居民休闲需求而建。从上个世纪翡翠项链的建设开始,公园常常与交通项目联系在一起。纽约卡尔·舒尔茨公园的散步道就悬挑于东河之上,并且它的下方建有一条公路。

[7] 许多城市公园的建立应归功于对健康和安全的考虑。修建费城费尔蒙特公园的一部分原因是为了保护城市供水,波士顿绿地系统的沼泽和河道在规划时是为了改善水质、减少水灾而设。圣·安东尼奥滨河道(1939—1941)也是为了应对水患而规划。

[8] 今天,城市公园的规模小到微型"袖珍"公园大到覆盖整个地区。纽约帕雷公园(1965—1968,泽恩和布林)只有一块建筑用地的大小,却是曼哈顿中心区的一块绿洲。费城国家独立历史公园仅仅包括历史建筑以及广场、景观区和花园。最近,为了经济复兴的需要,有的城市被整体设定为公园,比如马萨诸塞州洛厄尔。

[9] 公园已是一个内涵极其丰富的概念,它不断发展进化并且衍生出许多其他景观类型,例如公园道路、花园郊区、运动场、游乐园和工业园。如今,城市公园承载着许多关联功能:服务、健康、风景、运动、经济复兴和人

居环境。这些职责对于一个单一类型的开放空间来说是个复杂的重担,这也为美国城市公园在当前和未来所应担负的角色提出了挑战。

Reading Material A

Streetscape, Squares and Plazas[1]

Mark Chidister

街景和广场

[1]　Medieval and Renaissance European cities were the pattern for many colonial town plans and squares in America—a pattern freely adjusted to frontier realities and aspirations. European influences are still apparent in some town plans and squares. However, recent dramatic changes in urban growth and form have left an imprint of streets and plazas that are as different from their European counterparts as today's Americans are from their ancestors.

[2]　Closest to the planned towns of medieval Europe were the Spanish colonial settlements. These settlements were planned and organized on a grid around a regular, central plaza. The remnants of two such plazas remain in Santa Fe (1609) and Albuquerque[2] (1781). Like their medieval counterparts, such plazas were the spatial, political, commercial, social and often religious focus of the community-essential places, not amenities. In contrast to the uniformity of Spanish town plans, French English settlements varied in form. While gridiron plans predominated, a wide variety of street patterns were built. Invariably, a public green space such as the New Haven Green[3] (1638) (Fig. 1) in Connecticut was included. These commons[4] and residential squares[5] were used for grazing animals, parades, militia assemblies, ornamental gardens and settings for churches and town halls. Early public squares in the United States, in contrast to private residential squares of London and Paris, became symbols of a new egalitarian society. Yet none achieved the diversity and intensity of use found in Spanish colonial of European medieval squares. Rather, vital functions focused on main streets of small towns and districts of large cities.

Fig. 1　New Haven Green(1638) in Connecticut

[3] The rapid growth of cities during the late 19th and early 20th centuries was not accompanied by significant increases in central-city open space. Unlike European cities, where density was balanced by abundant plazas, American cities after World War II were complex, crowded places with little or no spatial relief. A half century earlier the World's Columbian Exposition in Chicago had projected the dream of cities with grand streets and generous squares—a dream largely unrealized.

[4] Faced with this lack of open space, city planners in the 1950s and 1960s devised incentive zoning[6] to encourage private development with plazas for public use. Despite the opportunities these new plazas afforded, there were costs: shaded streets from taller buildings, uncoordinated open space patterns and plazas sometimes catering more to private interests than public good.

[5] While corporate skyscraper plazas have dominated the scene since the 1960s, many civic and commercial plazas and streetscapes have been undertaken to make downtowns more habitable and desirable. Two adaptive use projects, Ghirardelli Square[7] (Fig. 2) (1962, Wurster, Bernardi and Emmons, with Lawrence Halprin and Associates) in San Francisco and Fanueil Hall Marketplace[8] (Fig. 3) (1976—1978, Benjamin Thompson Associates) in Boston are privately owned and managed but receive an intensity and diversity of use that few corporate plazas have generated. Recent civic plazas range from the brick expanse of Boston's City Hall Plaza[9] (Fig. 4) (1968, Kallman, McKinnell and Knowles) to the complexity of Minneapolis's Peavy Plaza[10] (Fig. 5) (1975, M. Paul Friedburg and Partners) and the Auditorium Forecourt Fountain[11] (Fig. 6) (1970, Lawrence Halprin) in Portland, Ore. On the whole, recent plazas have become dispersed local spaces centered around young office workers, weekday lunches and planned events[12]. Most streetscape projects serve as downtown open-air shopping malls to make working and shopping enjoyable. Many like Nicollet Mall[13] (Fig. 7) (1968—1981, Lawrence Halprin) in Minneapolis and the State Street Mall (1975—1981, M. Paul Friedburg) in Madison[14], Wis., have done well. Some less successful malls are being demolished and replaced with standard streets and sidewalks.

Fig. 2 Ghirardelli Square (1962, Wurster, Bernardi and Emmons, with Lawrence Halprin and Associates) in San Francisco

[6] Plazas and streets are being transformed. New "public" spaces are increasingly indoors and privately managed. The Ford Foundation atrium[15] (Fig. 8) (1967, Dan Kiley) in New York City was an oddity when built it is now the staple. Atriums, arcades, galleries, courts and skywalks abound, and traditional street life is moving indoors. Despite their pleasures and opportunities, they have not served as common spaces for the public. In a time when there is renewed interest in cities and public life, care must be taken to perpetuate and add places to our cities where access and use by all of the public are ensured.

UNIT 7 URBAN GREEN SPACE

Fig. 3 Fanueil Hall Marketplace (1976—1978, Benjamin Thompson Associates) in Boston

Fig. 4 Boston's City Hall Plaza   Fig. 5 Minneapolis's Peavy Plaza

Fig. 6 Auditorium Forecourt Fountain (1970, Lawrence Halprin) in Portland, Ore.

Fig. 7 Nicollet Mall (1968—1981, Lawrence Halprin) in Minneapolis

Fig. 8　Ford Foundation atrium (1967, Dan Kiley) in New York City

[NOTES]

[1]　squares and plazas 广场：square 指广场、方场、街区或四面有街道的房屋区；plaza 指城镇或城市中的公众广场或类似的开阔区域、集市广场，现代还指购物中心、综合大厦、公路旁的服务区。

[2]　Santa Fe and Albuquerque 圣塔菲和阿尔伯克基。

[3]　New Haven Green 纽黑文绿地：纽黑文市的规划被认为是美国第一个城市规划（1638），它按照格网布局进行规划，网格的中心是大型城市绿地，即纽黑文绿地。纽黑文绿地直到现在都是备受欢迎的城市中心开放空间。

[4]　commons 公共用地：指公共绿地或公共空间。

[5]　residential squares 居住绿地。

[6]　incentive zoning 奖励区划：是20世纪五六十年代城市规划师提出的鼓励私人建设广场供公众使用从而获得公共空间的方法，奖励区划规定开发者如果在高密度的商业区和住宅区内兴建一合乎规定的广场，则可获得增加20%的楼板面积的奖励；兴建步行走廊、拱廊也可得到一定的增建面积。奖励区划对美国公共事业作出了一定贡献，但也有很显著的弱点，就是容易使开发商利用奖励的机会不断突破建筑的高度控制，例如在西雅图和纽约，开发商通过公共空间的贡献得到建筑高度或容积率的提高，但最终的建设结果却与当初设定的城市形态相距甚远，高大建筑物在街道上的投下大量阴影，公共空间形式不协调。

[7]　Ghirardelli Square 旧金山吉拉德利广场：吉拉德利是有名的高级巧克力品牌，吉拉德利广场原为一栋制造巧克力的工厂，在1964年成功地再利用改造成为购物中心，并成为许多老建筑改造为小型购物中心的参考典范。

[8]　Fanueil Hall Marketplace 波士顿法尼尔厅市场：是美国波士顿一座历史建筑，靠近海滨和现在的政府中心，自从1742年起，法尼尔厅就是一个市场和会议厅，多个世纪以来都被作为城市的商业中心。

[9]　Boston's City Hall Plaza 波士顿市政厅广场。

[10]　Minneapolis's Peavy Plaza 明尼阿波利斯匹威广场。

[11]　Auditorium Forecourt Fountain 波特兰演讲堂前水景广场。

[12]　planned events 节庆：西方把不同类型的节庆统一称为 Event（事件），盖茨（Getz）把事先经过策划的事件（planned event）分为8个大类，即文化庆典（包括节日、狂欢节、宗教事件、大型展演、历史纪念活动）、文艺娱乐事件（音乐会、其他表演、文艺展览）、商贸及会展、体育赛事、教育科学事件、休闲事件、政治/政府事件、私人事件。

[13]　Nicollet Mall 尼考利特商业街。

[14]　State Street Mall in Madison 麦迪逊国家商业街。

[15]　Ford Foundation atrium 纽约福特基金会中庭。

[GLOSSARY]

atrium	中庭,天井	corporate plazas	公司广场	planned event	节庆活动
civic plaza	市民广场	incentive zoning	奖励区划	streetscape	街景
commercial plaza	商业广场	open-air shopping malls	户外商业街		

[NEW WORDS]

abound	vi.	富于,充满	gridiron	n.	烤架,格状物
adaptive	adj.	适合的,适应的	habitable	adj.	适于居住的,可居住的
adjust to	vt.	把……调节到,适应于	imprint	n.	印记,印象,印迹
arcade	n.	拱廊,拱街	incentive	adj.	激励的,鼓励性质的
aspiration	n.	热望,渴望,愿望	invariably	adv.	不变地,总是
atrium	n.	中庭,天井前庭	medieval	adj.	中世纪的
auditorium	n.	礼堂,会堂,演讲厅,音乐厅	militia	n.	民兵
cater	vi.	对……提供食品或服务,提供	oddity	n.	奇特,古怪,怪异的事物
colonial	adj.	殖民的,殖民地的	parade	n.	游行,阅兵,检阅
corporate	adj.	公司的,社团的,法人的	perpetuate	vt.	使永久化,使持久化,使持续
counterpart	n.	副本,相对物	plaza	n.	广场,露天广场,购物中心
court	n.	庭院,院子	remnant	n.	残余,剩余,残迹
demolish	vt.	拆毁,拆除	renaissance	n.	复兴,复活,文艺复兴
desirable	adj.	理想的,令人满意的,良好的	settlement	n.	殖民,殖民地
devise	vt.	设计,想出,创造,发明	skywalk	n.	行人天桥,高架步道
downtown	n.,adj.	市中心,闹市区,商业区	square	n.	广场
egalitarian	adj.	平等主义的	staple	n.	主要部分,重要内容
forecourt	n.	前院	streetscape	n.	街景
frontier	n.	国境,边境,最早的移民地	town hall	n.	市政厅
gallery	n.	走廊,画廊,美术馆	uncoordinated	adj.	不协调的,杂乱无章的,无计划的
graze	v.	放牧	uniformity	n.	一致,无差异,相同,统一
grid	n.	格子,网格,方格	zoning	n.	分区,区划,区域规划

[难句]

1. However, recent dramatic changes in urban growth and form have left an imprint of streets and plazas that are as different from their European counterparts as today's Americans are from their ancestors. 然而,近期城市发展和城市形态的显著变化给街道和广场留下的烙印,使得它们不同于欧洲的街道和广场,正如今天的美国人不同于他们的祖先一样。

2. In a time when there is renewed interest in cities and public life, care must be taken to perpetuate and add places to our cities where access and use by all of the public are ensured. 当城市和公众生活关心的东西更新时,必须要不断地关注这些变化,为我们的城市增加这些空间,并保证所有市民都能进入和使用。

[参考译文]

街景和广场
马克·齐迪斯特

[1] 中世纪和文艺复兴的欧洲城市为许多美国殖民城镇的规划和广场提供了模板——一种可以根据实际边界形态和人们意愿而自由调节的模式。欧洲影响在一些城镇规划和广场中依然明显。然而,近期城市发展和城市形态的显著变化给街道和广场留下的烙印,使得它们不同于欧洲的街道和广场,正如今天的美国人不同于他们的祖先一样。

[2] 与中世纪欧洲城镇规划格局最接近的是西班牙殖民地。这些殖民地的规划和组织是在围绕矩形中心广场的网格上布局的。现存的这类广场有两处,分别位于圣塔菲(1609)和阿尔伯克基(1781)。就像中世纪欧洲的这类广场那样,它们是社区空间、政治、商业和社会的焦点,往往也是宗教的焦点,它们是基本的场所,却不一定是宜人的地方。与西班牙城镇规划的整齐划一相反,法国和英国殖民地城镇形式却丰富多样。虽然网格状布局占据主导地位,却出现了各式各样的街道形式。然而不变的是城市中总有一处公共绿地,就像康涅狄格州的纽黑文绿地(1638)。这些公共用地和居住广场用于放牧、游行、民兵训练、观赏花园以及教堂和市政厅环境。美国早期的公共广场与伦敦和巴黎的私人居住广场不同,它们成为新型平等社会的象征。然而这类公共广场在使用上的多样性和密集性都比不上西班牙殖民时期欧洲中世纪广场。然而,它们更重要的功能是关注小城镇主要街道和大城市区域问题。

[3] 19世纪后期和20世纪早期的城市快速发展并没有伴随着城市中心公共空间的显著发展。不像欧洲城市那样,城市密度是通过大量广场来平衡,二战后的美国城市复杂、拥挤,缺乏开放空间。半个世纪前的芝加哥世界哥伦比亚博览会就曾梦想建设拥有宽阔街道和大型广场的城市——这个梦想基本上没有得到实现。

[4] 面对缺乏开放空间的问题,20世纪50年代和60年代的城市规划师提出奖励区划的方法鼓励私人建设广场供公众使用。尽管得到了一些新的广场,其代价却是高昂的:高大建筑物在街道投下大量阴影,不协调的公共空间形式,有时满足私人利益多过公共利益。

[5] 自从20世纪60年代,公司摩天大楼广场成为城市主要景观的同时,许多市民广场、商业广场和街道景观被建成,从而使市中心变得更加适合居住,也更加怡人。有两个项目可以证明,旧金山的吉拉德利广场(1962,伍斯特,贝尔纳迪和埃蒙斯,劳伦斯·哈普林事务所)和波士顿的法尼维厅市场(1976—1978,本杰明·汤普森事务所),他们归私人所有和并由私人管理,却具有极其丰富多样的用途,很少公司广场能够做到这一点。近期的城市广场从砌块砖铺设而成的单一形式的波士顿市政厅广场(1968,卡尔曼,麦金内尔和诺尔斯)到形态复杂的明尼阿波利斯匹威广场(1975,M.保罗·弗里德堡事务所)以及俄勒冈州波特兰市演讲堂前水景广场(1970,劳伦斯·哈普林)。总体来说,新广场往往散布在年轻的公司员工聚集、工作日午餐以及各种节庆活动的地点。大多数街景项目成为市中心户外商业街,使工作和购物更加愉快。许多像明尼阿波利斯市尼考利特商业街(1968—1981,劳伦斯·哈普林)和威斯康星州麦迪逊国家商业街(1975—1981,M.保罗·弗里德堡)都设计得很好。而一些不太成功的商业街则被拆除并以标准的街道和人行道取代。

[6] 广场和街道正发生着变化。新的"公共"空间逐渐走进室内并由私人管理。纽约福特基金会中庭(1967,丹·凯利)在建设之初是个令人感到奇怪的地方,但现在已成为当地的标志了。到处是中庭、拱廊、画廊、庭院和人行天桥,传统的街道生活移到了室内。尽管这些区域比较舒适,还为多种活动创造可能,它们还不是为公众服务的公共空间。当城市和公众生活关心的东西更新时,必须要不断地关注这些变化,并为我们的城市增加所有市民都能进入和使用的公共空间。

Reading Material B

Waterfronts

Roy B. Mann

滨水区

[1] During the second half of the 19th century, the park movement, which had demonstrated how recreational and civic needs could be met through the provision of parklands, grew and found expression on shore with designed public waterfront. Fredric Law Olmsted pioneered several forms of waterfront parks, among them a parkway-edged Muddy River segments of Boston's urban park, known as the Emerald Necklace, residential amenity shoreland in riverside, Ill (1868—1870), and, most spectacularly, waterfront exhibition grounds such as the 1893 World Columbia Exposition in Chicago (Fig. 1). The latter set the cornerstones for Chicago's quintessential lakefront park system and helped launch the City Beautiful movement that nurtured civic development well into the 20th century.

Fig. 1 waterfront exhibition of 1893 World Columbia Exposition in Chicago (Olmsted)

[2] Charles Eliot helped shape several recreational waterfronts, including the seaside Revere Beach Reservation[1] (Fig. 2) (1896) in Revere, Mass. Eliot and other landscape architects, including Olmsted, discovered that both coastal and river parkland could be imbued with naturalism in design, responding to the inherent qualities of the American waterfront environment. But the Beaux Arts[2] influence of the City Beautiful movement was active as well. It introduced Renaissance-inspired embellishment such as the Charles River[3] (Fig. 3) basin's balustraded

Fig. 2 Revere Beach Reservation (1896, Charles Eliot) in Revere, Mass

edgings and boat landings (1929, Arthur A. Shurtleff) in Boston; long formal promenades such as that of Manhattan's Riverside Park[4] (1868—1889, Olmsted and Vaux) and such classical structures as the Jefferson Memorial[5] (Fig. 4) (1943, John Russell Pope) in Washington D. C., where the Tidal Basin, on which the memorial is sited, was planted with Japanese flowering cherries and other landscape material in 1912 under the direction of landscape architect George E. Burnap.

Fig. 3   waterfront plan Charles River (1929, Arthur A. Shurtleff) in Boston

Fig. 4   Jefferson Memorial (1943, John Russell Pope) in Washington D. C.

[3]　After World War II, and particularly with the development of the interstate highway system in the 1950s, industry and warehousing abandoned many inner-city waterfronts, creating new potential for public access and leisure areas. At the same time, abandonment and industrial abuse produced serious aesthetic and environmental impact.

[4]　Facing new challenges posed by these transitional waterfronts and the need to accommodate the open space and recreation needs of growing urban and suburban populations, landscape architects and planners worked in the 1950s and early 1960s to replace blighted waterfronts with parks and other open space, some in conjunction with new civic facilities. Many broke with classical traditions and introduced contemporary design elements. Most sought to provide passive and quiet spaces as relief from the urban realm. Some attempted to emulate the waterfront of Europe, but the World War influence would not have a true impact until later in the 1960s.

[5]　In 1969 San Antonio's Hemisfair exposition[6] brought national attention to the city's Paseo del Rio[7], or riverwalk, a one-mile reach of open-air cafes, restaurants, shops and hotels conceived in the 1930s by architect Robert Huggeman as a simulation of a typical Spanish town form. With this example, with growing public support for historical preservation and adaptive use as seen in Boston's Faneuil Hall Marketplace (1967—1978, Benjamin Thompson Associates) and with growing public interest in social and cultural activity in public places, the corner was turned toward the more vibrant waterfront of the 1970s and 1980s.

[6]　A fascination with Europe's unique public spaces and waterfronts had been communicated to Americans by numerous observers in 1960s, including Lawrence Halprin[8] in his book *Cities* (1963) and Britain's Gordon Cullen[9] with his equality expressive *Townscape* (1961). In coastal cities, lakeshore communities and river towns, momentum grew for creating informal, marketplace-compatible and crowd accommodating spaces by the water. Designers emulate European and Japanese attention to details and architectural idioms. New paving features, marine bollards, historical sign motifs, port-styled lamps and other harbor-related accoutrements came into use. The simulation of European water-

fronts became not only desirable but feasible, while material and forms were used with growing originality to achieve an increasingly recognizable American waterfront idiom.

[7] long with Boston, other cities also reclaimed their historic waterfronts as public space opening the door to residential, retail and institutional activities. Baltimore's Inner Harbor restoration[10] (Fig. 5) (1974, Wallace, McHarg, Robers and Todd) included the market pavilions of Harborplace (1980, Benjamin Thomason Associates). Coastal areas were designed with the character of early American waterfronts, as for example, at Hilton Head Island's Harbour Town[11] (1970, Sasaki, Dawson, Domay Assoiciates). Some canal theme developments, Las Colinas in Dallas, for instance, echoed Venetian precedents.

[8] The search for contemporary water related forms has been carried out with increasing creativity. Riverbank Park (1979, CHNMB Associates) in Flint[12], Mich., introduced an Archimedes screw and flume as central elements in a geometric, sculptured scheme. In Tulsa, Okla, Athena Tacha's Blair Fountain[13] (Fig. 6) (1982) created a unique combination of dam, waterfalls and sculpture with light and water displays. Ottawa's Rideau, Canal, developed over many years, blended new park, restaurant and institutional edges with its traditional waterway and locks. Cincinnati's Serpentine[14] (Fig. 7) (1975, Zion and Ereen) created a dramatic water's edge highly compatible with the Ohio River's flood menace. The commitment to preserve a waterfront's environmental resources was dramatically expressed at the Sea Ranch residential project[15] (Fig. 8) in Sonoma County, Calif. (1967, Lawrence Halprin and Associates), which broadened public perception of the fragility of coastal landscapes and their site design considerations.

Fig. 5 Baltimore's Inner Harbor restoration
(1974, Wallace, McHarg, Robers and Todd)

Fig. 6 Blair Fountain (1982, Athena Tacha)
In Tulsa, Okla

Fig. 7 Cincinnati's Serpentine
(1975, Zion and Ereen)

Fig. 8 Sea Ranch residential project
(1967, Lawrence Halprin and Associates) in Sonoma County

[9] Today, increasing design innovation, preservation of historic features and social and cultural enrichment of public space have become new strengths as waterfronts are reclaimed. These qualities may have come to stay.

[NOTES]

[1] Revere Beach Reservation 若费尔海滩(1896):美国第一个公共海滩,由查尔斯·艾略特(Charles Eliot)主持开发与管理。

[2] Beaux Arts 布扎艺术:美术、艺术,富于装饰的艺术;尤指古典建筑风格。19世纪末起源于法国的建筑风格或与其相关的风格,特征为古典样式、对称、丰富多样的装饰以及宏大的规模。

[3] Charles River 波士顿查尔斯河滨水区(1929)。

[4] Manhattan's Riverside Park 曼哈顿滨水公园(1868—1889)。

[5] Jefferson Memorial 华盛顿特区杰斐逊纪念馆(1943):位于波多玛克河岸(Potomac River)的潮汐湖畔(Tidal Basin),由风景园林师乔治·E.柏耐普设计,种植大片日本樱花,杰斐逊纪念馆由建筑师约翰·罗塞尔·蒲柏设计,借用了罗马万神殿的构形,表达杰斐逊自由、独立和平等的理念。

[6] San Antonio's Hemisfair exposition 圣·安东尼奥博览会(1969)。

[7] Paseo del Rio:riverwalk,这里指圣·安东尼奥滨河步道。

[8] Lawrence Halprin 劳伦斯·哈普林(1916—2009):美国现代景观规划设计第二代的代表人物,美国战后最有影响力的景观设计师之一。主要作品有波特兰市的爱悦广场(Lovejoy Plaza)、柏蒂格罗夫公园(Pettygrove Park)和演讲堂前水景广场(Auditorium Forecourt Plaza)。他的著作《城市》(Cities)(1963)体现了他在城市设计中所涉及的各个方面。

[9] Gordon Cullen 高登·库伦(1914—1994):在他的著作《城镇景观》(Townscape)(1961)中,提出用创造"一系列视觉印象"(a serial vision)来构筑城镇景观,为后来的城市设计工作提供了一种有效的美学创作方法。

[10] Baltimore's Inner Harbor restoration 巴尔的摩内港保护区(1974)。

[11] Harbour Town 哈伯镇:位于南卡罗来纳州希尔顿头岛的度假圣地,1970由佐佐木英夫、唐森、道梅事务所设计。

[12] Riverbank Park in Flint 弗林特河滨公园(1979)。

[13] Athena Tacha's Blair Fountain 阿塞娜·塔哈的布莱尔喷泉(1982)。

[14] Cincinnati's Serpentine 辛辛那提的瑟潘汀(1975)。

[15] Sea Ranch residential project 滨海农场住宅开发项目(1967)。

[GLOSSARY]

Beaux Arts	艺术,美术	lakefront	滨湖	parkland	公园用地,公共用地	waterfront	滨水

[NEW WORDS]

abandon	vt.	放弃,遗弃,抛弃	blight	v.	毁坏,荒芜,衰退,枯萎
accoutrements	n.	配备,穿着	bollard	n.	系船柱
Archimedes	n.	阿基米德	coastal	adj.	海岸的,沿海的
balustrade	n.	栏杆,扶手	commitment	n.	承诺,保证,任务

continued

conjunction	n.	连接,关联,连接词	nurture	vt.	养育,培育,培养,滋长,助长
cornerstone	n.	奠基石,基石,基础	originality	n.	独创性,创造性,原创性,创意
echo	vt.	发出……的回声;重复,模仿	perception	n.	知觉,感知,洞察力,悟性,认识,理解,直觉,知觉过程,感性认识
embellishment	n.	装饰,修饰,布置,艺术加工			
enrichment	n.	丰富,充实,肥沃,浓缩	pose	vt.	提出,形成,引起,造成
feasible	adj.	可行的,切实可行的	promenade	n.	散步,漫步
flume	n.	水槽,引水沟	quintessential	adj.	精萃的,精髓的
fragility	n.	脆弱,虚弱	reclaim	vt.	收回,回收,改造
idiom	n.	风格,特色;习语,成语	retail	n.	零售业,零售
imbue	vt.	浸透,灌输,影响	scheme	n.	图,图表,图解;计划,方案
marine	adj.	海的,航海的,海运的,海产的	screw	n.	螺旋,螺孔,螺丝钉,螺旋桨
menace	n.	威胁,胁迫,危险	simulation	n.	模仿,模拟
momentum	n.	势头,动力,冲力	Venetian	adj.	威尼斯的
motif	n.	装饰图案,装饰图形;主题,主旨	vibrant	adj.	有活力的,充满生气的
naturalism	n.	自然主义	warehousing	n.	仓储

[难句]

A fascination with Europe's unique public spaces and waterfronts had been communicated to Americans by numerous observers in 1960s, including Lawrence Halprin in his book *Cities*(1963) and Britain's Gordon Cullen with his equality expressive *Townscape*(1961). 20世纪60年代大量到访欧洲的访客,对欧洲独特公共空间和滨水区产生的迷恋情结感染了美国人,其中包括劳伦斯·哈普林和他的《城市》(1963),以及英国戈登·库伦和他的《城镇景观》(1961)。

[参考译文]

滨水区
罗伊·B.曼

[1]　在19世纪下半叶,公园运动示范了如何通过提供公园用地来满足休闲和市民需求,进而发展出在水岸设计公共滨水区的方式。弗里德里克·劳·奥姆斯特德开创了几个滨水公园的形式,其中一个是沿公园道以"翡翠项链"著称的波士顿城市公园莫迪河段,这里是伊利诺斯州瑞弗里德滨水地居住条件宜人的地方(1868—1870),然而,最壮观的是滨水展览区,比如1893年芝加哥世界哥伦比亚博览会。后者成为芝加哥湖滨公园系统的精华区域,并促成了城市美化运动的发起,从而使城市更好地发展直至20世纪。

[2]　查尔斯·艾略特设计了几个休闲滨水区,包括位于马萨诸塞州若费尔海滨的若费尔海滩保护区。艾略特和其他包括奥姆斯特德在内的景观设计师发现,根据美国滨水环境的内在品质,不管是海岸还是河岸的公园用地,都能够在设计中渗透自然主义。但是,城市美化运动的布扎艺术影响也很活跃。布扎艺术引入了文艺复兴式的装饰艺术,比如波士顿查尔斯河滨水区水池边的栏杆和驳船码头(1929,阿瑟·A.舒克利夫);长长的规整式散步道,比如曼哈顿滨河公园(1868—1889,奥姆斯特德和沃克斯);还有一些传统结构,比如华盛顿特区的

杰斐逊纪念馆(1943,约翰·罗塞尔·蒲柏),1912在景观设计师乔治·E.柏耐普的指导下,在纪念馆所在的潮汐湖畔种植了日本樱花和其他园林植物。

[3] 二战后,尤其是20世纪50年代州际公路的发展,工业和仓库用地废弃后在内城留出了许多滨水区,为公众接近休闲区创造了新的潜在机会。同时,废弃地和废弃工业带来了严重的美学和环境问题。

[4] 面对这些因滨水区功能转变而带来的新的挑战和为不断增长的城市和郊区人口提供开放空间的需求,风景园林师和规划师在20世纪50年代和60年代早期致力于以公园和其他开放空间取代衰落的滨水区,一些地方还接入新的市政设施。许多地方打破了古典传统,引入当代设计要素。大多数都寻求创造安静闲散的空间作为城市中的放松之地。一些地方还试图仿效欧洲的滨水区,但是直到20世纪60年代,世界大战才真正影响到滨水区的设计。

[5] 1969年,圣·安东尼奥博览会将人们的注意力吸引到城市滨河步道,这里到达露天咖啡、饭店、商店和酒店仅1英里,是30年代由建筑师罗伯特·海格曼设计的模拟典型西班牙小镇模式的景观道。这个案例,以及波士顿法尼维厅广场(1967—1978,本杰明·托马斯事务所),使得公众对历史保护和适应性利用的支持不断增长,对公共场所社会和文化活动的兴趣也不断增长,这个被遗忘的角落变成了20世纪70年代和80年代具有活力的滨水区。

[6] 20世纪60年代大量到访欧洲的访客,对欧洲独特公共空间和滨水区产生的迷恋情结感染了美国人,其中包括劳伦斯·哈普林和他的《城市》(1963),以及英国戈登·库伦和他的《城镇景观》(1961)。在沿海城市、滨湖社区和河滨城镇,建造非正式的、有购物市场的、能够容纳大量人群的水边空间的势头渐渐增长。设计师效仿欧洲和日本,关注细节设计,运用建筑语言。其中运用了新的铺装形式、系船桩、历史符号、港口式样的灯具和其他与海港有关的设施。模仿欧洲滨水区不仅受人们欢迎,而且切实可行,但是材料和形式的使用愈发本土化,形成越来越易识别的美国滨水特色。

[7] 和波士顿一起,其他城市也将历史滨水区改造为公共空间,允许开发居住、零售和公共机构。巴尔的摩内港保护区(1974,华莱士、麦克哈格,罗伯茨和托德)拥有港口商业建筑。具有早期美国海滨特色设计的滨海区域,例如希尔顿海德岛的哈伯镇(1970,佐佐木英夫、唐森和道梅事务所)。一些运河开发区,例如达拉斯市的拉斯·科琳娜,则是模拟威尼斯模式。

[8] 目前展开的有关水体形式的探索越来越具有创造力。密歇根州弗林特河滨公园(1979,CHNMB事务所),引入了阿基米德螺旋和水槽,以几何的雕刻图案作为中心元素。在俄克拉荷马州塔尔萨,阿塞娜·塔哈的布莱尔喷泉(1982)创造了一个独特的结合水坝、瀑布和雕塑,以及灯光和水景展示的景观。渥太华的瑞多、科奈尔,经过多年的发展,将新型公园、饭店和公共设施与传统的水路和闸门结合。辛辛那提的瑟潘汀(1975,泽恩和厄瑞恩)创造了一个引人注目的水景,隐喻俄亥俄州水灾威胁。对于滨水环境资源保护的重视特别体现在加州索纳玛郡海滨农场居住项目(1967,劳伦斯·哈普林事务所),这扩大了人们对海岸景观脆弱性的感知以及他们对场地设计的思考。

[9] 今天,随着滨水区的改造,越来越多的设计变革,以及对公共空间历史特征和社会文化积淀的保护,已成为新的优势。这些特性也将永远留存下来。

[思考]
(1)美国最早的城市公园是哪一座?建于何时?出于什么目的而设?
(2)公园对城市起到哪些重要作用?
(3)美国的城市广场主要有哪些类型?有哪些实际案例?
(4)美国滨水景观发展的过程是怎样的?有哪些实际案例?

UNIT 8 RESIDENTIAL, INSTITUTIONAL AND CORPORATE LANDSCAPE

TEXT

Gardens
Jory Johnson

花　园

[1]　Native Americans made horticultural contributions to the early colonial settlements but had little influence on garden design. Until the late 18th century, most Americans could seldom afford time for anything more than utilitarian gardens and flower beds. In the South, plantation owners took advantage of commanding river vistas and slave labor to build expansive mansions and grounds; their central-axis gardens and long allees were influenced by English Renaissance gardens. Later, in the early 19th century, large country estates[1] (Fig. 1) in the rugged Hudson River topography favored the casual, picturesque style promoted by Andrew Jackson Downing.

[2]　In the latter part of the 19th and early 20th centuries, exotic plants and garden ornaments became widely available. Fashionable Victorian[2] embellishments-star-shaped flower beds and banks of the latest cultivars of lilacs, roses and hollyhocks-spread through the new "railroad suburbs[3]" of large cities. Frank Scott and other popular garden writers gave detailed recommendations for foundation plantings (over which there was wide disagreement on the necessity to conceal foundations with billowing shrubbery), front yards and other problems confronting suburban homeowners. Special garden features were emphasized such as sentimental fountains or colorful rings of red cannas, spirea, geraniums and other larger, showy flowers. Also popular were informal or wild gardens, which combined the latest horticultural introductions with a romantic love of nature.

Fig. 1　Gardens of country estates

[3]　The 1893 World's Columbian Exposition, with its Classical Revival architecture[4], stimulated interest in formal Italian and French gardens. The writer Edith Wharton and architect and landscape designer Charles A. Platt advocated architectonic elements, including pergolas, summerhouses and

balustrades, to achieve a progression of well-defined garden rooms. Formality long continued to be status symbol for many Americans. The Colonial Revival gardens[5] (1928—1937, Arthur A. Shurcliff) at Williamsburg, Va., with their small-scale borders and parterres, provided a popular model for the reduced expectations following the Depression, particularly in the South and the Mid-Atlantic states.

[4] California gardens from the late 19th to mid-20th centuries sought inspiration from Mediterranean countries. Irving Gill's unelaborated courtyard gardens, with drought-resistant plantings, were derived from Spanish missions[6]. California landscape architects such as Florence Yoch, Edward Huntsman-Trout, Ralph Cornell and Lockwood De-Forest were accomplished plantsmen and women who closely supervised the construction and maintenance of their gardens. Their projects were strongly influenced by formal Italian gardens and usually included elaborate series of formal lawns and architectonic planting designs. The Craftsman-style[7] bungalow, popularized by the brothers Greene, introduced outdoor living with sleeping porches and terraces. Bungalow gardens frequently were influenced by Japanese prototypes.

[5] The first significant influence of 20th-century art and architecture on American gardens can be found in the work of Fletcher Steele[8] (Fig. 2), and admirer of French Art Deco garden experiments[9] of the 1920s. But it was not until International Style[10] architecture arrived on the scene that American landscape architects began to reject formulaic classical plans. In the late 1930s, inspired by the ferment and excitement of the new architectural theories, James Rose, Garrett Eckbo and Dan Kiley[11] began exploring appropriate garden forms and ecological strategies for the modern era.

Fig. 2 Fletcher Steele and his garden

Fig. 3 Miller garden

[6] Kiley extended the rectilinearity and open plan of modern architecture into the garden(Fig. 3). For Rose and Eckbo, the most important concepts were embodied in Eckbo's book *Landscapes for living*[12] (1950). Many Americans in the post-Korean War building boom believed that elegant gardens were a luxury or indulgence. Eckbo's home landscaping book showed how such quotidian necessities of suburban life as laundry areas, children's sandboxes and barbecue grills could be part of a sophisticated garden design. Homeowners also were encouraged to use industrial products such as poured concrete and other inexpensive materials(Fig. 3).

UNIT 8　RESIDENTIAL, INSTITUTIONAL AND CORPORATE LANDSCAPE

[7]　Many of these ideas were first developed in California by Lawrence Halprin, Robert Royston, Douglas Baylis and others. With its year-round temperate climate, California became synonymous with redwood decks, swimming pools and strong paving and ground plane patterns. The "California style[13]" (Fig. 4), promised low maintenance and relaxed entertaining. In addition, the incorporation of Japanese garden motifs such as picturesque moss beds, evergreens, ferns and artfully placed stones[14] were considered an ideal response to the cool refinement of contemporary-style homes.

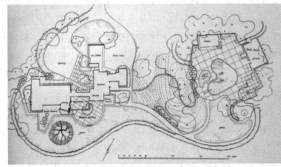

Fig. 4　Tomas Church and his "California style" garden-Donnel Garden

[8]　By the 1970s various economic and resource pressures reduced the average size of residential lots. In addition, smaller families had less need for all-purpose yards. Many homeowners also were growing tired of large suburban lawns and the blandness of low-maintenance gardens. The environmental movement helped evoke an interest in natural landscapes, including the aesthetic pleasures of perennial gardens, ornamental grasses and garden architecture. Recent garden design has been influenced by the historic preservation movement[15], which has increased public awareness of America's garden history. It is doubtful, however, that Americans will return en masse to historical revival styles.

[9]　The history of the garden in America could be described as a movement from revivals or copies of European styles to an exploration of America's social, cultural and geographical diversity. While American gardens have not achieved a singular style, the latter half of the 20th century represents the first consistent break with classical ideals and European models. As new styles and directions are developed, and as American gardens begin to receive the same attention now given to European and Asian gardens their rich heritage will achieve deserved recognition.

[NOTES]

[1]　country estates 乡村庄园:美国的乡村地产源自法国文艺复兴城堡、意大利别墅以及英国乡村庄园,这些庄园是财富的象征。华盛顿的佛农山(Mount Vernon)(1799)和杰斐逊的蒙地梭罗(Monticello)(1768—1809),一方面保留了英国文艺复兴的一些特色,另一方面又将18世纪英国景观园艺家 Batty Langley 和 Humphry Repton 的浪漫主义园林融入美国独特的景观中。19世纪初富有的工业家远离城市,到森林中寻找安静的住所,这种占地广阔的地产首先出现在波士顿外围及美丽的哈德逊河谷。马萨诸塞州 Philips 庄园(1881)和北卡罗来纳州的 Biltmore 庄园(1888—1895),是乡村地产设计浪漫时期的巅峰之作。1880年到二战,乡村庄园设计受到美术学院的影响,将建筑上均衡、对称和空间等级的观念直接移转到花园和景观设计上。Ferruccio Vitale, Fletcher Steele, Olmsted 兄弟,Warren H. Manning 和 Jensen 等人设计了许多庄园。二战以后,较前卫的设计有加利福尼亚州的 Donnel 住宅(1948年,Thomas Church),印第安纳州 Miller 住宅(1955,Dan Kiley),这些设计将法国传统的规则式和美术花园融入现代设计中。美国经济的繁荣使乡村地产的设计一直是景观设计师主要的

实践领域。

[2] Victorian 维多利亚风格：是19世纪英国维多利亚女王在位期间（1837—1901）形成的艺术复兴风格，它重新诠释了古典的意义，扬弃机械理性的美学，以奢华、装饰、绚丽为特征，视觉设计上矫揉造作，繁琐装饰、异国风格占据重要地位。在建筑上的表现则是历史上各种建筑式样的复兴，它并非简单复制，而是加入了更多现代元素，并运用新的建筑材料，是多种风格相融合的混合风格。

[3] railroad suburbs 铁路郊区：以铁路与主城联系的郊区城镇或卫星城。

[4] Classical Revival architecture 古典复兴式建筑：又称新古典主义建筑，18世纪60年代到19世纪流行于欧美一些国家，一方面起于对巴洛克（Baroque）和洛可可（Rococo）艺术的反对，一方面受启蒙运动思想的影响，崇尚古代希腊、古罗马的文化艺术。在建筑方面，古罗马的广场、凯旋门和记功柱等纪念性建筑成为效法的榜样。采用古典复兴建筑风格的主要是国会、法院、银行、交易所、博物馆、剧院等公共建筑和一些纪念性建筑，对一般的住宅、教堂、学校等影响不大。美国独立以后，在摆脱殖民统治的同时，希望借助希腊、罗马的古典建筑来表现民主、自由、光荣和独立，典型的例子如华盛顿国会大厦（The Capital of Washington D.C）和林肯纪念堂（Lincoln Memorial）。

[5] Colonial Revival gardens 殖民复兴式花园：早期北美洲的殖民者来自于欧洲几个航海国家——英国、法国、荷兰、西班牙和葡萄牙，加利福尼亚州受西班牙传教士的影响大于文艺复兴的影响，法国殖民者热衷于17世纪的巴洛克式花园，而英国和荷兰则钟情于威廉和玛丽风格。威廉斯伯格（Williamsburg）作为弗吉尼亚州的州府，同时也是早期美国筑园的中心，最早的新教徒们因为反对独裁统治而移居到了美洲这个蛮荒之地，法式花园或者奥古斯都式花园都不适合当时的条件，但是质朴、自然的荷兰式花园受到了青睐。弗吉尼亚州威廉堡殖民复兴花园的小规模装饰花坛，成为小型规则花园设计所喜爱的模式，特别盛行于美国南部和大西洋沿岸中部各州。

[6] Spanish missions 西班牙传教士风格：是指从1769年到1823年，西班牙对北美洲进行拓荒的殖民计划，当时的西班牙国王Charles三世指示神父Serra带领传教士与军队，建立教堂，传经布道，驯化印第安原住民，他们在圣地亚哥建立的第一个Mission，既是任务，也是传教堂，更是西班牙征服者的殖民中心。西班牙传教运动对美国西部海岸的地中海风格的形成有着深远影响，并留下大量西班牙传教士风格的建筑与花园。

[7] Craftsman-style 工匠风格：美国原创建筑风格，强调和表现木质构件的结构美，精湛以及淳厚，自然而朴实，建筑以木、砖、石等自然元素为主。

[8] Fletcher Steele 弗莱彻·斯蒂里（1885—1971）：最早将欧洲前卫花园引入美国的风景园林师之一。

[9] French Art Deco garden experiments 法国艺术博览会。

[10] International Style 国际风格：最初由格罗皮乌斯（Walter Gropius，1883—1969）提倡，在斯图加特的德意志制造联盟住宅展（1927）中展出的住宅样式，20—30年代，国际风格在欧美广泛流传，后流行于全世界，成为最重要的现代建筑风格之一。最主要的特点是以钢、水泥、玻璃为主要材料，呈方形或长方形外观，不加繁琐装饰，讲究功能运用。代表人物与作品有格罗皮乌斯的法古斯工厂（Fagus Factory）与包豪斯（Bauhaus）校舍，密斯（Mies van der Rohe）的湖滨公寓（Lake Shore Drive Apartments）与西格拉姆大厦（Seagram Building）、柯布西耶（Le Corbusier）的萨伏依别墅（Villa Savoye）等。

[11] James Rose, Garrett Eckbo and Dan Kiley 詹姆斯·罗斯（1913—1991）、盖瑞特·埃克博（1913—1991）和丹·凯利（1912—2004）：美国现代主义景观运动的引领者，哈佛革命（Harvard Revolution）的三位发起人。

[12] Landscapes for living《生活景观》（1950）。

[13] California style 加州花园风格：20世纪40年代在美国西海岸，由托马斯·丘奇（Thomas Church，1902—1978）开创的一种不同于以往的私人花园风格，这种带有露天木制平台、游泳池、不规则种植区域和动态平面的小花园为人们创造了户外生活的新方式，不仅受到中产阶层的喜爱，在美国风景园林行业中也引起了强烈的反响，成为当时现代园林的代表。

[14] artfully placed stones 置石艺术。

[15] historic preservation movement 历史保护运动。

UNIT 8　RESIDENTIAL, INSTITUTIONAL AND CORPORATE LANDSCAPE

[GLOSSARY]

California style	加州(花园)风格	drought-resistant plant	抗旱植物
Classical Revival	古典复兴式,新古典风格	foundation planting	基础种植
Colonial Revival	殖民复兴式	historic preservation movement	历史保护运动
Craftsman-style	工匠风格	International Style	国际风格

[NEW WORDS]

allee	n.	(林荫)小径	lilac	n.	丁香,紫丁香,淡紫色
architectonic	adj.	建筑式样的	maintenance	n.	维持,维护,保养
balustrade	n.	栏杆	Mediterranean	adj.	地中海的
barbecue	n.	烧烤野餐,烤架	moss	n.	苔藓,地衣,沼泽
billow	n.	巨浪,波涛	ornament	n.	装饰,点缀,美化
blandness	n.	温柔,爽快	parterre	n.	花坛,花圃
bungalow	n.	单层小屋,(有凉台的)平房,别墅	perennial	adj.	(植物)多年生的,终年的,长期的
canna	n.	美人蕉	pergola	n.	藤架,花架,廊
casual	adj.	偶然的;不经意的,随意的	picturesque	adj.	风景如画的,独特的,别致的
cultivar	n.	栽培变种,(栽培)品种	porch	n.	门廊,走廊
elaborate	adj.	精心制作的,复杂的	progression	n.	连续,一系列
elegant	adj.	优美的,优雅的,漂亮的	prototype	n.	原型
embellishment	n.	装饰,修饰,润色	quotidian	adj.	每日的,每日发生的,平凡的
en masse	v.	全体地,一同地	rectilinearity	n.	直线性
estate	n.	庄园,住宅区,房地产,产业	redwood	n.	红杉,红木,红色的木材
evoke	vt.	唤起,引起	rugged	adj.	崎岖的,多岩石的,粗犷的
exotic	adj.	异国情调的,奇异的,醒目的	sentimental	adj.	感情的,感伤的,多愁善感的
fashionable	adj.	流行的,时髦的	showy	adj.	华丽的,艳丽的,引人注目的
ferment	n.	激动,动荡	shrubbery	n.	灌木,灌木林
fern	n.	蕨类植物	singular	adj.	单独的,非凡的,奇特的
formulaic	adj.	公式化的,千篇一律的,刻板的	spirea	n.	绣线菊
geranium	n.	天竺葵	summerhouse	n.	凉亭,亭子
hollyhock	n.	蜀葵	synonymous	adj.	同义的
homeowner	n.	房主,屋主	terrace	n.	台地,台阶;露台,平台
horticultural	adj.	园艺的	utilitarian	adj.	实用的,有效用的,功利的
indulgence	n.	放任,沉迷,爱好,享受	Victorian	adj.	维多利亚时代的,维多利亚式
informal	adj.	不规则的,非正式的	year-round	adj.	全年的,全年不变的

[参考译文]

花 园

乔里·约翰

［1］ 美国原住民对早期殖民地的园艺贡献很大，却对花园设计的影响很小。直到18世纪末，除了实用性花园和花坛，大多数美国人很少在园林上花费时间。在南部，种植园主利用居高临下的河流景色，奴役奴隶们为他们建造奢华的府邸和花园，而这些中轴对称的花园和长长的林荫道受到了英国文艺复兴式园林的影响。后来，在19世纪早期，哈德逊河两岸崇山峻岭中的大型乡村庄园，开始偏爱安德鲁·杰克逊·唐宁所宣扬的那种随意自然、风景如画般的风格。

［2］ 在19世纪后期和20世纪早期，具有异国情调的植物和花园装饰物被广泛采用。流行的维多利亚式星形装饰花坛和栽有新培育的丁香、蔷薇和蜀葵的堤岸遍及大城市的新"铁路郊区"。弗兰克·斯科特和其他一些受欢迎的花园作家对基础种植（关于使用大面积的灌木丛覆盖建筑基础的必要性，很多人也提出了反对意见）、前院以及郊区房主所面临的其他问题都提出了详尽的建议。花园中的一些特色内容常被强调，例如浪漫的喷泉或由红花美人蕉、绣线菊、天竺葵和其他大型艳丽花卉所组成的色环等。此时还流行不规则的或具有野趣的花园，它们结合了最新的园艺知识并展现出对自然的浪漫之爱。

［3］ 1893年世界哥伦比亚博览会古典复兴式建筑的展出，激起人们对意大利和法国规整式花园的兴趣。作家伊迪斯·沃顿和建筑师兼景观设计师查尔斯·A.布拉特倡导建筑式的景观元素，包括廊架、凉亭和栏杆，从而获得一系列精美的园林空间。规整式长期以来被许多美国人当作地位的象征。弗吉尼亚州威廉堡的殖民复兴式花园（1828—1837，阿瑟·A.舒克利夫）以较小尺度的花园和花坛，成为经济萧条时期对花园要求不高的人们所喜爱的一种模式，尤其盛行于南部和沿大西洋中部各州。

［4］ 加利福尼亚州的花园从19世纪末至20世纪中不断从地中海国家寻找灵感。欧文·吉尔种植着抗旱植物的粗放式花园，就来源于西班牙传教士风格。加州风景园林设计师，如弗洛伦斯·夭，爱德华·亨斯曼-特劳特，拉尔夫·科奈尔和洛克伍德·德福里斯特，很多都是很有成就的园丁，他们精心建造和管护他们的花园。他们的设计受意大利规整式花园影响很大，常常在花园里布置一系列精致的规整式草坪和建筑式整形植物。由格林兄弟推广的工匠风格别墅，以寝廊和露台将户外生活理念引入花园。这种花园的设计常常受到其原型——日本园林的影响。

［5］ 20世纪的艺术和建筑对美国花园最早最显著的影响，可以从弗莱彻·斯蒂里以及20世纪20年代法国艺术博览会的钦慕者的作品中找到。但是，直到国际风格式建筑出现时，美国风景园林师才开始摒弃刻板的古典设计。在20世纪30年代后期，受到纷杂激荡的新建筑理论的激发，詹姆斯·罗斯、盖瑞特·埃克博和丹·凯利开始探索适于新时代的花园形式和生态策略。

［6］ 凯利将现代建筑的直线性和开放式规划引入了花园。罗斯和埃克博最重要的观念就体现在埃克博的《生活景观》一书（1950）中。在朝鲜战争后建筑大发展时期，许多美国人认为优雅的花园是一种奢侈品或享受品。埃克博的家庭园林手册解释了郊外生活的日常必需内容，如洗衣区域、儿童沙池和野餐烤架，是如何成为复杂园林设计的一部分。他还鼓励房主使用工业产品，例如现浇混凝土和其他廉价材料。

［7］ 这些观点很多最早是在加州由劳伦斯·哈普林、罗伯特·罗伊斯顿、道格拉斯·贝利斯等人提出的。因其全年气候温和，加州成为红木平台、游泳池、坚实的铺装路面和地面铺纹的同义词。"加州风格"体现了较低的维护需求和舒适的休闲娱乐。此外，"加州风格"还结合了日本园林的要素，如生动的苔藓植床、常绿植物、蕨类植物和置石艺术，相对于精致冰冷的现代住宅来说这是一种理想的搭配。

［8］ 到20世纪70年代各种经济和资源压力使住宅的平均占地面积减小。此外，规模变小了的家庭对多用途院子的需求也在减少。许多房主也逐渐厌倦了大面积的郊外草坪和低维护花园的平淡。环境运动唤起人们对自然景观的兴趣，包括能给人带来审美享受的多年生花卉花园、观赏草和花园建筑。近期的花园设计受到历史保护运动的影响，该运动增加了公众对美国花园历史的了解。然而令人怀疑的是，美国人是否又将完全回归到历史复兴风格中去。

［9］ 美国花园的历史可以说是一个起源于欧洲风格的拷贝与复兴，并通过探索适应美国社会、文化和地理多

样性的过程。尽管美国花园还没有形成一种独立的风格,但在20世纪后半期却开始持续不断地打破传统理念和欧洲模式。随着新风格和新趋势的不断发展,美国花园开始受到与欧洲和亚洲花园同样的关注,其创造的丰富遗产必将得到赞誉。

Reading Material A

Housing Environments

Michael Laurie[1]

居住环境

[1] The house represents shelter, privacy and security for a family or group, and a cluster of housed-of families-constitutes a community. The quality of our communities and thus the lifestyles they support is a function of many factors, including provision for streets, spacing of dwellings, availability of private gardens, distribution of public open space and relationship to the work place. Site planning for housing is a major contribution of the modern landscape architect.

[2] Historically, the Jeffersonian dream of single-family farms[2] equidistantly spaced in an agricultural landscape and William Penn's compact plan for Philadelphia (1683) were two contrasting alternatives for community form in 18th-century America. Later, the desire to get away from the squalor and congestion of industrialized 19th-century cities resulted in the development of suburbs. Connected by trolley and railroad services, suburban communities combined the pleasures of the countryside with the economic and cultural opportunities of the city. H. W. S. Cleveland, Frederick Law Olmsted and other pioneer landscape architects designed such suburbs with winding, tree-planted streets, single-family homes and surrounding lawns. Natural features were emphasized, and public greens and small parks provided a romantic landscape setting for the villas. A good example of the romantic suburb is Riverside, Ⅲ. (1868-1870), near Chicago, designed by Olmsted and Vaux.

[3] Although this housing type was largely for the middle class and, therefore, not widely affordable, it did provide a model for those concerned with a general improvement in housing conditions. The garden city[3] ideal, defined in England by Ebenezer Howard in 1898, resulted in complete towns of limited size for lower-income families such as Letchworth (1908) and Welwyn (1924) (Fig.1).

Fig. 1 Ebenezer Howard and his first "garden city" Letchworth in England

[4] Inspired by these examples, the socially conscious architects Clarence Stein and Henry Wright designed Radburn, N.J.[4] (1928-1929) (Fig.2). There, a new factor, the automobile, was incorporated into the design. A system of 40-acre superblocks with cul-de-sac access to the houses retained a central green space with pedestrian paths from every street at grade. Six superblocks made up a neighborhood with approximately 960 dwellings at an overall density of four per acre.

Fig. 2　Radburn, N. J. (1928—1929)

[5]　At Baldwin Hills (1941), Los Angeles, the same architects, with landscape architect Fred Barlow, further developed this model for rental units. One large superblock of 80 acres accommodated 627 units, with parking provided at a ratio of three to one. Private patios opened onto pedestrian ways giving access to a central village green, one-half mile long and varying from 50 to 250 feet in width.

[6]　In spite of the apparent advantages of these communities, America's private-sector housing built after World War II was dominated by the subdivision concept epitomized by Levittown[5] (Fig. 3), N. J. (1955). Exceptions occurred in the public and corporate sectors, where, following European practice, often massive housing blocks were surrounded by green space and playgrounds. Stuyvesant Town (1947) in New York City is a notable example. In other situations, such as the new towns of Irvine, Calif. (1964), Reston, Va. (1964), and Columbia, Md. (1965), a variety of housing types from highrise apartments[6] and cluster housing[7] to conventional subdivisions[8] were combined into planned environments, including spacious and accessible green areas for recreation while achieving an economical overall density[9].

Fig. 3　Levittown, N. J. (1955)

[7]　In the future the planned unit, condominium and rental development and the typical subdivision with single-family homes are likely to undergo change as new concerns-social cohesion, affordable housing[10], resource conservation, water quality, solar orientation[11] and travel time to work-are taken seriously. The Village Homes development[12] (1972) of Michael Corbett in Davis, Calif., with solar orientation for all houses, narrow streets, bicycle and pedestrian paths, community gardens and facilities, and drought-tolerant planting is a recent example responding to many of these issues.

[NOTES]

[1]　Michael Laurie 迈克尔·劳瑞(1931—2003):加州大学伯克利分校教授,风景园林教育家,曾是宾夕法尼亚大学伊恩·麦克哈格的学生,著有《风景园林学导论》(*An introduction to landscape architecture*),是西方风景园林教学经典入门书籍。

UNIT 8　RESIDENTIAL, INSTITUTIONAL AND CORPORATE LANDSCAPE

[2]　single-family farms 独立家庭农场。

[3]　garden city 田园城市：埃比尼泽·霍华德（Ebenezer Howard, 1850—1928）在他的著作《明日的田园城市》（*Garden Cities of Tomorrow*）中提出的理想城市模式，一种兼具城市和乡村特点的新型居住形式，并针对当时英国大城市所面临的问题，提出用实现土地社区所有制、建设田园城市的方法逐步消灭土地私有制。《明日的田园城市》作为一本公认的最具有经典型的城市规划专著，它导致了城市规划专业的建立，霍华德也因此成为城市规划界的开山鼻祖。除了英国建设的莱奇沃思（Letchworth）和韦林（Welwyn）两座田园城市以外，在奥地利、澳大利亚、比利时、法国、德国、荷兰、波兰、俄国、西班牙和美国都建设了"田园城市"或类似称呼的示范性城市。

[4]　Radburn, N. J. 新泽西州雷德伯恩：克拉伦斯·斯坦（Clarence Stein）和亨利·怀特（Henry Wright）对雷德伯恩（Radburn）小镇的规划中提出人车分流系统，由步行和车行两套独立的道路系统组成，在人、车发生冲突的地方设置简易立交，以避免机动车交通干扰居民的居住环境，这种人车分流的交通系统至今仍为许多城市住区规划所采用。Radburn还将居民日常活动场地用步行系统连接起来，车行系统沿外围设置成尽端路（cul-de-sac access）形式。Radburn的互相连接的绿色空间网络及绿道（interconnected networks of greenspaces and greenways）被认为是对风景园林职业的独特贡献。

[5]　Levittown 利维顿：是美国二战后重要的郊区住宅开发模式，一种装配式的、可大规模生产的郊区住宅生产方式，对美国社区发展和郊区化产生了深远的影响。利维顿镇产生的背景是战后退役士兵对住宅的大量需求，以及1946—1947年的婴儿生育高峰，使城市中可出售的住宅远远不能满足这些需求，利维特（Levitt）父子公司创造的这种低成本、模式化郊区城镇模式迅速风靡。

[6]　highrise apartments 高层公寓住宅。

[7]　cluster housing 集聚住宅。

[8]　conventional subdivisions 传统住宅。

[9]　overall density 总建筑密度。

[10]　affordable housing 可负担住宅或经济适用房。

[11]　solar orientation 朝阳或朝阳方向。

[12]　The Village Homes development 乡村住宅发展规划。

[GLOSSARY]

drought-tolerant planting	耐旱植物	social cohesion	社会凝聚力
garden city	田园城市，花园城市	socially conscious	社会意识

[NEW WORDS]

affordable	adj.	担负得起的	neighborhood	n.	社区，街区，邻里，邻里单位
cohesion	n.	粘连，团结，凝聚力	patios	n.	内院，庭院，露台
condominium	n.	分契式公寓，共管公寓	privacy	n.	（不受干扰的）独处，隐私
congestion	n.	拥挤，堵车，（人口）稠密	shelter	n.	庇护所，居所，住处
cul-de-sac	n.	尽端路，死胡同，绝境	squalor	n.	肮脏，贫穷
dwelling	n.	住宅，寓所，居住	subdivision	n.	细分；供出卖而分成的小块土地
epitomize	vt.	使成……缩影，集中体现	superblocks	n.	车辆禁行区
equidistant	adj.	距离相等的，等距的	trolley	n.	电车
housing	n.	房屋，住宅群，住房供给	undergo	vt.	经受，经历，遭受

[难句]

1. The quality of our communities and thus the lifestyles they support is a function of many factors, including provision for streets, spacing of dwellings, availability of private gardens, distribution of public open space and relationship to the work place. 我们社区的质量和由其支撑的生活方式是许多因素的功能所在,包括提供街道、分隔住宅、留出私人花园、布置公共开放空间以及与工作地点的联系等。

2. Six superblocks made up a neighborhood with approximately 960 dwellings at an overall density of four per acre. 6个车辆禁行区组成一个约有960套住宅的邻里单位,其平均密度为每英亩4户住宅。

[参考译文]

居住环境

迈克尔·劳瑞

[1] 房屋对于一个家庭或团体来说代表着庇护所、私密和安全,住宅房屋的集合则组成一个社区。我们社区的质量和由其支撑的生活方式是许多因素的功能所在,包括提供街道、分隔住宅、留出私人花园、布置公共开放空间以及与工作地点的联系等。对住宅区进行场地规划是现代风景园林师的一个主要贡献。

[2] 在历史上,杰斐逊的在农业景观空间中等距离布置独立家庭农场的梦想和威廉姆·佩恩紧凑的费城规划(1683),是美国18世纪社区模式的两个对比明显的模式。后来,远离肮脏拥挤的19世纪工业城市使郊区得到发展。通过有轨电车和铁路联系,郊区社区将乡村生活的愉悦和城市经济和文化要素结合在一起。H. W. S. 克里夫兰、弗里德里克·劳·奥姆斯特德和其他先驱者,设计了这类拥有蜿蜒林荫道、独立住宅及周边草坪的郊区。规划强调自然特性,公共绿地和小公园为别墅提供了浪漫的景观环境。这种浪漫式郊区的一个很好的案例是由奥姆斯特德和沃克斯设计的位于芝加哥附近的伊利诺斯州瑞弗塞得住宅区(1868—1870)。

[3] 虽然这种住宅形式主要是为中产阶级考虑,多数人很难负担得起,但它确实为那些关注住房条件总体改善的人提供了一种模式。1898年由英国埃比尼泽·霍华德提出的花园城市理念,为低收入家庭创造了由小规模住宅组成的完善的城镇,如莱奇沃思(1908)和韦林(1924)。

[4] 受到这些例子的激发,一些有社会意识的建筑师克拉伦斯·斯坦和亨利·怀特设计了雷德伯恩(1928—1929)。在他们的设计中,结合了汽车这个新的要素。这个40英亩带有尽端式入户路的车辆禁行区系统保留了中心绿地,从每一条街都有人行道接入中心绿地。6个车辆禁行区组成一个约有960套住宅的邻里单位,平均密度为每英亩4户。

[5] 在洛杉矶鲍德温山(1941),还是同样的建筑师,与风景园林师弗雷德·巴洛一起,在为出租房住宅区的设计中进一步发展了这种模式。这个面积80英亩的大型住宅区包含627套住宅,停车率为每3户一个停车位。私人小院通向步行道,使住户能够到达1.5英里长,50到250英尺宽的小区中心绿地。

[6] 尽管这些社区有着明显的优点,美国二战后建造的私人住宅形式还是以新泽西利维顿城为代表独户住宅模式占主导地位。然而,在公共和公司住宅中也有一些特例,他们借鉴欧洲的案例,常常以绿地和运动场包围大型住宅区。纽约斯泰弗森特镇(1947)就是一个有名的例子。其他还有一些案例,例如加州尔湾(1964)、弗吉尼亚州雷斯顿(1964)和马里兰州哥伦比亚(1965),在进行环境规划时,综合考虑了各种类型的住宅形式,如高层公寓和集聚住宅到传统独户住宅,以及供人们休闲的大规模可进入绿地,同时获得经济节约的总建筑密度。

[7] 将来,随着人们对社会凝聚力、可负担住宅、资源保护、水质、房屋朝向和上班时间等因素的关注,规划单体、共管公寓和出租住宅区,以及典型的独户住宅将会经历新的变化。迈克尔·考伯特为加州戴维斯做的乡村住宅发展规划(1972)就是致力于解决这些问题的近期案例,如所有房屋均朝阳、狭窄的街道、自行车道和散步道、社区花园和各类设施、耐旱植物的运用等。

Reading Material B

Institutional and Corporate Landscapes[1]

Cheryl Barton

社会机构及公司景观

[1]　In its concern with providing meaningful open space, landscape architecture has always related fundamentally to human welfare. Institutional environments have benefited from the expertise of landscape architects for more than a century, but the corporate landscape is a more recent phenomenon. In the 18th and early 19th centuries, institutions existed primarily to take care of problems that the family was unable to handle—care of the mentally retarded, orphaned children, the sick and the handicapped. Many of these places were little more than warehouses. A similar fate met workers during the Industrial Revolution[2] who fled their farms to grasp opportunities in the cities. There, too, they were confined in warehouse-type environments for long, exhausting hours.

[2]　Reflecting post-Civil War reform movements, American landscape architects demonstrated innovative and practical responses to the social problems of alcoholism, mental illness and disease in the design of new institutional settings. Pure air and water, sunlight and other natural amenities were important factors in planning hospitals, asylums and sanatoriums. Spacious, well-designed grounds reflected advances in medicine. These projects could involve many disciplines and were often coordinated by landscape architects whose knowledge of environmental factors provided the requisite overview.

[3]　Frederick Law Olmsted, once secretary of the U. S. Sanitary Commission[3] (later the American Red Cross), planned several significant institutions, including Columbia Institution for the Deaf[4] (1866) (Fig. 1) in Washington D. C.. The plan for the Iowa Hospital for the Insane[5] (1871, H. W. S. Cleveland), Mt. Pleasant, involved patients in the therapeutic planting of native trees and shrubs on the asylum grounds. Scientific evidence has indicated that direct contact with nature produces positive effects on healing. Today, site and building design for contemporary institutions place a premium on exterior views for patients as well as easy accessibility to courtyards and gardens.

Fig. 1　Columbia Institution for the Deaf
(1866) in Washington D. C.

[4]   The corporate landscape evolved after World War II because of rapid growth in industry and the advent of suburbia. Here, the symbolism of wealth and power was expressed in attractive, parklike locations. People now spend considerable time in corporate settings, which in many places have replaced Main Street as a community's social crossroads. Affinities with early country villas and manor houses have also been suggested, as these buildings and their elaborate grounds once housed a large population of family and servants, curiously analogous to 20th century managerial, clerical and maintenance staff.

[5]   Today's corporate headquarters landscapes, such as those for Deere and Company[6] (Fig. 2) (1963, Sasaki Associates[7]), Moline, Ill. an important precedent; Carlson Center[8] (Fig. 3) (1988, EDAW Design Group), Minnetonka, Minn. ; and Pacific Bell[9] (1987, MPA Design), San Ramon, Calif. , contain extensive open space and recreational resources for their work forces, including tennis and volleyball courts, jogging paths, sculpture gardens, lakes and other facilities to ensure more loyal and productive workers and to add value to the corporate real estate portfolio. Furthermore, such features convey image, status, power, prestige and product excellence—critical ingredients in a competitive corporate culture.

Fig. 2 Deere and Company (1963, Sasaki Associates)

Fig. 3 Carlson Center[8] (1988, EDAW Design Group)

[6]   For many suburbs, the corporation has become a community center. In its commitment to employees and the public, PepsiCo[10] (Fig. 4) (1965, Edward D. Stone, Jr. Associates)—an integration of landscape architecture, architecture and sculpture in Purchase, n. Y. —exemplified a new creativity in the corporate landscape of its time. Its sculpture collection is displayed in garden terraces, becoming an outdoor art museum for the public and a showpiece landscape for the corporate landlord. Many corporations have purchased former estates and preserved their historic landscapes. At Corporate Woods[11] (1976, The SWA Group[12]), Overland Park, Kans. , and the TRW World Headquarters[13] (1985, Sasaki Associates), Lyndhurst, Ohio, landscape architects have effectively contrasted contemporary architecture with wooded and pastoral settings. Codex's world headquarters[14] (1986, Hanna/Olin), Canton, Mass, incorporates an elegant interior atrium at the heart of its complex of buildings to recall the outdoors year-round.

UNIT 8　RESIDENTIAL, INSTITUTIONAL AND CORPORATE LANDSCAPE

Fig. 4　PepsiCo (1965, Edward D. Stone, Jr. Associates)

[7]　Urban corporate landscapes, which were the earliest examples, also have thrived. One innovation was the Kaiser roof garden [15] (Fig. 5) (1960, Osmundson and Stanley), Oakland, Calif., a three-acre urban open space 23 stories above the street. While 90 percent of the site is covered by the building, 60 percent actually becomes semipublic open space. Dallas's Fountain Place [16] (Fig. 6) (1985, Dan Kiley), with a similar open space ratio, is a shimmering, monumental water garden in sharp contrast to the arid Texas landscape.

Fig. 5　Kaiser roof garden　　　　　　　Fig. 6　Dallas's Fountain Place (1985, Dan Kiley)
(1960, Osmundson and Stanley)

Fig. 7　Christian Science Church's world　　　　Fig. 8　Williams Square
headquarters (1973, Sasaki Associates)　　　　　(1984, The SWA Group)

[8]　Increasingly, site plans have been developed to express particular corporate and institutional values. The Christian Science Church's world headquarters [17] (1973, Sasaki Associates) in Boston is a powerful, ceremonial space, personifying the church's social stature and wealth. Landscape architects and their clients have approached corporate site design as an art form, creating dramatic tension between geometric and biomorphic forms. Rich materials and sculpture abound in Williams Square [18]

· 147 ·

(1984, The SWA Group), Las Colinas, Tex., where regional land forms are borrowed and reinterpreted on the site. In Englewood, Colo., the mission of Chevron Geothermal[19] (1985, Hargreaves Associate) is depicted metaphorically by its landscape of exfoliating layers of sedimentary rock and simulated steam. The desired effect is a more memorable and enduring image.

[9]　The significance of the institutional and corporate landscape types today is that they have demonstrated that both quality open space and memorable imagery are vital to the human experience.

[NOTES]

[1]　institutional and corporate landscapes 社会机构及公司园区景观：institutional 指事业机构的或社会公共机构的、慈善机构的；corporate 指公司的、企业的。
[2]　Industrial Revolution 工业革命或产业革命。
[3]　U. S. Sanitary Commission 美国卫生局：后来的美国红十字会（American Red Cross）。
[4]　Columbia Institution for the Deaf 哥伦比亚聋人学院。
[5]　Iowa Hospital for the Insane 爱荷华州精神病院。
[6]　Deere and Company 迪尔公司：也称作约翰迪尔公司，美国历史悠久的以农业设备为主的跨国公司，建筑由沙里宁（Saarinen）设计，环境景观由佐佐木英夫事务所设计。
[7]　Sasaki Associates 佐佐木英夫事务所：佐佐木英夫（Sasaki）日裔美国人，美国 SWA 集团创始人之一，Sasaki 事务所创始人，在各类城市公共领域的景观规划与设计方面有众多实践，是美国近现代城市空间的塑造者之一。
[8]　Carlson Center 卡尔森中心：位于明尼唐卡的国际会议中心，其环境景观由易道公司（EDAW Design Group）设计。
[9]　Pacific Bell 太平洋贝尔公司：美国最老牌的电话公司。
[10]　PepsiCo 百事可乐公司园区：其环境景观由 EDSA 公司（Edward D. Stone, Jr. Associates）设计。
[11]　Corporate Woods 科珀里特伍兹办公区。
[12]　The SWA Group：SWA 设计公司，其前身是 Sasaki, Walker and Associates，由佐佐木英夫与彼得·沃克创建，是美国近现代景观的重要缔造者。
[13]　TRW World Headquarters 汤普森·拉莫·伍尔德里奇公司世界总部：TRW 是美国以宇航业和飞机制造业为主的公司。
[14]　Codex's world headquarters 科戴克斯公司世界总部。
[15]　Kaiser roof garden 凯撒屋顶花园。
[16]　Dallas's Fountain Place 达拉斯喷泉广场。
[17]　Christian Science Church's world headquarters 基督教科学会世界总部。
[18]　Williams Square 威廉姆斯广场。
[19]　Chevron Geothermal 雪弗龙地热公司：全球地热能最大生产商，其环境景观由哈格里夫斯事务所设计。

[GLOSSARY]

corporate landscape	公司景观	institutional environment	社会公共机构环境

UNIT 8　RESIDENTIAL, INSTITUTIONAL AND CORPORATE LANDSCAPE

[NEW WORDS]

accessibility	n.	易接近，可到达	memorable	adj.	值得纪念的；难忘的
affinity	n.	近似，相像，喜爱，亲近	mentally	adv.	心理上，精神上，智力上
alcoholism	n.	酒精中毒	metaphorically	adv.	隐喻地，比喻地
analogous	adj.	类似的，相似的	monumental	adj.	纪念碑的，雄伟的，巨大的
arid	adj.	干旱的，贫瘠的，无生气的	orphaned	adj.	成为了孤儿的
asylum	n.	精神病院，避难所，庇护所	overview	n.	概述，概况，总的看法
biomorphic	adj.	生物形态的	personify	vt.	象征，体现，拟人化
clerical	adj.	职员的，办公室工作的	portfolio	n.	投资组合，总资产
confine	vt.	限制，局限于	premium	n.	重视，非常珍贵
depict	vt.	描绘，刻画，描述，叙述	reinterpret	vt.	使再生，使化身
discipline	n.	学科	requisite	adj.	需要的，必要的，必不可少的
exfoliate	vt.	片状剥落	retarded	adj.	智力上迟钝的
exhausting	vt.	用尽，耗尽，使筋疲力尽	sanatorium	n.	疗养院，疗养所
flee	vi.	避开，逃避	sanitary	adj.	卫生的，公共卫生的，保健的
geometric	adj.	几何图案的，几的	sedimentary	adj.	沉积的，冲积成的
grasp	vt.	抓住，抓紧	semipublic	adj.	半公共的，半开放的
handicapped	adj.	残疾的，弱智的	shimmering	adj.	闪烁的，闪闪发光的
headquarter	n.	总部，设总部	showpiece	n.	展出品，陈列品，样板
ingredient	n.	（构成）要素，因素	suburbia	n.	郊区及其居民，郊区居民
insane	adj.	（患）精神病的，精神失常的	therapeutic	adj.	疗法的，对身心健康有益的
landlord	n.	房东，地主，店主	thrive	vi.	兴盛，兴旺，繁荣
managerial	adj.	管理（上）的，经营（上）的	warehouse	n.	仓库，货栈
manor	n.	庄园，领地，庄园大厦	welfare	n.	福利，健康

[难句]

Affinities with early country villas and manor houses have also been suggested, as these buildings and their elaborate grounds once housed a large population of family and servants, curiously analogous to 20th century managerial, clerical and maintenance staff. 公司园区被认为与早期的乡村别墅和庄园住宅非常接近，因为这些建筑和精致的室外场地曾住过的大量家人与佣人，与20世纪的公司管理、杂务与维护人员有着奇怪的相似。

[参考译文]

社会机构及公司景观

谢丽尔·巴顿

［1］ 由于风景园林师致力于为人们提供有意义的开放空间，他们总是与人类福利息息相关。社会公共机构环境受益于风景园林师的专业技术已有一个多世纪，但是公司景观却是一种比较新的现象。在18世纪和19世纪早期，社会公共机构的工作主要是处理那些家庭无法处理的问题——照顾智障儿童、孤儿、病人和残疾人。但是这些地方的环境很多都只是比仓库稍微好一点。工业革命时期，那些离开农场到城市寻求机遇的工人们也遭遇了相似的命运。在那里，他们也长期被限制在仓库一般的环境里，度过漫长而疲惫的时光。

［2］ 回顾内战后的改革运动，美国的风景园林师们在设计新的社会公共机构环境时针对酒精中毒、精神病及疾病等社会问题提出了创新、实用的方法。纯净的空气和水、阳光及其他自然事物对于医院、精神病院和疗养院的规划来说是很重要的因素。宽敞的、良好设计的场所反映了医疗条件的改进。这些项目可能有许多学科参与，但常常由风景园林师协调，他们运用对各种环境要素的知识做出必不可少的项目总体规划。

［3］ 曾任美国卫生局（后来的美国红十字会）官员的弗雷德里克·劳·奥姆斯特德，规划了几个重要的公共机构，包括华盛顿特区的哥伦比亚聋人学院（1866）。芒特普莱森特城的爱荷华精神病院规划（1871年，H. W. S. 克里夫兰），在精神病院的场地内种满有助于治疗的乡土树木和灌木丛将病人包围。科学证据已经表明，直接接触自然对治疗能产生积极的影响。当代公共机构的景观和建筑设计重视病人的室外视觉感受以及与庭院、花园的紧密连接。

［4］ 二战后，由于工业的快速增长和郊区的出现，公司园区逐渐发展起来。这时，财富和权利的象征体现在具有吸引力、公园一般的位置。现在，人们有相当多的时间是待在公司环境中，在很多地方，公司已经取代主街变成社区的社交中心。公司园区被认为与早期的乡村别墅和庄园住宅非常相似，因为这些建筑和精致的室外场地曾住着大量家人与佣人，与20世纪的公司管理、杂务与维护人员有着奇怪的相似。

［5］ 现在的公司总部景观，比如伊利诺斯州莫林的迪尔公司（1963年，佐佐木英夫事务所），一个重要的先例；明尼苏达州明尼唐卡的卡尔森中心（1988年，易道设计公司）和加州圣拉蒙的太平洋贝尔公司（1987年，MPA设计公司），为了舒缓员工的工作强度而设计了大量的开放空间和娱乐资源，包括网球场、排球场、慢跑道、雕塑花园、湖面和其他设施，来确保员工更加忠诚和高产，同时为公司的地产投资组合增值。而且，这些特色还反映了公司形象、地位、实力、声望和产品的卓越特性——一个竞争性公司文化中关键要素。

［6］ 对于很多郊区来说，公司已成为社区中心。百事公司（1965年，EDSA设计事务所）——一个位于纽约珀彻斯的集风景园林、建筑和雕塑为一体的公司园区——在实现它对公司职员和公众的承诺中，形成当时公司景观中具有创造性的典范。他的雕塑藏品陈列在公园的台地上，对公众来说它是一个室外艺术博物馆，对公司老板来说它是一个成功的景观展品。很多公司购买过去的地产并保护其历史性景观。在堪萨斯州奥弗兰公园的科珀里特伍兹办公区（1976年，SWA公司）及俄亥俄州林德赫斯特的汤普森·拉莫·伍尔德里奇公司世界总部（1985年，佐佐木英夫事务所），风景园林师们使当代建筑与树林和田园风格的环境形成鲜明对比。位于马萨诸塞州坎顿的科戴克斯公司世界总部（1986年，汉纳/奥林设计事务所），在它建筑综合体的中心设计了一个精美的室内中庭，来唤起人们全年的室外感受。

［7］ 城市公司景观，作为公司景观的最早范例，也兴盛起来。加州奥克兰的凯撒屋顶花园（1960年，奥斯芒德森和斯坦利）是一个创新，它是一个三英亩大的城市开放空间，位于高于街道23层楼的屋顶。尽管90%的场地被建筑遮挡，实际上却有60%成为半公共的开放空间。达拉斯水景广场（1985年，丹·凯利），开放空间比例与前者类似，是一个与干旱的德克萨斯州景色形成鲜明对比的闪闪发光的巨大水景园。

［8］ 逐渐的，越来越多的场地规划成为体现公司和社会机构特殊价值观的载体。位于波士顿的基督教科学会的世界总部（1973年，佐佐木英夫事务所）是一个强有力的、礼仪性空间，来体现教会的社会声望和财富。风景园林师以及他们的客户把公司场地设计作为一种艺术形式来处理，在几何形态和生物形态之间创造富于戏

UNIT 8　RESIDENTIAL, INSTITUTIONAL AND CORPORATE LANDSCAPE

剧性的张力。德克萨斯州拉斯科林纳斯的威廉姆斯广场(1984年,SWA公司),拥有丰富的材料和大量的雕塑,借鉴当地的区域地貌并进行重新诠释。在科罗拉多州恩格尔伍德的雪弗龙地热公司(1985年,哈格里夫斯设计事务所)景观,是通过沉积岩的剥落片岩和模拟雾气以隐喻的手法来表现的。希望达到的效果是一个更加令人难忘和永久的形象。

[9]　现在,社会机构及公司景观类型的重要性在于,它们证明了开放空间的质量和令人难忘的形象对人们的体验来说至关重要。

[思考]
(1)美国花园的发展主要受到哪些风格和思潮的影响?代表性案例有哪些?
(2)美国居住环境规划经历了哪些模式?这些模式产生的背景是什么?
(3)美国近现代社会机构及公司景观有哪些代表性案例?

PART IV

Characters, ideas and works of Landscape Architecture in modern time

第四部分 近现代风景园林人物、思潮及作品

UNIT 9 LANDSCAPE PEOPLE AND THOUGHTS I

TEXT

Frederick Law Olmsted[1]
弗雷德里克·劳·奥姆斯特德

[1] Frederick Law Olmsted (April 25, 1822-August 28, 1903) (Fig. 1) was an American landscape designer and father of American landscape architecture, famous for designing many well-known urban parks, including Central Park[2] (Fig. 2) and Prospect Park[3] in New York City. Other projects include the country's oldest coordinated system of public parks and parkways in Buffalo[4] (Fig. 3), New York, the country's oldest state park, the Niagara Reservation[5] in Niagara Falls, New York, Mount Royal Park[6] in Montreal, the Emerald

Fig. 1 Frederick Law Olmsted

Fig. 2 Central Park

Necklace[7] in Boston, Massachusetts, the Belle Isle Park[8], in Detroit, Michigan, the Grand Necklace of Parks[9] including Washington Park in Milwaukee, Wisconsin, the Cherokee Park (and the entire parks and parkway system) in Louisville, Kentucky, the Jackson Park, the Washington Park, the Midway Plaisance in Chicago for the World's Columbian Exposition, the landscape surrounding the United States Capitol building, George Washington Vanderbilt II's Biltmore Estate in Asheville, and Montebello Park in St. Catharines, Ontario.

Fig. 3 Buffalo, New York's park and parkway system

[2] Olmsted was born in Hartford, Connecticut. His father,

John Olmsted, a prosperous merchant, took a lively interest in nature, people, and places, which was inherited by both Frederick Law and his younger brother, John Hull. After working as a seaman, merchant, and journalist, Olmsted settled on a farm on the south shore of Staten Island that his father helped him to acquire in January 1848. The house, in which Olmsted lived, still stands today at 4515 Hylan Blvd, near Woods of Arden Road.

[3]　Olmsted also had a significant career in journalism. In 1850, he traveled to England to visit public gardens, where he was greatly impressed by Joseph Paxton's Birkenhead Park[10], and subsequently published *Walks and Talks of an American Farmer* in England in 1852. Interested in the slave economy, he was commissioned by *the New York Daily Times* (now *The New York Times*) to embark on an extensive research journey through the American South and Texas from 1852 to 1857. The Texas trip produced his narrative account published as *A Journey Through Texas* (1857) and recognized as the work of an astute observer of the land and lifestyles of Texas. Olmsted took the view that the practice of slavery was not only morally odious, but expensive and economically inefficient. His dispatches were collected into multiple volumes which remain vivid first-person social documents of the pre-war South. The last of these, *Journeys and Explorations in the Cotton Kingdom* (1861), published during the first six months of the American Civil War, helped inform and galvanize antislavery sentiment in New England.

[4]　Olmsted's friend and mentor, Andrew Jackson Downing[11], the charismatic landscape architect from Newburgh, New York, first proposed the development of New York's Central Park as publisher of The *Horticulturist* magazine. It was Downing who introduced Olmsted to the English-born architect Calvert Vaux[12], whom Downing had personally brought back from England as his architect-collaborator. After Downing died in a widely publicized steamboat explosion on the Hudson River in July 1852, in his honor Olmsted and Vaux entered the Central Park design competition together—and won (1858). On his return from the South, Olmsted began executing the plan almost immediately. Olmsted and Vaux continued their informal partnership to design Prospect Park in Brooklyn from 1865 to 1873, and other projects. Vaux remained in the shadow of Olmsted's grand public personality and social connections. The design of Central Park embodies Olmsted's social consciousness and commitment to egalitarian ideals. Influenced by Downing and by his own observations regarding social class in England, China and the American South, Olmsted believed that the common green space must always be equally accessible to all citizens. This principle is now so fundamental to the idea of a "public park" as to seem self-evident, but it was not so then. Olmsted's tenure as park commissioner can be described as one long struggle to preserve that idea.

[5]　In 1863, he went west to become the manager of the Mariposa mining estate in the Sierra Nevada Mountains in California. For his early work in Yosemite Valley[13], Olmsted Point near Tenaya Lake is named after him. In 1865 Vaux and Olmsted formed Olmsted, Vaux and Company. When Olmsted returned to New York, he and Vaux designed Prospect Park; suburban Chicago's Riverside; Buffalo, New York's park system[14]; Milwaukee, Wisconsin's grand necklace of parks; and the Niagara Reservation at Niagara Falls.

[6] Olmsted not only created city parks in many cities around the country, he also conceived of entire systems of parks and interconnecting parkways which connected certain cities to green spaces. Two of the best examples of the scale on which Olmsted worked are one of the largest pieces of his work, the park system designed for Buffalo, New York and the system he designed for Milwaukee, Wisconsin.

[7] Olmsted was a frequent collaborator with Henry Hobson Richardson[15] for whom he devised the landscaping schemes for half a dozen projects, including Richardson's commission for the Buffalo State Asylum.

[8] In 1883 Olmsted established what is considered to be the first full-time landscape architecture firm in Brookline, Massachusetts. He called the home and office compound "Fairsted", which today is the recently-restored Frederick Law Olmsted National Historic Site. From there Olmsted designed Boston's Emerald Necklace, the campuses of Stanford University and the University of Chicago, as well as the 1893 World's Fair in Chicago among many other projects.

[9] In 1895, senility forced Olmsted to retire. In 1898 he moved to Belmont, Massachusetts and took up residence as a resident patient at McLean Hospital, whose grounds he had designed several years before. He remained there until his death in 1903, and was buried in the Old North Cemetery, Hartford, Connecticut.

[10] After Olmsted's retirement and death, his sons John Charles Olmsted[16] and Frederick Law Olmsted, Jr.[17] continued the work of their firm, doing business as the Olmsted Brothers. The firm lasted until 1950. A quotation from Olmsted's friend and colleague architect Daniel Burnham could well serve as his epitaph. Referring to Olmsted in March, 1893, Burnham said, "An artist, he paints with lakes and wooded slopes; with lawns and banks and forest covered hills; with mountain sides and ocean views."

[NOTES]

[1] Frederick Law Olmsted 弗雷德里克·劳·奥姆斯特德(1822—1903):奥姆斯特德是一位充满传奇色彩的人物,因设计纽约中央公园而声名显赫,并因创办美国风景园林专业、美国风景园林师协会及设计了大量亲近普通市民的风景园林作品而被人们尊称为"美国风景园林之父"。据统计,奥姆斯特德和他的公司一生承接了大约500个项目,其中包括100多个公园和娱乐场、200个私人庄园、50个居住区设计以及40余所校园设计,对美国20世纪风景园林的发展产生了重大影响。奥姆斯特德是一位理论著述良多的作家,在他的职业生涯中亲手书写了600多封信函和报告,涉及300个设计项目,其著作的完整名录中包括描述他的南方之旅的信函,以及由美国卫生委员会出版的各种文件,一共有300多项。当他早年游历英国时期,于1852年完成了他的第一本专著——《美国农夫在英国的游历和谈话》,对英国的乡村风光和伯肯黑德公园大加赞赏,称它为"人民的花园"。后来,他以美国北方报纸《纽约每日时报》通讯员记者的身份对美国南部进行研究,期间创作了三卷本《我们的努力国家》,于1861年出版。在写作上的成功,使他受邀成为《普特南月刊》及《国家》杂志的主编,另外,他还在《园艺师》等其他报纸、期刊上发表了多篇关于公园及景观对于满足胜利、精神需求的文章,并在1868年,出版了一本关于城市规划方面的书籍——《先驱的条件与美国文明的倾向》。在书中他批判了在综合规划中只顾眼前利益而忽视长远利益的设计,产生了积极而又深远的影响。奥姆斯特德以丰富的人生阅历、综合的理论知识记录、评述了对风景园林的看法,对后世产生了很大的影响。奥姆斯特德还以行动促进了美国风

景园林教育进步。从1860年到1900年,奥姆斯特德及其他风景园林师在城市公园、绿地、广场、校园、居住区及自然保护地等方面所做的规划设计奠定了风景园林学科的发展。1900年,奥姆斯特德之子F. L. Olmsted. Jr和A. A. Sharcliff首次在哈佛大学开设了风景园林专业课程,并在美国首创了4年制的风景园林专业学士学位。由于奥姆斯特德及其合作者的实践以及专业教育在哈佛大学的确立,使美国的风景园林一开始便定位在一个很大的活动范围内,包括城市公园和绿地系统、城乡风景道路系统规划设计,居住区、校园、地产开发、农产和国家公园的规划设计和管理,随后又进一步扩展到主题公园和高速公路系统的景观设计,这使风景园林师成为人居环境的主要规划设计师和创造者。

[2] Central Park 中央公园:是美国纽约市曼哈顿区大型都市公园,面积843英亩(3.41平方千米),长4千米,宽800米,是曼哈顿中心区的一片绿洲。

[3] Prospect Park 希望公园:由奥姆斯特德和沃克斯设计,1866年动工,历时两年完成,拥有茂密的森林、广阔的绿地以及60英亩的人工湖,成为布鲁克林的珍宝。它见证了工业时代的来临,人口转移都市,以及城市的整体更新。

[4] system of public parks and parkways in Buffalo 纽约州布法罗公园及公园道系统:是美国最早建成的具有真正意义的公园系统。1868年,奥姆斯特德在布法罗市放射型道路系统形态的基础上,规划了61米宽的公园路连接三个不同功能的公园,组成一个较完整的公园系统,最北面的是特拉华公园(Delaware),西面的是弗兰特公园(Front),位于东部的是巴拉德公园(Parade)。

[5] Niagara Reservation 尼亚加拉瀑布自然保护区。

[6] Mount Royal Park 皇家山公园:位于加拿大蒙特利尔,由奥姆斯特德设计,建于1876年。

[7] Emerald Necklace "翡翠项链":这一公园系统被公认为是美国最早规划的真正意义上的绿色通道。

[8] Belle Isle Park 贝尔岛公园。

[9] Grand Necklace of Parks 大公园群。

[10] Birkenhead Park 伯肯黑德公园:是英国历史上第一个公共筹款建立的城市公园。

[11] Andrew Jackson Downing 安德鲁·杰克逊·唐宁(1815—1852):风景园林师、园艺师、作家。唐宁在19世纪中叶为美国创造了风景园林艺术,成为简洁、自然、永恒的自然主义风格流派的伟大代表,对美国社会生活的许多方面都带来了深远的影响,被誉为美国风景园林的鼻祖。

[12] Calvert Vaux 卡尔沃特·沃克斯:美国近代建筑师及风景园林师,纽约中央公园的设计者之一。

[13] Yosemite Valley 约瑟米蒂河谷:美国加利福尼亚州中东部峡谷。

[14] Buffalo, New York's park system 纽约州布法罗公园系统:是美国最早的游憩空间的配套系统,由奥姆斯特德和卡尔沃特·沃克斯在1868—1896年设计。

[15] Henry Hobson Richardson 亨利·霍布森·查理森:被认为是美国历史上最伟大的建筑师之一,曾和弗雷德里克·劳·奥姆斯特德合作过一些项目。

[16] John Charles Olmsted 约翰·查尔斯·奥姆斯特德:奥姆斯特德之子。

[17] Frederick Law Olmsted, Jr. 小弗雷德里克·劳·奥姆斯特德:奥姆斯特德之子,父子同名。

[GLOSSARY]

National Historic Site	国家历史遗址	public park	公共公园,公立公园
parkway	公园道路,景观道路,风景车道	World's Fair	世界博览会

[NEW WORDS]

account	n.	描述,描写;叙述,记述;报道	embark on		从事,着手,开始做某事
antislavery	adj.	反对奴隶制度的	epitaph	n.	墓志铭,纪念死者的文字
astute	adj.	敏锐的,机敏的,精明的	galvanize	vt.	刺激,使震惊,惊起,使兴奋
asylum	n.	精神病院,避难所,庇护	horticulturist	n.	园艺师
charismatic		有魅力的;有感召力的	mentor		指导者
commission	vt.	委任,委托,使服役	odious	adj.	可憎的,可厌恶的,令人作呕的
	n.	授权,委托,委员会	self-evident	adj.	不证自明的,不言而喻的
conceived of	vt.	想象,设想	senility	n.	高龄,老迈,老年期
dispatch	n.	新闻报道,通讯;电讯,电文	steamboat		汽艇,汽船
egalitarian	adj.	平等主义的	tenure	n.	任职期

[参考译文]

弗雷德里克·劳·奥姆斯特德

[1] 弗雷德里克·劳·奥姆斯特德(1822.4.25—1903.8.28)是一位美国景观设计大师,同时也是闻名于多项著名城市公园设计的美国风景园林之父,作品包括纽约市的中央公园和希望公园。奥姆斯特德的其他设计项目还包括位于纽约州布法罗(有时也译作水牛城)的美国最古老的公共公园系统和公园道路;美国最古老的州立公园——坐落于纽约州尼亚加拉瀑布的尼亚加拉自然保护区;蒙特利尔的皇家山公园;马萨诸塞州波士顿的"翡翠项链";密歇根州底特律的贝尔岛公园;大公园群,包括威斯康辛州密尔沃基市的华盛顿公园;肯塔基州路易斯维尔市的切诺基公园(以及整个公园和公园道路系统);芝加哥哥伦比亚世界博览会中的杰克逊公园、华盛顿公园和大道公园;美国国会大厦周边景观;阿什维尔的乔治·华盛顿·范德比尔特二世的比尔特摩庄园;以及加拿大安大略省圣凯瑟琳的蒙特贝罗公园。

[2] 奥姆斯特德出生于美国康涅狄格州哈特福德。他的父亲,约翰·奥姆斯特德是一个成功的商人,他对自然、人和场地有着浓厚的兴趣,这些兴趣都被弗雷德里克·劳和他的弟弟约翰·赫尔继承了下来。奥姆斯特德当过海员、商人和记者,之后在父亲的帮助下,他在1848年1月斯塔腾岛南岸的一个农场定居了下来。而这座奥姆斯特德曾居住过的房子,至今仍伫立在雅顿路树林旁的海兰大道4515号。

[3] 奥姆斯特德在其新闻行业的职业生涯中也取得了非凡的成就。1850年,他前往英国参观公共公园时,被约瑟夫·帕克斯顿设计的伯肯黑德公园深深打动,随后于1852年发表了《美国农夫在英国的游历和谈话》一书。出于对奴隶制经济的兴趣,他受《纽约每日时报》(现在的《纽约时报》)委托,为进行研究于1852到1857年间穿梭于美国南部和德克萨斯州。他的德克萨斯州之旅促成了他叙述性作品——《穿越德克萨斯州的旅行》(1857年)的发表,而这部作品也被公认为是出自一个敏锐观察者对德克萨斯州土地和生活方式的记录。奥姆斯特德认为奴隶制不仅在道德上是可恶的,在经济上也昂贵且低效。他的报道被收录成册,一直被作为以生动的第一人称记录的战前南部地区社会性文件。他的最后一篇文章《在棉花王国中的旅程和探索》(1861年),出版于美国内战初期的6个月,在新英格兰地区影响和激励了反对奴隶制度的情绪。

[4] 奥姆斯特德的朋友兼导师,安德鲁·杰克逊·唐宁,是一位来自纽约州纽堡的极具魅力的风景园林师。作为《园艺师》杂志的出版商,他首次提出了对纽约中央公园的开发。也是唐宁将奥姆斯特德介绍给了他亲自从英格兰带来作为建筑师合作者、出生于英国的建筑师卡尔沃特·沃克斯。1852年7月,在唐宁死于一个广为报道的哈得逊河汽船爆炸事件后,为了纪念他,奥姆斯特德和沃克斯一起参加了中央公园设计竞赛,并且获胜

157

(1858年)。在从南方回来的路上,奥姆斯特德几乎是立即开始执行这项计划。1865年至1873年期间,奥姆斯特德和沃克斯继续以非正式伙伴关系设计了位于布鲁克林的希望公园和其他项目。而沃克斯仍然处于奥姆斯特德强大的外向型人格和社会关系阴影中。中央公园的设计体现了奥姆斯特德对于平等主义理想的社会意识和承诺。由于受到唐宁和其自身对于英国、中国和美国南部社会阶层观察的影响,奥姆斯特德认为,公共的绿色空间必须对所有市民平等开放。这一原则作为当代"公共公园"理论的基础似乎是不言自明的,然而在当时却并非如此。奥姆斯特德担任公园专员的任职期可以被描述为一个为维护这一想法而进行长期斗争的过程。

[5] 1863年,奥姆斯特德去西部担任加利福尼亚州内华达山脉的美利坡撒矿业地产的经理。特尼亚湖畔的奥姆斯特德点就是为了纪念他在约塞米蒂河谷早期所做的工作而以他的名字命名的。1865年,沃克斯和奥姆斯特德成立了"奥姆斯特德-沃克斯"公司。在奥姆斯特德回到纽约后,他和沃克斯设计了希望公园;芝加哥里弗塞得郊区社区、布法罗的纽约公园系统;威斯康辛州密尔沃基的大公园群,以及尼亚加拉瀑布的尼亚加拉自然保护区。

[6] 奥姆斯特德不仅在国内许多城市设计建造城市公园,同时他也在构思整体的公园系统和将城市与绿色空间相连的公园道路网络。奥姆斯特德在这种尺度上最好的两个作品,一个是其最大的项目之一,为纽约布法罗设计的公园系统,另一个是他为威斯康辛州密尔沃基所设计的公园道路系统。

[7] 奥姆斯特德是亨利·霍布森·查理森频繁的合作者,奥姆斯特德为其设计了六个项目的景观方案,其中就包括委任于理查德森的布法罗州立精神病中心(现为布法罗州立医院)。

[8] 1883年,奥姆斯特德在马萨诸塞州的布鲁克林成立第一个人们认为的全职景观设计公司。他将家庭和办公室称作复合的"费尔斯特德",也就是现在刚被修复的奥姆斯特德国家历史遗址。在那里奥姆斯特德设计了波士顿"翡翠项链",斯坦福大学和芝加哥大学校园景观,以及1893年芝加哥世界博览会等众多项目。

[9] 1895年,奥姆斯特德因年老体衰不得不被迫退休。1898年他移居马萨诸塞州的贝尔蒙,作为住院病人生活在麦克里恩医院,而那里的场地是他多年前设计的。他一直在那里生活直到1903年去世,死后被埋在康涅狄格州哈特福德的老北坟场。

[10] 在奥姆斯特德退休和去世后,他的儿子约翰·查尔斯·奥姆斯特德和小弗雷德里克·劳·奥姆斯特德,以奥姆斯特德兄弟的名字继续经营他们的公司,一直持续到1950年。奥姆斯特德的朋友和建筑师同事丹尼尔·伯纳姆的描述可以被恰当地引作他的墓志铭,1893年3月在谈到奥姆斯特德时,伯纳姆说"奥姆斯特德是一位艺术家,他用湖泊和树木繁茂的山脉来作画,用草坪、河岸和森林覆盖的丘陵来作画,用山峰和海景来作画。"

Reading Material A

Ian McHarg[1]

伊恩·麦克哈格

[1] Ian McHarg (1920-2001) (Fig. 1) was one of the true pioneers of the environmental movement. Born near the gritty, industrial Scottish city of Glasgow, he gained an early appreciation of the need for cities to better accommodate the qualities of the natural environment that until then had largely been shunned. After serving in World War II, McHarg went to the United States to attend Harvard University[2], where he picked up degrees in landscape architecture and city planning. He was responsible for the creation of the Department of Landscape Architecture at the University of Pennsylvania[3]. McHarg, however, would not be confined to the halls of academia. In 1960, McHarg hosted his own show, "The House We Live In" on the CBS television network[4], an early effort to publicize

discussion about humans and their environment. The show, along with a later PBS[5] documentary, helped make McHarg a household name when he published his landmark book, *Design With Nature*[6] (Fig. 2), in 1969. In it, McHarg spelled out the need for urban planners to consider an environmentally conscious approach to land use, and provided a new method for evaluating and implementing it. Today *Design With Nature* is considered one of the landmark publications in the environmental movement, helping make McHarg arguably the most important landscape architect since Frederick Law Olmsted.

 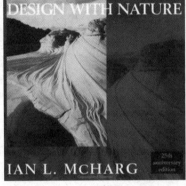

Fig. 1 Ian McHarg(1920-2001)　　　Fig. 2 Design with Nature (1969)

[2]　During the 1950s and 1960s, the nascent American Interstate Highway System[7] began to spread its tentacles around ever-increasing swaths of land. In its wake, suburban sprawl increased in scope and intensity. Highway planners and subdividers focused almost exclusively on narrow cost-benefit and efficiency considerations in choosing how to implement their ideas. According to Ian McHarg, "the task (of design) was given to those who, by instinct and training, were especially suited to gouge and scar landscape and city without remorse—the engineers."

[3]　McHarg's quote came from Design With Nature, in which he laid forth the argument that form must follow more than just function; it must also respect the natural environment in which it is placed. "(The engineer's) competence is not the design of highways," McHarg explained, "merely of the structures that compose them—but only after they have been designed by persons more knowing of man and the land." While this might seem a fairly obvious concept in the twenty-first century, in the late 1960s this was cutting-edge thought.

[4]　An important reason why the environment played such a small role in planning and design stemmed from the lack of a method to quantify and display information about the natural environment in any meaningful way. In the days before advanced computer technology, there was no way to store, process, or present large amounts of spatial data.

[5]　McHarg's way around this limitation was through the use of map overlays. As the addage goes, a picture is worth a thousand words, and McHarg felt that visually displaying spatial data could convey large amounts of information in a concise manner. McHarg demonstrated his approach with the example of a highly controversial road construction project in Staten Island[8] (Fig. 3), New York, known

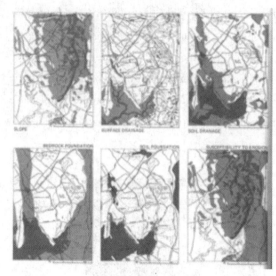

Fig. 3 Richmond Parkway in Staten Island

as the Richmond Parkway[9]. Engineers had proposed that the most cost-efficient route for the parkway lay along a five-mile stretch of scenic Greenbelt parkland[10], which would connect the two ends of the island along the straightest path. Local protests arose by 1968, and McHarg's firm was asked to review the various proposed routes.

[6] McHarg analyzed the situation with respect to "social values," or benefits and costs to society caused by the construction of a multipurpose facility such as a major traffic artery. There were many factors that went into the broad field of social values, including historic, water, forest, wildlife, scenic, recreation, residential, institutional, and land values. A map transparency was created for each factor, with the darkest gradations of tones representing areas with the greatest value, and the lightest tones associated with the least significant value. All of the transparencies were then superimposed upon one another over the original map. The darkest areas showed the areas with the greatest overall social values, and the lightest with the least, following the format of each individual layer. The social value composite map was then compared with similar maps constructed for geologic and hazard considerations, and the result was a clear picture of where to situate the controversial stretch of road. The road, McHarg concluded, should be built in an area west of the Greenbelt, thus saving the socially valuable forest and parks from the bulldozer.

[7] In the end, neither his nor any other proposal was ever acted upon, nor the Richmond Parkway, later renamed the Korean War Veterans Parkway[11], remains unfinished. McHarg's use of the overlay method, while not resolving this thorny issue, was a forerunner to later forms of complex analysis in multi-criteria route evaluation and decision making.

[NOTES]

[1] Ian McHarg 伊恩·麦克哈格(1920—2001):伊恩·伦诺克斯·麦克哈格(Ian Lennox McHarg),《设计结合自然》(Design with Nature)的作者。由于他在运用生态学原理处理人类生存环境方面做出的特殊贡献,曾多次获得荣誉,包括1972年美国建筑师学会联合专业奖章、1990年乔治·布什总统颁发的全美艺术奖章及日本城市设计奖。

[2] Harvard University 美国哈佛大学:位于马萨诸塞州剑桥市。著名的常春藤盟校成员。培养出了8位美利坚合众国总统与上百位诺贝尔获得者。在多个领域拥有崇高的学术地位及广泛的影响力,被公认为是当今世界最顶尖的高等教育机构之一。

[3] University of Pennsylvania 美国宾夕法尼亚大学:位于宾夕法尼亚州费城。著名的常春藤盟校成员之一。培养出多位诺贝尔奖获得者以及许多美国国家院士。

[4] CBS television network 美国哥伦比亚电视公司:CBS 全称为 Columbia Broadcasting System。美国三大全国性商业广播电视网之一。1927年成立,总部设在纽约。

[5] PBS 美国公共广播公司:全称为 Public Broadcasting Service。美国的一个公共电视机构,由 354 个加盟电视台组成,成立于 1969 年,总部位于维吉尼亚州阿灵顿县。

[6] *Design with Nature*《设计结合自然》:该书是英国著名环境设计师麦克哈格的代表作,1971 年获全美图书奖,是一本具有里程碑意义的专著,以丰富的资料、精辟的论断,阐述了人与自然环境之间不可分割的依赖关系、大自然演进的规律和人类认识的深化。作者提出以生态原理进行规划操作和分析的方法,使理论与实践紧密结合。

[7] American Interstate Highway System 美国州际高速公路系统。

[8] Staten Island 斯塔腾岛:美国纽约湾对面的岛屿。

[9] Richmond Parkway 里士满公园道。

[10] Greenbelt parkland:绿带公园。

[11] Korean War Veterans Parkway 朝鲜战争退伍老兵大道:一条穿过美国纽约斯塔滕岛南岸的公园大道。原里士满大道,1997 年正式更名为朝鲜战争退伍老兵大道。

[GLOSSARY]

composite map	综合图	map overlay / overlay method	叠图/分析法
environmental movement	环境运动	spatial data	空间数据

[NEW WORDS]

academia	n.	学术界,学术环境	nascent	adj.	初期的;发生中的,开始存在的
artery	n.	干线,要道	parkland	n.	公园用地,适于辟建公园的土地
bulldozer	n.	推土机	publicize	vt.	宣传(某事物)
competence	n.	能力,技能	remorse	n.	懊悔,悔恨
concise	adj.	简明的,简要的	scar	vt.	使留下伤痕,损害外观
controversial	adj.	有争议的,引起争议的	Scottish	adj.	苏格兰的,苏格兰式的
cutting-edge	n.	(刀片的)刃口,刀刃	shun	vt.	避开,规避,避免
forerunner	n.	先驱	sprawl	vi.	蔓生,蔓延
Glasgow	n.	格拉斯哥(英国城市)	stem from	vi.	源于,基于,出于
gouge	vt.	凿,挖	subdivider	n.	地块划分商
gradation	n.	阶段,等级;渐变	superimpose	vt.	使重叠,使叠加
gritty	adj.	有砂砾的,多沙的;刚强的	swath	n.	收割的刈痕,细长的列
host	vt.	主办,主持	tentacle	n.	触手,触角,触须
instinct	n.	本能,天性,直觉	thorny	adj.	棘手的,引起争议的,多刺的
multi-criteria	n.	多准则	transparency	n.	幻灯片,透明的东西,透明度

[参考译文]

伊恩·麦克哈格

[1] 伊恩·麦克哈格(1920—2001)是环境运动当之无愧的先驱者。麦克哈格出生于苏格兰多风沙的工业城

市格拉斯哥,他很早就理解到城市最好与自然环境本质相适应的需求。这种看法在当时很大程度上被忽视了。在参加了第二次世界大战后,他进入美国哈佛大学深造,在那里他获得了风景园林与城市规划的学位。随后,他负责了宾夕法尼亚风景园林系的创建。然而,麦克哈格并没有仅专注于学术教育。1960年,麦克哈格在CBS主持了他自己的电视节目"我们所居住的房屋",这是他早期对人类与所生存的环境的宣传普及所付出的努力。这个节目与之后PBS推出的纪录片,让他在1969年出版代表作《设计结合自然》的时候,已是家喻户晓。在这本书中,麦克哈格阐释了城市规划师对于土地利用要考虑采用环境保护的方法的需求,并提供了评估和实施的新方法。如今,《设计结合自然》被认为是环境保护运动里程碑似的一本著作,也让麦克哈格成为继弗雷德里克·劳·奥姆斯特德之后最重要的风景园林师。

[2] 在1950年代至1960年代之间,新兴的美国州际高速公路系统把它的"触须"延伸到不断扩大的土地裂痕周边。随之而来的是郊区在范围和密度上的扩张。高速公路的规划师与设计师几乎都关注于怎样能缩减成本效益与高效率的方法来实施他们的想法。如伊恩·麦克哈格所言,设计项目都给了那些天性使然或经过训练,只专注于挖掘、破坏景观与城市而丝毫不感到羞愧的工程师们。

[3] 麦克哈格《设计结合自然》中的观点之一,提出形式不仅仅只遵从于功能,它同样也需要注重它所处的自然环境。"(工程师的)能力不是设计高速公路,"麦克哈格解释到,"这些设计仅是由构筑物完成组合而已,只有由真正了解人类与土地的人经手,才能被称作设计。虽然这种思想在21世纪看起来是非常显而易见,然而在1960年这可谓是一种前卫的思想。"

[4] 环境在规划与设计中只扮演一个小角色,一个重要的原因在于缺乏有意义的自然环境量化和展示信息的方法。在先进的计算机技术普及之前,没有办法储存,处理或者展示大量的空间数据。

[5] 麦克哈格通过利用叠图分析法从而解决了这个限制。正如格言所道:一图道千言,麦克哈格认为用图像描述空间数据能够以非常简洁的形式传递大量的信息。他结合了处在纽约斯塔滕岛,名为里士满公园道这样一个备受争议的项目来示范了他的方法。工程师们提出建设这条公园道最具成本效益的路线是沿风景优美的绿带公园修建五英里,这条道路以最笔直的路径连通小岛的两端。当地的抗议者在1968年集结起来,麦克哈格的公司被请求重新审提议的各种路线。

[6] 麦克哈格结合"社会价值"对现场进行了分析,或者说建设多功能设施例如交通主干道给社会带来的利益与代价。在社会价值这个宽泛的命题里有许多因素,包括历史的、水、森林、野生动物、风景、休憩娱乐、住宅、公共机构和土地价值。每个因素都创建了一张透明的图,颜色最深的地方代表着该因素最有价值的区域,最浅与最不重要的价值关联。所有的透明图层叠加在一起放在原始的地图上。颜色最深的地方显示着最重要的总的社会价值,而最浅的则最少,其余的是不同的单独图层的形式。然后再将这张复合的社会价值地图与为地质和灾害而绘制的相似地图比较,最后得到一张清晰的有争议的道路的修建地点图。据麦克哈格推断,这条大道应该修建在绿带公园的西边,由此从推土机下保留下来了最具社会价值的森林与公园。

[7] 最终,麦克哈格和其他方案都没有被实施,里士满公园道路也没被采用,后来被改名为朝鲜老兵退伍大道,依旧没有完工。虽然麦克哈格的叠图分析法没有解决这个棘手的项目,但这却是之后多指标路径评估与决策中的综合分析法的先驱。

Reading Material B

Thomas Church[1]
托马斯·丘奇

"Any tendency to design for design's sake, to create a pattern within which the owner must live according to rules set by the designer, is headed for frustration, if not disaster."

UNIT 9　LANDSCAPE PEOPLE AND THOUGHTS Ⅰ

[1]　The name of Thomas Church (1902-1978) (Fig. 1) is synonymous with the California school of garden[2] design. After studying landscape architecture at Berkeley[3], he graduated from Harvard in 1926, grounded in the Beaux-Arts tradition.

[2]　His early work was confined to modest suburban gardens in and around San Francisco[4], where steep slopes shaped his design approach by preventing him from relying too much on symmetry and formal arrangement. As the winner of a traveling scholarship, he studied garden and landscape design in France, Spain, and Italy in the late 1920s, and found great similarities between these Mediterranean countries and California.

Fig. 1　Thomas Church (1902-1978)

Fig. 2　Sullivan garden

[3]　The Sullivan garden[5] (Fig. 2), in San Francisco, is remarkable for its use of a bold diagonal geometry that creates a dynamic sense of perspective as it cuts across the site. This technique echoed the work of André and Paul Véra and Gabriel Guevrekian[6] in France, but Church also appeared able to accentuate length and increase the sense of space. In other gardens he had started to use timber decking, an appropriate material for the Californian slopes, together with the built-in benches that would be found in gardens later in the twentieth century.

[4]　As the Bauhaus[7] influence started to permeate design circles in the mid-1930s, Church found himself drawn to Europe again. He became fascinated by Cubism and also met, in Finland, the architect Alvar Aalto[8], whose work was much more organic and sensuous than that of the Bauhaus. These two influences seemed to set him on a completely new trajectory, transforming his work.

[5]　The concept of Cubism was applied by including in the design a range of important views rather than employing a fixed or central viewing position. This increases the dynamic of the garden, creating a sensation of movement and spatial complexity, based on asymmetry. The garden that illustrates this style is El Novillero (Fig. 3), created by Church for Mr. and Mrs. Dewey Donnell in the Sonoma Valley[9], California, and seen by many as the icon of twentieth-century garden design.

Fig. 3　El Novillero (Donnel Garden)

[6]　Completed in 1948, this garden blends function, modern materials, Cubist theory, and sensitivity to location. It sits on the valley side, embracing the view of the marshes beyond, through which the river meanders in elegant curves. These soft lines are echoed in the swimming pool, sited on the main garden terrace, the last a regular grid of in situ concrete, based on the architectural grid of the

house. The contrast between this grid and the curves of the pool terrace are redwood decks, carefully arranged around the oaks already there, which spread their mature canopies across the garden, framing the view and decorating the ground with complex shadow patterns. The boundary between garden and landscape is almost non-existent, allowing the eye to run on to the distant Pacific. A sense of space is this garden's overwhelming and exhilarating characteristic.

[7] In 1948 Church created a garden for a beach house at Aptos[10], south of San Francisco. This is a composition of deck, sand, and simple planting, again combining sinuous curves and a controlling grid. The design expressed its function as a space for leisure and pleasure in which gardening was not a central issue.

[8] Church published his first book, *Gardens Are for People* [11] —now considered a classic—in 1955. He writes almost as if in conversation with a client, looking at how they live their life, how best a garden might suit them, how much time they might have for gardening, and how they might bring their entertaining out of doors. Although nowadays we tend to take these considerations for granted, in the 1950s the idea that the garden could be used for anything other than plants was a huge departure from the accepted view of its role.

[9] In many of the case studies that he includes in *Gardens Are for People*, Church not only explains how design solutions have been achieved but how topography might have affected the layout of a design. The tone is friendly, inviting the reader to understand his outlook on design and avoiding a dictatorial stance. In particular, Church acknowledged that the domestic scene was changing dramatically and that the garden was becoming smaller and more democratic. He was the first designer to remark on these changes in society and to address them for a wider clientele.

[10] Many of Church's gardens use expansive areas of paving, partly for functional reasons and partly to reduce maintenance for the American middle classes. Paths rarely figure in his gardens, and, as a result, the whole ground plan is freed to allow hard and soft materials to flow together. This informality and ease expressed through a curvilinear geometry relates back to Aalto, who created a famous swimming pool for the Villa Mairea[12] (Fig. 4), Finland, in 1938; it perhaps also links to the work of Roberto Burle Marx[13] (Fig. 5), whose work in Brazil in the 1930s had probably been publicized.

Fig. 4　Swimming pool in the Villa Mairea

Fig. 5　Roberto Burle Marx's Work

[11] Church produced more than two thousand commissions in his long and productive career as a

garden designer and landscape architect. Not all of these exhibited the modernist principles for which Church is justly famous, for he was not a purist, preferring to listen to his clients and to work with them rather than against them. At the same time those gardens that allowed Church to explore modernism have lived on, making him one of the most revered designers of the twentieth century.

[NOTES]

[1]　本文节选自 *Influential Gardeners* 书中"Thomas Church"一文。托马斯·丘奇(Thomas Church, 1902—1978)是美国现代园林的开拓者,他从 20 世纪 30 年代后期开始,开创了被称为"加州花园"的美国西海岸现代园林风格,丘奇等加州现代园林设计师群体被称为"加利福尼亚学派",其设计思想和手法对今天美国和世界的风景园林设计有深远的影响。丘奇的主要作品有唐纳花园(Donnel Garden, 1948)、阿普托斯(Aptos, 1948)花园和旧金山瓦伦西亚公共住宅工程(Valencia Public Housing, 1939—1943)。1951 年,丘奇获得美国建筑学会(AIA)艺术奖章,1976 年获得美国风景园林学会金奖,1955 年,出版著作《园林是为人的》(*Gardens Are for People*)。他的事务所培养了一系列成功的风景园林师,如罗斯坦(R. Royston)、贝里斯(D. Baylis)、奥斯芒德森(T. Osmundson)和哈普林(L. Harplin)等。

[2]　California school of garden 加州花园学派:托马斯·丘奇等加州现代园林设计师群体被称为加利福尼亚学派,其设计思想和手法对今天美国和世界的风景园林设计有深远的影响。

[3]　Berkeley 伯克利:加利福尼亚大学伯克利分校。

[4]　San Francisco 旧金山:美国加利福尼亚西部港市。

[5]　Sullivan garden 沙利文花园:位于美国加利福利亚州。

[6]　Gabriel Guevrekian 加布里埃尔·古埃瑞克安:20 世纪初景观设计师,以 1925 年在巴黎装饰工艺博览会上设计的光与水的花园及 1926 年设计的三角形花园闻名。

[7]　Bauhaus 包豪斯建筑学派:1919 年由德国建筑师瓦尔特·格罗皮乌斯(WalterGropius, 1883—1969)创立,他将当时的德国魏玛艺术学校发展为融建筑、雕刻、绘画与一体的"包豪斯"学校,并发扬德意志制造联盟的理想,从美术结合工业探索新建筑的精神。包豪斯强调自由创作、反对墨守成规,将工艺同机器生产相结合,强调各门艺术间的交流,因此成为 20 年代最激进的艺术汇集地之一,包豪斯学校在现代建筑、工业设计、工艺美术史上具有极为重要的地位。

[8]　Alvar Aalto 阿尔瓦·阿尔托:芬兰现代建筑大师,是现代建筑的重要奠基人之一,也是现代城市规划、工业产品设计的代表人物。

[9]　Sonoma Valley 索诺玛谷:美国加州最受瞩目的两大葡萄产区之一。

[10]　Aptos 阿普托斯:是旧金山一个未合并的地区,圣克鲁斯县,其中包括几个小的社区。

[11]　*Gardens Are for People*《花园为人们而建》:这本书发行于 1955 年,包含了托马斯·丘奇的设计哲学和精髓,同时包含很多实用的建议。

[12]　Villa Mairea 玛利亚别墅:阿尔托(Alvar Aalto)的建筑作品之一。

[13]　Roberto Burle Marx 布雷·马克斯:他被公认为是 20 世纪最有天赋的景观设计师之一,他同时也是画家、雕塑家和植物学家。他的作品体现了巴西的文化传统与发源于欧洲的现代艺术思想的结合,创造了适合当地气候特点和植物材料的崭新风格,开辟了景观设计的新天地,并与巴西的现代建筑运动相呼应。

[GLOSSARY]

Beaux-Arts	美术,美术学院	Cubism	立体主义	modernism	现代主义	modernist	现代主义者

[NEW WORDS]

accentuate	vt.	使突出,强调	marsh	n.	沼泽,湿地
asymmetry	n.	不对称	meander	vi.	蜿蜒而流
blend	vt.	混合	oak	n.	橡树
built-in	adj.	嵌入的,内装的,内部的,内置	overwhelming	adj.	势不可挡的,压倒一切的,巨大的
canopy	n.	天篷,遮篷,苍穹	permeate	vi.	弥漫,渗透,普及
clientele	n.	委托人,顾客,客户	revered	adj.	受人尊敬的
curvilinear	adj.	曲线的,由曲线组成的	sensuous	adj.	刺激感官的,感觉官能的
diagonal	adj.	对角线的,斜的,斜纹的	similarity	n.	相似,类似,相像性,相似之处
dictatorial	adj.	独裁的,专制的,霸道的,专横的	sinuous	adj.	弯弯曲曲的,迂回的,乖僻的
dynamic	adj.	动态的,有动力的,有力的	stance	n.	姿态,态度,立场
exhilarating	adj.	使人愉快的,令人振奋的	symmetry	n.	对称(性),匀称,整齐
frustration	n.	挫折,失败,沮丧	synonymous	adj.	同义词的,同义的,含义相同的
grid	n.	网格,格子	timber	n.	木材,林木,用材林
informality	n.	非正式,非公式,简略	topography	n.	地形学,地形,地貌,地势
in-situ	n.	就地,在原处,自然(环境)	trajectory	n.	轨道,轨迹,弹道

[参考译文]

托马斯·丘奇

"任何为了设计而设计的意向,去创造一个使用者必须按照设计师所设定的规则去生活的模式,如果不是灾难,那么必然走向失败。"

[1] 托马斯·丘奇(1902—1978)(图1)的名字是"加州花园学派设计"的代名词,在加利福尼亚大学伯克利分校学习景观设计后,他于1926从哈佛大学毕业,成长于传统美术学院派中。

[2] 他的早期作品局限于旧金山及其周围的中型郊区花园。陡峭的坡度塑造了他的设计方式,使他不过度依赖对称和正规的布局。作为一个留学奖学金的获得者,在19世纪20年代末期,他在法国、西班牙和意大利都学习过景观设计,并发现了这些地中海国家和加利福尼亚之间存在极大的相似性。

[3] 在旧金山的沙利文花园(图2),因大胆的对角形的运用而著名,在切割场地的同时创造了一个充满活力的景象。这个技巧仿效了法国的安德烈、保罗·维拉和加布里埃尔·古埃瑞克安的作品,但丘奇还展现了强调长度和增加空间感的能力。在其他的花园中他开始运用木平台,一种适合加利福尼亚坡地的材料,同时开始运用在20世纪晚期可见于花园中的嵌入式长凳。

[4] 在19世纪30年代中期,当包豪斯建筑学派的影响开始渗透设计圈,丘奇发现自己再次被欧洲所吸引。他变得着迷于立体主义,还在芬兰遇见了建筑师阿尔瓦·阿尔托,他的作品比包豪斯建筑学派的更加有机和具有美感。这两个影响因素似乎促使他走上了一条全新的轨道,转变了他的作品。

[5] 立体主义的概念被用于将一系列重要的视角融入设计而不是使用一个固定的或中心的视点位置。这增加了庭院的活力,基于不对称性创造了一种动感和空间的复杂性。展现这种风格的花园是丘奇为杜威·唐纳夫妇创建的位于加利福尼亚索诺玛谷的唐纳花园(图3),被很多人视作20世纪园林设计的代表。

[6] 这个于1948年被建造完成的花园融合了功能、现代材料、立体主义理论和地域敏感性。它坐落于山谷旁,环抱着前方沼泽的风景,其间河流迂回曲折成优美的弧度。柔性的线条被重复用于主花园平台上的游泳池。主花园平台是场地中最后一块基于房屋的建筑轴网的规则的现浇混凝土块,形成强烈对比的混凝土平台

与泳池曲线之间构筑了一处红木平台，被巧妙地设置在已有的橡树周围，在整个花园伸展着它们成熟的树冠，框景的同时用复杂的阴影图案装饰地面。花园和风景的界限几乎不存在，使人们的视线可以到达遥远的太平洋。空间感是这个花园最强烈、让人振奋的特点。

［7］ 1948年在旧金山南部的阿普托斯，丘奇为一个沙滩房屋设计了花园。这是一个平台、沙地和简单种植的组合，又一次结合了蜿蜒的曲线和一个控制性的网格。这个设计表现了它作为一个休闲和娱乐空间的功能，而园艺并非一个核心问题。

［8］ 丘奇在1955年出版了他的第一本书《花园为人们而建》——现在被认为是一部经典。他的写作几乎像是在与委托人的谈话中，查看他们如何生活，一个花园如何最适合他们，他们愿意为园艺付出多少时间，他们将如何在户外进行娱乐。尽管现在我们把这些考虑当作理所当然，在19世纪50年代花园能被用于除种植以外其他任何事情的想法，与它所被公认的作用有巨大的背离。

［9］ 在《花园为人们而建》一书所包含的诸多案例中，丘奇不单阐释了设计解决方案如何获得，并且解释了地形可能会如何影响一个设计的布局。写作语气友好，吸引读者去理解他在设计上的见解，从而避免一个独裁专横的态度。值得一提的是，丘奇认可国内的景象正在发生急剧的变化，花园变得越来越小，越来越民主化。他是第一个评论这些社会变化的设计师，并为更广范围的客户作出解释。

［10］ 很多丘奇的花园运用大量面积的铺装，部分原因是为了功能，部分是为了减少美国中产阶级的维护。道路在他的花园中很少突显，因而整个地面的设计可以自由地允许硬质和软质材料相互流动。这种通过曲线形状轻松表达的随意和舒缓来源于阿尔托，他在1983年为芬兰的玛利亚别墅设计了一个著名的游泳池（图4）；它或许与布雷·马克斯的作品也有关联（图5），其作品在19世纪30年代巴西被广泛宣传。

［11］ 作为一个园艺设计师和风景园林师，丘奇在他漫长且高产的职业生涯中完成了超过2 000个设计委托。这些设计不全都表现了丘奇所为之著名的现代主义原则，因为他不是一个纯粹主义者。他乐意倾听他的客户，并和他们一同工作，而不是与他们对抗。同时，那些允许丘奇去探索现代主义的花园继续存在着，使他成为20世纪最受尊敬的设计师之一。

［思考］
（1）奥姆斯特德对美国乃至世界风景园林的发展有哪些重要贡献？
（2）奥姆斯特德在各个设计领域最有代表性的作品是哪些？
（3）麦克·哈格的生态思想有什么重要意义？他所开创的生态设计方法是什么？
（4）托马斯·丘奇开创的"加州花园风格"主要特点有哪些？

UNIT 10 LANDSCAPE PEOPLE AND THOUGHTS II

TEXT

Dan Kiley[1]
丹·凯利

"The thing that's important is not something called design, it's how you live, it's life itself. Design really comes from that. You can not separate what you do from your life."

[1] Dan Kiley (b. 1912) (Fig. 1, Fig. 2), is a truly celebrated landscape designer, an example to countless designers in the second half of the twentieth century and, at the age of ninety, still hard at work. Kiley was born in Boston, Massachusetts, and feels fortunate to have been poor as a child, suggesting that this broadened his understanding. He enjoyed his sense of freedom but also subconsciously read the structure of his home territory. Part of this was the Arnold Arboretum[2], through which he would walk home from school. He was keenly aware of his environment and the power of the landscape as a backdrop to his life, noting in detail the golf course where he caddied as a teenager or the woodlands that surrounded his grandmother's farm. The landscape was always a part of him, and in consequence his designs seem to be part of the landscape.

Fig. 1 Dan Kiley relief in Jefferson National Expansion Memorial

Fig. 2 Dan Kiley

Fig. 3 Warren Manning

[2] In the early 1930s, Kiley was offered a job by Warren Manning[3] (Fig. 3), and then considered the top plants man in American, for whom Fletcher Steele[4] had also worked. Later, he enrolled as a special student at the Harvard Graduate School, while retaining his position in Manning's office.

In this way he enjoyed some detachment from his studies and developed an independent outlook. This became an important factor in the famous rebellion at Harvard[5] in which Kiley, James Rose[6], and Garrett Eckbo[7] reacted against the drudgery of Beaux-Arts pattern-making by rote. Kiley feels that it was Eckbo who led the way into modernism, but the sense of release in exploring space and three-dimensional structure proved both welcome and timely for all three designers. Kiley left Harvard in 1938, without a degree but desperate to explore the field of landscape design.

[3] In the early years of his career Kiley was introduced to Louis Kahn[8] (Fig. 4), a talented modern architect whose work Kiley considered to be inspired. They worked together on a number of schemes for the United States Housing Authority, and Kiley was to learn a lot of from this experience. He also absorbed modern design principles, taking particular note of Margaret Goldsmith's *Design for Outdoor Living*[9], published in 1942. Her principle of breaking down the barrier between indoor and outdoor space related closely to work that Kiley had carried out in 1941 at the Collier Residence[10], Virginia. In this garden, a series of terraces provided space for outdoor entertainment, and the "room outside" was effectively born.

Fig. 4　Louis Kahn

[4] During the Second World War, Kiley turned to architectural design before being drafted into the army. He was sent to Germany, where he was involved in the refurbishment of the palace at Nuremberg[11] in preparation for the trials of Nazi war criminals. While traveling, he discovered the work of André Le Nôtre, whose clarity of line was to leave an indelible impression on him.

[5] Later Kiley met the designer Eero Saarinen[12], with Kevin Roche[13], of the Miller house[14] (Fig. 5, Fig. 6), in Columbus, Indiana. Saarinen invited Kiley to design the gardens, and he created what he regards as his finest work, "The Irwin Miller garden… it seems a miracle. I can't quite understand how I did it."

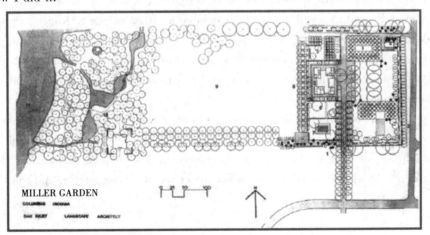

Fig. 5　Plan view of Miller house

Fig. 6 Miller house garden

[6]　Architect, garden designer, and client were all thinking in the same way, an unusual convergence in any design commission. The geometry of the house, completed in 1955, is taken out into the landscape as a grid. The glass walls allowed this transition to be almost seamless, and the sense of space is exhilarating. The simplicity of treatment also contributes to the timeless elegance of the garden. Lawn, graveled allées planted with gleditsia, and low decks of paving are separated by blocks of ground cover, allowing space and light to flow through the entire design. Borrowing from both modernist and classical themes, this abstract composition bridges the divide between the two to create a timeless work.

[7]　Forty years later, in the Kimmel residence[15], Connecticut Kiley was still able to bring out these qualities of space and light, making the gardens emerge from and relate to the surrounding landscape. The main lawn is edged with a wide channel paved in slate to create a shining strip that reflects the broad sky. Kiley's work is characterized by an understatement that makes it belong to its site and compels other designers to say "of course." He manages to bridge the two disciplines of architecture and landscape design as a result of successful synthesis and sensitivity.

[8]　In addition to garden designs, Kiley is also noted for his handling of urban space, producing schemes such as the Henry Moore Sculpture Garden[16], in Kansas City, in which he explored the development of landscape design in relation to works of art. Fountain Place[17], in Dallas, is perhaps his most famous work, a celebration of water in the searing heat of Texas, in which pools planted with trees cover most of the urban plaza. Every day more than 2.4 million litres (635,000 US gallons) of water is pumped through the pools.

[9]　What comes across in both Kiley's gardens and his extensive writing is his great passion for the landscape, for people's interaction with their environment, and a sense that he feels privileged to have been involved in this work. His influence on other designers has been immense, both in the United States and around the world. Few designers have enjoyed a career as long as Kiley's, in which he has created work that in almost every respect represents the aspirations of the twentieth century. He is the acceptable face of modernism, able to create in spatial terms a voluptuous elegance that is both thrilling and sublime, when his fellow modernists seemed to have become obsessed with the international style and the horrors of graying concrete.

[10]　Kiley's designs remain as fresh and inspiring in the early twenty-first century as they did at the start of his career. He maintains that a design emerges as a concept from the nature of the problem set, after which the functions are analyzed, and synthesized. This is pure modernism, and it underlines the difference between decorative pattern-making or stylistic interpretation and true design.

[NOTES]

[1]　Dan Kiley 丹·凯利(1912—2004):丹·凯利是"哈佛革命"的发起者之一,美国现代风景园林的奠基人之一。凯利的引路人曼宁(Warren Manning)关注大尺度的土地利用,他告诉凯利要从个人的直觉出发,从个体的体验和经验中去寻找解决基地问题的办法,而不要去模仿某种形式。在他的事务所,凯利避免了进入行业之初

受到当时各种保守或激进思想的影响和禁锢,而是埋头于实际的工程实践中,学到了大量关于植物的知识,积累了许多工程的经验,并且对于什么是景观设计中最重要的问题有自己独到的理解。凯利的设计通常从基地和功能出发,确定空间的类型,然后用轴线、绿篱、整齐的树阵和树列、方形的水池、树池和平台等古典语言来塑造空间,注重结构的清晰性和空间连续性。空间微妙变化主要体现在材料的质感色彩、植物的季相变化和水的灵活运用。凯利的作品通常使用古典的要素,但他的空间是现代的、流动的。他从基地的情况、客户的要求以及建筑师的建议出发,寻找解决这块基地功能最恰当的图解,将其转化为一个个的功能空间,然后用几何的方式将其组织起来,着重处理空间的尺度、空间的区分和联系。他认为,"对基地和功能直接而简单的反映是最有效的方式之一,一个好的设计师,是用生动的想象力来寻找问题的症结所在,并使问题简化,这是解决问题最经济的方式,也是所有艺术的基本原则"。半个多世纪以来,凯利的作品不计其数,虽然他的设计语言可以归纳为古典的,他的风格可视为现代主义的,但他的作品从来没有一种特定的模式。凯利经常从建筑出发,将建筑的空间延伸到周围环境中,他的几何空间构图与现代建筑看起来非常协调,许多建筑师欣赏他的风格选择凯利作为合作伙伴,如路易斯·康、小沙里宁、贝聿铭、凯文·罗奇等,成为二战后美国最重要的一些公共建筑环境的缔造者。他曾获得过各种组织的60多个奖项。1992年获得哈佛大学杰出终身成就奖。

[2]　Arnold Arboretum 阿诺德植物园。

[3]　Warren Manning 沃伦·曼宁(1860—1938):继承和发展了奥姆斯特德(Olmsted)特别是艾略特(Eliot)的景观系统思想,而成为景观分析研究和实践的领头人。他是有文字可循的最早使用地图叠加技术进行景观规划的风景园林师。

[4]　Fletcher Steele 弗莱彻·斯蒂里(1885—1971):是最早将欧洲前卫花园思想带到美国的设计师,斯蒂里多数著作的主题都是郊区住宅的布局思想,绝大多数作品都是庄园尺度的。从在波士顿独立开业后,他穿行于整个新英格兰区,一直到北部的纽约州,伊利诺伊州莱克福里斯特地区,底特律以及北卡莱罗那州底阿什维尔,最后作品是为在离他家几英里远处设计的一个异常安静但又很有影响力的花园,尚未完工,斯蒂里就去世了,享年86岁。

[5]　rebellion at Harvard 哈佛革命:20世纪30—40年代,由于第二次世界大战,欧洲不少有影响力的艺术家和建筑大师来到美国寻找安身之地,主要的艺术中心从巴黎转移到了纽约。1937年,德国现代建筑师、建筑教育家格罗皮乌斯来到美国,担任哈佛大学设计研究生院院长,他将包豪斯的办学精神带到哈佛,彻底改变了哈佛建筑专业的"学院派"传统。然而风景园林的教授们试图忽视这些,他们认为园林的革新无非是规则式和不规则式之间的微妙平衡。渴求新思想的学生们不愿意接受这样的观点,他们通过研究现代艺术和现代建筑的作品和理论,探索它们在景观上的可能的应用。这些学生中的最突出的是凯利、罗斯和埃克博。他们成为美国景观界最负盛名三位现代主义大师,掀起了现代主义的潮流,动摇并最终导致哈佛风景园林系"巴黎学院派"教条解体和现代设计思想的确立。这就是著名的"哈佛革命"。其他还有斯蒂里和英国建筑师唐纳德,他们与风景园林学科的旧的守护者之间展开了论战。

[6]　James Rose 詹姆斯·罗斯(1913—1991):是美国现代主义园林设计大师。在20世纪现代艺术和现代建筑蓬勃发展的背景下,詹姆士·罗斯反对学院派的传统方式,探索现代园林的形式、空间、材料,其作品具有简洁、灵活、实用的特点,被誉为美国现代园林的先驱。

[7]　Garrett Eckbo 盖瑞特·埃克博(1910—2000):是美国著名的景观规划师,著有《生活景观》(*Landscape for Living*)。盖瑞特·埃克博阐述了一个重要思想,即艺术与规划是等同的,只不过它们采用了不同的表达手段而已。

[8]　Louis Kahn 路易·康(1901—1974):是美国现代建筑师。路易·康发展了建筑设计中的哲学概念,认为盲目崇拜技术和程式化设计会使建筑缺乏立面特征,主张每个建筑题目必须有特殊的约束性。他的代表作品为宾西法尼亚大学理查德医学研究中心、耶鲁大学美术馆、索克大学研究所、爱塞特图书馆、孟加拉国达卡国民议会厅、艾哈迈德巴德的印度管理学院等。著作有《建筑是富于空间想象的创造》《建筑·寂静和光线》《人与建筑的和谐》等。

[9]　Margaret Goldsmith's *Design for Outdoor Living* 玛格丽特·史密斯1942年出版的《户外生活设计》。

[10]　Collier Residence 科利尔住宅。

[11]　palace at Nuremberg 纽伦堡宫。

[12] Eero Saarinen 埃罗·沙里宁(1910—1961):沙里宁是美国著名建筑设计师和工业设计师,他在建筑上的代表作有杰佛逊纪念碑、纽约肯尼迪国际机场、美国杜勒斯国际机场等。

[13] Kevin Roche 凯文·罗奇(1922—1961):罗奇1922年生于爱尔兰都柏林,1948年移民美国,1964年成为美国公民。1951年,他加入位于密歇根州 Bloomfield Hills 的沙里宁事务所。自1954年至沙里宁1961年辞世,凯文·罗奇是其最主要的设计成员。代表作品为 Deere West 办公大楼、福特基金会大楼、联邦广场饭店和奥克兰博物馆。

[14] Miller house 米勒庄园(1955):庄园内建筑由沙里宁(Eero Saarinen)设计,其花园(Miller Garden)是丹·凯利(Dan kiley)的第一个现代主义设计作品,也是现代景观设计最具影响的作品之一,凯利利用树干作为结构,利用绿篱进行围合,塑造了一系列室外的功能空间。米勒庄园的成功是建筑师与风景园林师无间合作的成果。

[15] Kimmel residence 凯莫尔住宅。

[16] Henry Moore Sculpture Garden 亨利·摩尔雕塑公园:亨利·摩尔(1898—1986,Henry Moore),英国雕塑家。摩尔以他的大型铸铜雕塑和大理石雕塑而闻名,受到英国艺术圈的推崇,他的创作为英国在现代主义艺术中占据了一席之地。

[17] Fountain Place 喷泉广场:位于德州达拉斯市,是丹·凯利代表作品之一,凯利成功地运用结构主义的手法,将水、植物和喷泉等园林要素都作了有序的组织。

[GLOSSARY]

glass wall	玻璃幕墙	rebellion at Harvard	哈佛革命	the geometry of the house	房屋几何学
ground cover	地被植物	room outside	外部居室	three-dimensional structure	三维结构

[NEW WORDS]

arboretum	n.	植物园	outlook	n.	观点,见解,展望,前景
backdrop	n.	背景幕,(事件的)背景	privileged	adj.	有特权的,特许的
barrier	n.	分界线,隔阂,障碍	react	vi.	反对,对抗;作出反应,回应
caddy	vt.	供差遣,当球童	rebellion	n.	谋反,叛乱,反抗
convergence	n.	集中,会聚,聚焦	refurbishment	n.	整修,翻新,刷新,翻新工程
desperate	adj.	不顾一切的,拼命的	rote	n.	死记硬背,机械的做法,生搬硬套
detachment	n.	分开,分离;超然,超脱	searing	adj.	烧灼的,灼热的,炽热的
discipline	n.	学科;训练;纪律	slate	n.	板岩,石板,石片
drudgery	n.	苦工,单调沉闷的工作	stylistic	adj.	风格上的,格式上的,体裁上的
effectively	adv.	实际上,事实上;有效地	subconsciously	adv.	潜意识地
geometry	n.	几何(学),几何图形,几何体	sublime	adj.	崇高的,壮丽的,宏伟的
gleditsia	n.	皂荚	synthesized	adj.	合成的,综合整理出,综合
immense	adj.	极大的,巨大的,无边的	territory	n.	地盘,领域,范围,领土,地区
in relation to		与……有关	trials	n.	审判,审讯,试用,试验;选拔赛
indelible	adj.	去不掉的,不能拭除的,擦不去	underline	vt.	加强,强调;在……下面画线
keenly	adv.	敏锐地,敏感地,强烈地,热心地	understatement	n.	有节制的陈述或表达,轻描淡写
Nazi	n.	(德国的)纳粹党员,纳粹主义	voluptuous	adj.	撩人的,艳丽的,奢侈逸乐的

UNIT 10　LANDSCAPE PEOPLE AND THOUGHTS Ⅱ

[参考译文]

丹·凯利

"重要的事情,不是所谓的设计,而是你如何生活以及生活本身。
设计真正来源于此。你无法将你的所作所为从生活中分离出来。"

[1]　丹·凯利(生于1912年)是一位名副其实的景观设计师,在90岁高龄仍坚持努力工作,为20世纪后半叶的诸多设计师树立了榜样。凯利出生在马萨诸塞州的波士顿,他觉得作为一个穷人的孩子是很幸运的,认为这拓展了他的见解。他享受自由的感觉,而且还下意识地勘查了他家周边区域的景观结构。其中的一部分是他从家走到学校途中需要经过的阿诺德植物园。他敏锐地感知他周边的环境和作为其生活背景的景观的力量,在做少年球童时他就详细记录了高尔夫球场地的每个细节和围绕在祖母农场周围的林地。景观一直是他的一部分,因此他的设计似乎也是景观的一部分。

[2]　在20世纪30年代初,凯利得到了沃伦·曼宁提供的工作机会,曼宁被认为是全美最优秀的植物种植专家,弗莱彻·斯蒂里也曾为其工作。随后,他作为特招生进入哈佛大学研究生院学习,同时保留其在曼宁事物所的工作职务。他因此从理论学习中脱离出来并形成独立的见解。这也成为他后来参与发起著名的"哈佛革命"的重要因素,在"哈佛革命"中,凯利、詹姆斯·罗斯和盖瑞特·埃克博反对学院派程式化且机械刻板的设计风格。凯利认为,是埃克博率先进入现代主义的,但在对空间及三维结构的探索上,三个设计师都达成了共识。凯利在1938年没有拿到学位就离开了哈佛,急切地开始了他在景观设计领域的探索。

[3]　在最初几年的职业生涯里,凯利被引见给了路易斯·康,一位才华横溢的现代建筑师。凯利认为他的作品富有创造力。他们一同工作,为美国住宅局提供了诸多设计方案,在这段经历中,凯利学到了很多。他还掌握了现代设计的原则,尤其关注玛格丽特·史密斯于1942年出版的《户外生活设计》一书。她主张消除室内外空间界限的原则被凯利应用于1941年弗吉尼亚的科利尔住宅设计中。在这个庭院中,一系列平台为户外娱乐活动提供了空间,"外部居室"就这样产生了。

[4]　在第二次世界大战期间,凯利在被征入伍前,转向了建筑设计。入伍后他被派遣到德国,参与重建纽伦堡宫作为审判纳粹战犯的法庭。在旅行期间,他参观了安德烈·勒·诺特的作品,其清晰的直线形式给他留下了不可磨灭的印象。

[5]　后来,凯利在印第安纳州哥伦布斯的米勒庄园项目中遇到了设计师埃罗·沙里宁和凯文·罗奇。沙里宁邀请凯利来设计花园,在那里他创造出了自己认为最杰出的作品:"埃尔文·米勒花园……这简直就是个奇迹,我都不知道我是怎么做到的。"

[6]　建筑师、园林设计师和客户意见一致,这种不寻常的统一,在任何设计委托中都是难以实现的。1955年出现的房屋几何学被作为框架应用于景观中。玻璃幕墙使得这一转变近乎无懈可击,营造出的空间令人倍感愉悦。这种简约的处理方法也有助于创造出永恒雅致的庭院。草坪、碎石小径旁种植着皂荚树;低矮平台的铺设被块状的地被植物隔断;让空间和光线在整个设计中流动。米勒花园作为一个抽象的设计,同时借鉴了现代主义和古典主义,并将两者有机地结合起来,是不朽的佳作。

[7]　四十年后,在康涅狄格州的凯莫尔住宅设计中,凯利仍然创造出这些空间和光线效果,使得庭院既能从周边环境中脱颖而出又不显突兀。主草坪由一条铺设着板岩的宽大水渠限定,形成一条能映射出广阔天空的光带。凯利作品的特点是在不经意间与场地相契合,使得其他的设计师不得不赞同。基于他优秀的综合能力和对设计的敏感性,凯利成功地融合了建筑设计和风景园林两门学科。

[8]　除了庭院设计,凯利对城市空间的处理也值得称道,如在堪萨斯城的亨利摩尔雕塑公园的方案中,他探索发展了景观设计和艺术作品之间的关系;还有达拉斯的喷泉广场,这也许是他最著名的作品,在这个以水作为礼物送给炎热德州的项目中,种有树木的水池覆盖了大部分的城市广场,每天有240多万公升(635 000美国加仑)的水灌入水池。

[9]　凯利的庭院和他大量的文章展现出他对景观的无限激情,他对人们与其生存环境间相互影响的热忱理解,以及他对他所从事工作的荣誉感。无论是美国还是世界各地,他对其他设计师的影响都是巨大的。极少有

173

设计师能拥有像凯利这样长的职业生涯,并且在这期间他创造出了几乎在各方面代表20世纪人们期望的作品。他是当之无愧的现代主义者,在其他现代主义者们痴迷于国际风格和单一的灰色混凝土时,他能运用空间语言来创造出令人惊艳的舒适和高雅。

[10] 凯利在21世纪初期的作品,也同其职业生涯初期一样,鲜活生动而富于启发性。他坚持认为设计是一种概念,是从一系列问题中提炼出来的本质,而后才进行其功能的分析与整合。这便是纯粹的现代主义,它强调了装饰性图案或者风格诠释与真正设计的区别。

Reading Material A

Peter Walker[1]
彼得·沃克

"More than ever, we need…spaces for discovery, repose, and privacy in our increasingly bewildering, spiritually impoverished, overstuffed, and under maintained garden Earth."

Fig. 1 Peter Walker

[1] It would be easy to label Peter Walker (b. 1932) (Fig. 1) as a minimalist, but when we use any tag, the detail and message is often lost or diluted, and in Walker's case such a description would not do justice to the spiritual significance of his work. His designs display a strong sense of space, so it comes as no surprise to find that the work of André Le Nôtre has been a huge influence.

[2] Walker delights in the contrast of formality with nature, bringing sharply defined shapes into competition with softer vegetation or arranging plants into regimented lines or patterns. Many of his larger schemes use shapes, forms, and patterns that work well from above, creating parterre designs for the twentieth and twenty-first centuries. The garden for the Hotel Kempinski[2] (Fig. 2), in Munich, is a grid-based parterre garden with a secondary, overlapping, angled grid that creates a complex pattern of angles and a sense of rhythm.

[3] At one time Walker worked with Martha Schwartz, and they shared a penchant for land art[3]. Walker goes further and sees himself making marks on the landscape that relate to an ancient tradition, the results of which can be seen across the world.

[4] Themes and imagery play a part in linking ideas to their specific locations in Walker's designs. The gardens that make up the Centre for Advanced Science and Technology[4], Japan, are redolent of ancient Japanese gardens translated to the twentieth century. Walker uses stone,

Fig. 2 Garden of Hotel Kempinski, Munich

moss and grass to create "mountains"—cones that are monumental and provide a link with the gardens of the Muromachi period[5] (1338—1573) in Japan.

[5]　The gardens include raked gravel setts with highly polished "piers" of granite, mist fountains to swirl around tall bamboos, and huge burnt logs to provide counterpoint. The use of the various gardens for quiet contemplation and the careful placement of stepping stones are also both Japanese in origin, and lend the whole complex an air of sophisticated thoughtfulness and calm.

[6]　Walker engages in the production of large public landscapes. Their scale allows dramatic concepts to emerge, often combined with clever and witty details or installations. The Tanner Fountain[6] (Fig. 3), Massachusetts, is a perfect circle of stones from which emerges a fine mist that sits over the rugged boulders like a ghostly shroud. In the Toyota Municipal Museum of Art[7] in Japan the lake contains a huge, circular air fountain, producing a ring of white water that disturbs the otherwise calm, reflective surface. Only visible from a distance, the circle appears both mysterious and fascinating.

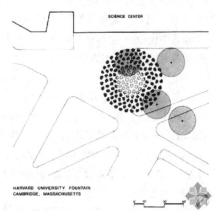

Fig. 3　View of Tanner Fountain and the plan (right)

[7]　Perhaps this designer's most serene and technically accomplished design, however, is that for the Plaza Tower and Town Center Park, Costa Mesa[8] (Fig. 4), California. The whole complex consists of stylized landscapes, but at the base of the main tower is a pool made up of concentric rings of reflecting water, retained by a skin of stainless steel to produce razor-sharp edges. The water shimmers, moving only

Fig. 4　Plaza Tower and Park, Costa Mesa

to slip over the edges into deep rills. The pattern of the water continues through the abstract patterns of paving stones, selected and placed to give varying degrees of light reflection.

[8]　Walker includes among his more recent influences Garrett Eckbo, Thomas Church, and Lawrence Halprin[9]. However, it is Dan Kiley—a designer also deeply influenced by Le Nôtre—who is pre-eminent for him. There is certainly a classical quality evident in the work of both Walker and Ki-

ley that seems to sit happily alongside their modernist credentials.

[9]   Recently Walker has become fascinated by the gardens of the modern age, researching and discovering the story of modernism in the United Sates through the production of the book *Invisible Gardens*[10]: The Search for Modernism in the American Landscape. He feels somehow cheated of a full training in design history, and the project allowed him to fill in some of the gaps. More than anything, Walker's commissions represent a hybridization of philosophies, with influence and inspiration coming from land artists such as Michael Heizer[11] and Robert Smithson[12], notably the latter's Spiral Jetty; artists such as Jasper Johns[13] and Yves Klein[14]; and sculptors such as Andy Goldsworthy[15] and Richard Serra[16]. In his writing, Walker talks of the loss of expression in the process of modern abstraction and the resultant loss of faith. In his own work, he strives to restore expression, spirituality, and a sense of profound contact with the natural world.

[NOTES]

[1]   Peter Walker 彼得·沃克(1932—   ):"极简主义"景观设计风格的代表人物,美国风景园林师协会(ASLA)理事,美国设计师学院荣誉奖获得者,美国风景园林师协会城市设计与规划奖获得者。彼得·沃克有着丰富的从业和教学经验,一直活跃在风景园林教育领域,1978—1981年曾担任哈佛大学设计研究生院风景园林系主任。他最著名的著作是与梅拉尼·西蒙合作完成的《看不见的花园:寻找美国景观的现代主义》。1958年彼得·沃克与哈佛大学设计研究生院佐佐木英夫教授(Hideo Sasaki,1919—2000)共同创立了SWA(Sasaki Walker Associates)景观设计事务所,其逐渐发展壮大,后成为美国当代最著名的景观设计公司之一。1983年于加利福尼亚州伯克利市成立了彼得·沃克景观设计事务所(Peter Walker and Partners,简称PWP)。

[2]   Hotel Kempinski 凯宾斯基酒店:位于德国慕尼黑,是彼得·沃克早期的极简主义设计作品之一。

[3]   land art:大地艺术:20世纪60—70年代,许多极简主义雕塑家开始走出画廊和社会,来到遥远的牧场和荒漠,创造出一种巨大的超人尺度的雕塑,即大地艺术。大地艺术由雕塑发展而来,但与雕塑不同的是,大地艺术与环境结合更加紧密,是雕塑与景观设计的交叉艺术。大地艺术的叙述性、象征性、人造与自然的关系,以及表现出来的自然的神秘,对当代景观规划设计的发展中起到了不可忽视的作用。随着废弃地成为大地艺术家的创作舞台后,在土地漫长的生态恢复过程中,它以艺术的主题提升了景观的质量,改善了环境的视觉价值。因此大地艺术也成为各种废弃地的更新、恢复、再利用的有效手段之一。利用工业废弃地建造公园最有代表性的案例是德国景观设计师拉茨设计的杜伊斯堡风景公园。

[4]   Centre for Advanced Science and Technology 日本京都高科技中心。

[5]   Muromachi period 室町时代(1333—1568):是日本史中世时代的一个划分。

[6]   Tanner Fountain 哈佛大学泰纳喷泉:是第一个"景观艺术行为"项目,位于一个交叉路口,是由159块巨石组成的圆形石阵,所有石块都镶嵌于草地之中,呈不规则排列状。

[7]   Toyota Municipal Museum of Art 丰田市美术馆:该美术馆位于知名汽车生产厂商丰田汽车所在地的丰田市的一个山上,这座由日本建筑师谷口吉生设计的美术馆由两栋建筑构成,一栋用来展览当地一位陶艺家的作品,而另一栋则用于陈列丰田不断增长的当代艺术收藏品。

[8]   Plaza Tower and Town Center Park 广场大厦及市中心公园:科斯塔梅萨,加利福利亚州,彼得·沃克的一个设计项目,曾获得美国风景园林师协会的设计奖。

[9]   Lawrence Halprin 劳伦斯·哈普林:美国当代风景园林设计大师、理论家。他的作品体现了现代主义风景园林进展的各个方面,包括设计的社会作用、对适应自然系统的强调,以及功能和过程对形式产生的重要性等。他的一系列以自然作为戏剧化景观场所规划灵感来源的城市公共景观设计,不仅是优美的城市风景而且更是人们游憩的场所,从而成为城市中人性化的开放空间。

[10] *Invisible Gardens*《看不见的花园》:是一部关于在1925—1975年曾为美国景观学做出过重要贡献的一些个人和公司发展历程的著作。书中讨论的项目,既有现代景观中的经典作品,也有鲜为人知、但实际上却应该拥有同那些经典作品相同地位的作品,因此该书成为美国景观学在战后巨大的社会变革中为争取行业地位和为追求现代审美理想所做出的文化贡献的一种记载,并成为西方园林景观方面的经典理论书籍。

[11] Michael Heizer 迈克尔·海泽(1944—):是现代艺术家,大地艺术的先驱,声称"艺术是激进的,美国的"。他认为雕塑应该体现美国西部雄浑的自然地理特征,并且迥异于惯用的欧洲模式,他在美国西部荒漠地带完成了"孤立的垃圾""双重否定""复合体1号"等数十件庞大的作品。

[12] Robert Smithson 罗伯特·史密森(1938—1973):是著名的美国大地艺术家,其最有名的作品为螺旋状的防波堤(Spiral Jetty)。

[13] Jasper Johns 贾斯伯·琼斯(1930—):是美国波普艺术的艺术家,他的著名画册有"美国国旗""虚伪的起点""数字2""走廊"等。

[14] Yves Klein 伊夫斯·克莱因(1928—1962):是法国艺术家,反传统艺术和后现代主义的拥护者。他的作品主要关于战后的欧洲,自认为是在创作最后的物质性绘画作品,此后的绘画艺术必将演化成纯精神的产物。

[15] Andy Goldsworthy 安迪·格兹乌斯(1956—):是美国著名的大地艺术家,摄影师,环保主义者。他分别把石材和木材的应用推向高潮。

[16] Richard Serra 理查德·塞拉(1939—):是美国极简抽象派艺术家,影像导演,其艺术作品是大规模组装的金属片,参加过概念艺术运动。

[GLOSSARY]

land art	大地艺术	stainless steel	不锈钢
minimalist	极简主义者	stepping stones	汀步

[NEW WORDS]

abstraction	n.	抽象化,提炼,抽象派作品	pier	n.	水上平台,码头,柱子
boulder	n.	卵石,巨石,巨砾	polished	adj.	精致的,完美的,优雅的
concentric	adj.	同一中心的,同轴的	pre-eminent	adj.	卓越的,优秀的
contemplation	n.	注视,凝视,沉思,冥想,期望	rake	vt.	以耙子耙平,搜索,梳理
credentials	n.	资格证书,资格;证明,证书	razor-sharp	adj.	锋利的,犀利的
diluted	adj.	无力的,冲淡的	redolent	adj.	使人联想或回想起某事物的
granite	n.	花岗岩,花岗石	regimented	adj.	严格控制的
gravel	n.	沙砾,砾石,石子	rill	n.	小河,小溪
hybridization	n.	杂交,杂种培植,配种	serene	adj.	宁静的,安详的
monumental	adj.	雄伟的,巨大的,不朽的	shimmer	vi.	闪闪发光,发微光
moss	n.	苔藓,地衣,泥灰沼,沼泽	shroud	n.	裹尸布,寿衣;遮蔽物,覆盖物
overlapping	n.	重叠,搭接	stainless	adj.	不会染污的,不生锈的,无瑕疵的
parterre	n.	(庭院中的)花坛,花圃	swirl	vi.	旋转,打旋
penchant	n.	(强烈的)倾向,爱好	witty	adj.	机智的,情趣横生的

[参考译文]

彼得·沃克

"在我们愈加使人迷惑的,精神贫穷的,过度增生的,在维持下的地球花园中,我们比任何时候都需要发现,休息和隐私的空间。"

[1] 人们很容易给彼得·沃克贴上一个极简主义者的标签,但我们运用任何标签时,细节和信息往往丢失或被稀释。根据沃克的情况,这样的描述对他的作品在精神上的重要性来说并不公正。他的设计表现了一种强烈的空间感,因此认为安德烈·勒诺特尔的作品对其有巨大的影响力并不惊奇。

[2] 沃克喜欢形式和自然的对比,将清晰设定的形状与植被软柔形成对照,或者将植被安置成严格控制的线条或图案。他的很多大型方案将形状、形式和图案良好结合,创造了20世纪和21世纪的现代花坛设计。慕尼黑凯宾斯基酒店的花园是一个基于网格的花圃,和一个成一定角度的网格与之重叠创造了一个复合的有角度的图案和一种韵律感。

[3] 沃克一度和玛莎·舒瓦茨合作,他们都有一种对大地艺术的强烈爱好。沃克在这条路上走得更远,他认为自己在大地上创作一种与古老的传统有关联的印记,其成果可以让全世界都可以看到。

[4] 在沃克的设计中主题和意象扮演着将创意与特定场地相联系的角色。为日本京都高科技中心设计的花园,被认为是古老的日本园林在20世纪的诠释。沃克运用石头、苔藓和草地创造"山体"——纪念碑似的圆锥体,提供了与日本室町时代(1338—1573)园林的联系。

[5] 花园包含用耙子耙过的砾石和高度抛光的花岗石墩,雾喷泉萦绕在高大的竹子周围,与烧制过的巨大原木形成对比。为安静冥想营造不同的花园及小心放置的汀步皆源于日本,给予整体园林意蕴丰富和静谧的氛围。

[6] 沃克曾参与多项大型公共景观设计。它们的规模使设计概念得以戏剧化地呈现,并时常结合着聪明且妙趣横生的细节或小品。马萨诸塞州的泰纳喷泉是一个完整的圆形石阵,其中喷出薄薄的雾霭,如同幽灵般弥漫在凹凸不平的大石块上。日本丰田市美术馆的湖中有一个巨大的圆形空气喷泉,制造出一圈白色水花打破原本平静、反光的湖面。当从一定距离欣赏时,圆环既显得神秘,又使人着迷。

[7] 加利福尼亚州科斯塔梅萨市的广场大厦及市中心公园或许是他最静谧且富于技巧的作品。整个复杂的景观群由多种风格的景观组成,大厦主体的基部是一个同心圆组成的镜面水池,采用不锈钢表皮支撑,从而产生锋利的边缘。水体闪闪发光、从边沿跌落至深深的水槽。水体继续通过被选择放置的抽象的石头铺装图案,提供不同程度的反光。

[8] 沃克认为他更近一些时期受影响于埃克博、托马斯·丘奇和劳伦斯·哈普林。然而,丹·凯利,一个同样被勒·诺特尔深深影响的设计师,才是他心中真正卓越的设计师。无疑的,在沃克和凯利的作品中都具有很明显的古典特质,且似乎乐于与他们作为现代主义者的凭证相伴。

[9] 沃克近一段时期被现代花园所吸引,其《看不见的花园》一书研究和探索美国现代主义的故事:在美国景观中的搜寻现代主义。他感觉到被设计历史的完整教育所欺骗,这促使他去填补一些缺口。此外,沃克的任务代表各种哲学的一种混合,受到一些艺术家的影响和启发,如大地艺术家迈克尔·海泽和罗伯特·史密森。值得一提的是后者设计的螺旋防波堤,又如艺术家贾斯伯·琼斯和伊夫斯·克莱因;雕塑家安迪·格兹乌斯和理查德·塞拉。在他的著作中,沃克提及现代抽象过程中表现力的丢失和发生的信仰丢失。在他自己的作品中,他力求修复表现力、精神性和与自然产生深刻联系的感觉。

Reading Material B

Martha Schwartz[1]
玛萨·舒瓦茨

"I like to see and understand a space, to hear everyone's point of view in order to respond to many different needs."

[1] Martha Schwartz (b. 1950) (Fig. 1) began her career in fine art, studying printmaking in Michigan, but found herself fascinated by land art. The work of practitioners such as Robert Smithson, Nancy Holt[2], and Michael Heizer[3] jumped from the confines of the studio and into the landscape on a large scale. Schwartz was keen to apply her artistic training in this way.

[2] At that time landscape architecture was skill-based and regarded as technical rather than artistic. A generally conservative alliance of practitioners appeared to be oblivious to the potential for the expansion of their work and, in a characteristically humorous aside, Schwartz suggested that her individualistic approach "was like farting in church".

Fig. 1 Martha Schwartz

[3] Much of her work may be described as installation, concepts combining artistic expression with functional reality. Her most famous work is the Bagel Garden[4] (Fig. 2), in Boston, Massachusetts, created as a joky home-coming present for a friend. Containing a hint of the surreal[5], it sent ripples through the landscape-architectural establishment. She had the garden photographed, using the powerful images and the opportunity to write about her work and made her mark.

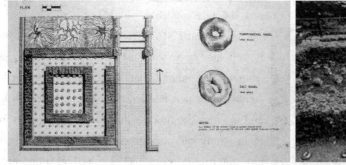

Fig. 2 View of Bagel Garden and the plan view (left)

[4] Schwartz takes time to understand the space that is to be transformed. The factors that have brought the space to its current status are all considered, and she sees herself dealing with a specific point in time in a site's history. She brainstorms with her design staff, allowing a range of ideas to be thrown together. She describes the conceptual phase as like dreaming, with the design process and the realization proving greatly thrilling.

[5] She admires the work of Peter Walker for its minimal serenity and that of Andrè Le Nôtre for his

outstanding spatial arrangement, which she would be happy indeed to achieve.

[6] In her teaching she tells her students to work intuitively. The designer must consider a huge amount of information, she says, but there is more, and the intuitive must at least be acknowledged. A strong sense of Schwartz's identity is evident in her work, and, coming from a family background of art and architecture, she sees her artistic sense as its primary engine. She senses that historically Jews were not allowed to follow a profession, and creativity was their response.

[7] Schwartz is very interested in how people use the space she designs, how they interact with them, and what they gain from their experience. Some are intimate, such as the Splice Garden[6] (Fig. 3), in Cambridge, Massachusetts—a tiny rooftop area that gives the impression of two quite different gardens sharing the same space. The installation was commissioned for a biomedical research centre, the title and concept relating to the genetic engineering carried out there. Other works form huge landscapes, usually urban in character, such as a new pedestrian square in Manchester, England. The Federal Court Plaza (Fig. 4) in Minneapolis shares this scale, completed with soft green drumlins and stark paving patterns and textures. All of these designs are alive with vigour and excitement, but each tells another story for those who care enough to reflect on the work in greater depth.

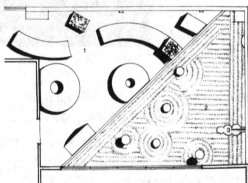

Fig. 3　View of Splice Garden and the plan (left)

Fig. 4　Federal Court Plaza in Minneapolis

[8] Schwartz strives to make both herself and her clients happy with her concepts and their realization. She admits that her clients must be farsighted and open-minded enough to travel with her and to enjoy the process too. Her ideas need to remain strong and clear to succeed, and she regards any watering-down of concepts as both unacceptable and unworkable. The results are sometimes dramatic, often humorous, but always imaginative and dynamic. They remain the breath of fresh air that jolted a whole profession into a more artistic and creative mode.

UNIT 10　LANDSCAPE PEOPLE AND THOUGHTS Ⅱ

[NOTES]

[1]　本文节选自 *Influential Gardeners* 书中"Martha Schwartz"一文。玛莎·舒瓦茨(1950—　)美国哈佛大学研究生设计学院终身教授,有着超过 30 年的从业经历,在美国马塞诸塞州的剑桥和英国伦敦拥有自己的设计事务所。玛莎·舒瓦茨的作品充满着独特的艺术气息,在全球享有盛誉,并且已经赢得众多的国际设计大奖。她的作品极为多样化,包括从大型新城规划、城市复兴总体规划到城市公共开放空间设计、居住社区和商业地产的景观规划以及现代艺术展设计等城市规划、景观设计和艺术设计的各个领域。

[2]　Nancy Holt 南希·霍尔特(1938—　):罗伯特·史密森(Robert Smithson)的妻子,是一名著名的美国大地艺术家,作品为太阳隧道(Sun Tunnels)。

[3]　Michael Heizer 迈克尔·海泽(1944—　):是现代艺术家,大地艺术的先驱,声称"艺术是激进的,美国的"。他认为雕塑应该体现美国西部雄浑的自然地理特征,并且迥异于惯用的欧洲模式,他在美国西部荒漠地带完成了"孤立的垃圾""双重否定""复合体 1 号"等数十件庞大的作品。

[4]　Bagel Garden 面包圈花园:1979 年舒瓦茨为自己家设计的面包圈花园。1980 年玛萨舒瓦茨《风景园林》杂志第一期上发表的面包圈花园,在美国风景园林领域引起了对后现代主义的广泛讨论,它被认为是美国风景园林师在现代风景园林中进行后现代主义尝试的第一例。

[5]　surreal 超现实主义:超现实主义是在法国开始的文学艺术流派,它的主要特征是以所谓"超现实""超理智"的梦境、幻觉等作为艺术创作的源泉,认为只有这种超越现实的"无意识"世界才能摆脱一切束缚,最真实地显示客观事实的真面目。

[6]　Splice Garden 拼合园:舒瓦茨与彼得·沃克(Peter Walker)合作把法国的巴洛克花园和日本古典园林"拼接"在一起,并"种"上了修剪过的塑料植物。

[GLOSSARY]

| Installation（Art） | 装置（艺术） | spatial arrangement | 空间布局 |

[NEW WORDS]

bagel	n.	硬面包圈	intuitive	adj.	直觉的,直观的,凭直觉获知的
biomedical	adj.	生物医学的	jolted	adj.	颠簸的
brainstorm	n.	灵机一动,突来的灵感,集体讨论	oblivious	adj.	未察觉,不注意,忘记
conservative	adj.	保守的,守旧的	pedestrian	n.	步行者,人行道
drumlin	n.	鼓丘,(冰河漂积成的)小丘	printmaking	n.	版画复制(术)
dynamic	adj.	动态的,充满活力的,强有力的	ripples	n.	波纹,涟漪
fart	vi.	〈讳〉放屁,(尤指)放响屁	serenity	n.	安详,宁静
hint	n.	暗示,提示,线索,心得	stark	adj.	僵硬的,光秃秃的,荒凉的
humorous	adj.	滑稽有趣的,有幽默感的	surreal	adj.	超现实的,犹如梦幻的
individualistic	adj.	个人主义者的,利己主义者的	vigour	n.	精力,活力,力量,效力,气势
intimate	adj.	亲密的,密切的,私人的,个人的	watering-down		放宽标准

[参考译文]

玛萨·舒瓦茨

"我喜欢去看和理解一个空间,去为了回应许多不同的需求而倾听每个人的观点。"

[1] 玛萨·舒瓦茨(生于1950年)(图1)的职业生涯始于艺术。在密歇根学习版画时,她发现自己着迷于大地艺术。大地艺术家如罗伯特·史密森,南希·霍尔特和迈克尔·海泽的作品,从工作室的局限跳出到大尺度的景观。舒瓦茨热衷于以这种方式运用她的艺术知识。

[2] 那个时代风景园林是基于技巧的并被认为是一项技术而不是艺术。一个通常保守的从业者同盟似乎注意不到他们工作扩展的潜力,以别具一格的幽默方式来说,舒瓦茨提到她的个人主义的方式"像在教堂里放屁"。

[3] 很多她的作品可能被描述为装置艺术,其概念结合艺术表达和实际功能。她最著名的作品是在马萨诸塞州波士顿的面包圈花园(图2),被当作给朋友的一个玩笑式的归家礼物而创造。它包含超现实主义的暗示,在景观建筑圈的权威派中激起了涟漪。她为花园拍了照片,运用有力的图像和撰写她作品的机会显露头角。

[4] 舒瓦茨花费时间去理解将被改变的空间。那些使空间变成它现在的情形的因素均被考虑,她理解自己正在应对一个场地历史中的一个特别的时刻。她和她的设计团队进行头脑风暴,让一系列想法被扔在一起。她形容概念设计的阶段像是在做梦,通过设计过程和实现被证实非常令人激动。

[5] 她欣赏彼得·沃克的作品,因其极简的宁静,以及安德烈·勒·诺特尔的作品,因他出众的空间布局。她对真正实现那样的作品充满热情。

[6] 在她的授课中,她告诉学生凭直觉工作。她说设计者必须考虑大量信息,但不光这样,直觉无论如何必须至少被认可。强烈的舒瓦茨的特性在她的作品中很明显,来自一个艺术和建筑背景的家庭,她将自己的艺术天赋看作主要的引擎。她感到从历史的角度来说犹太人不被允许追随一种职业,从而,创造性便是他们对此的回应。

[7] 舒瓦茨对人们如何使用她所设计的空间非常感兴趣,人们如何与之互动,会从体验中获得什么。有一些空间是私人的,像是在马萨诸塞州剑桥市的拼合园(图3):一个极小的屋顶,给予两个相当不同的花园共享一个空间的效果。这个装置艺术是被一个生物医学研究中心所委托,与基因工程学有关的名称和概念被运用于此。其他的作品往往形成大尺度的景观通常是城市特性的,如英国曼彻斯特市新的人行广场。在明尼阿波利斯城的联邦法院广场也是这种规模,运用柔软的绿色鼓丘和质朴的铺装图案和材质完成。所有这些设计因充满活力和令人激动而鲜活,但每个作品却都在向那些足够在意作品本身反映更深层次深度的人们讲述着另一个故事。

[8] 舒瓦茨致力于使她本人和委托人都满足于她理念的实现。她承认她的客户必须具有远见、思想开阔,才能够和她一起在过程中旅行并享受。她需要保持强烈和清晰的想法才能取得预期效果,她认为任何放宽标准的概念都难以接受。有时候,结果充满戏剧性或滑稽有趣,但充满想象力和活力。它们保持"呼吸"新鲜空气,从而撼动整个职业到达一个更加具有艺术性和创造性的模式。

[思考]

(1) 丹·凯利的设计风格是什么?主要作品有哪些?

(2) 丹·凯利的设计思想对现代景观设计有哪些启发?

(3) 彼得·沃克的景观设计有哪些特点?主要作品有哪些?

(4) 彼得·沃克的设计思想对现代景观设计有哪些启发?

(5) 玛莎·舒瓦茨的设计风格是什么?主要作品有哪些?

(6) 玛莎·舒瓦茨众多的设计作品分别受到哪些艺术风格的影响?

PART V

Theories and methods of planning and design
第五部分 规划设计理论与方法

UNIT 11 ECOLOGY AND ENVIRONMENT PROTECTION

TEXT

Environmental Movement
环境运动

Blue Marble composite images generated
by NASA in 2001 (left) and 2002 (right).

[1] The environmental movement, a term that includes the conservation and green movements[1], is a diverse scientific, social, and political movement for addressing environmental issues.

[2] Environmentalists advocate the sustainable management of resources and stewardship of the environment through changes in public policy and individual behavior. In its recognition of humanity as a participant in (not enemy of) ecosystems, the movement is centered on ecology, health, and human rights.

[3] The environmental movement is represented by a range of organizations, from the large to grassroots. Due to its large membership, varying and strong beliefs, and occasionally speculative nature, the environmental movement is not always united in its goals. At its broadest, the movement includes private citizens, professionals, religious devotees, politicians, and extremists. Environmentalists are also often linked with other social movements, such as human and animal rights and pacifism.

Introduction

[4] The environmental movement in the United States can be traced back to the early conservation movement[2] and the establishment of Hot Springs National Park[3] in 1832. Two early conservationists stood out as leaders in the movement; they were Henry David Thoreau[4] and George Perkins Marsh[5]. Thoreau was concerned about the wildlife in Massachusetts; he wrote *Walden; or, Life in*

the Woods as he studied the wildlife from a cabin. Marsh was influential with regards to the need for resource conservation.

[5]　The roots of the modern environmental movement can be traced to attempts in nineteenth-century Europe and North America to expose the costs of environmental negligence, notably disease, as well as widespread air and water pollution, but only after the Second World War did a wider awareness begin to emerge.

[6]　During the 1950s, 1960s, and 1970s, several events illustrated the magnitude of environmental damage caused by humans. In 1954, the 23 man crew of the Japanese fishing vessel Lucky Dragon 5 was exposed to radioactive fallout from a hydrogen bomb test at Bikini Atoll[6]. The publication of the book *Silent Spring* (1962) by Rachel Carson[7] drew attention to the impact of chemicals on the natural environment. In 1967, the oil tanker Torrey Canyon went aground off the southwest coast of England, and in 1969 oil spilled from an offshore well in California's Santa Barbara Channel. In 1971, the conclusion of a law suit in Japan drew international attention to the effects of decades of mercury poisoning on the people of Minamata.

[7]　At the same time, emerging scientific research drew new attention to existing and hypothetical threats to the environment and humanity. Among them were Paul R. Ehrlich, whose book *The Population Bomb*[8] (1968) revived concerns about the impact of exponential population growth. Biologist Barry Commoner generated a debate about growth, affluence and "flawed technology." Additionally, an association of scientists and political leaders known as the Club of Rome published their report *The Limits to Growth*[9] in 1972, and drew attention to the growing pressure on natural resources from human activities.

[8]　Meanwhile, nuclear proliferation and photos of Earth from outer space emphasized the consequences of technological accomplishments, as well as Earth's truly small place in the universe.

[9]　In 1972, the United Nations Conference on the Human Environment[10] was held in Stockholm, and for the first time united the representatives of multiple governments in discussion relating to the state of the global environment. This conference led directly to the creation of government environmental agencies and the UN Environment Program[11]. The United States also passed new legislation such as the Clean Water Act[12], the Clean Air Act[13], the Endangered Species Act[14], and the National Environmental Policy Act[15], the foundations for current environmental standards.

[10]　Since the 1970s, public awareness, environmental sciences, ecology, and technology have advanced to include modern focus points like ozone depletion, global climate change, acid rain, and the harmful potential of genetically modified organisms (GMOs).

Scope of the movement

Before flue gas desulfurization was installed, the air-polluting emissions from this power plant in New Mexico contained excessive amounts of sulfur dioxide.

[11]　Biological studies
- Environmental science is the study of the interactions among the physical, chemical and biological components

of the environment;
- Ecology, or ecological science, is the scientific study of the distribution and abundance of living organisms and how these properties are affected by interactions between the organisms and their environment.

[12] Primary focus points
- The environmental movement is broad in scope and can include any topic related to the environment, conservation, and biology, as well as preservation of landscapes, flora, and fauna for a variety of purposes and uses.
- The Conservation movement seeks to protect natural areas for sustainable consumption, as well as traditional (hunting, fishing, trapping) and spiritual use.

[13] Other focus points
- Environmental health movement[16] dates at least to Progressive Era[17], and focuses on urban standards like clean water, efficient sewage handling, and stable population growth. Environmental health could also deal with nutrition, preventive medicine, aging, and other concerns specific to human well-being. Environmental health is also seen as an indicator for the state of the environment, or an early warning system for what may happen to humans.
- Environmental Justice is a movement that began in the U.S. in the 1980s and seeks an end to environmental racism and prevent low-income and minority communities from an unbalanced exposure to highways, garbage dumps, and factories. The Environmental Justice movement seeks to link "social" and "ecological" environmental concerns, while at the same time preventing de facto racism, and classism.
- Ecology movement could involve the Gaia Theory[18], as well as Value of Earth and other interactions between humans, science, and responsibility.
- Deep Ecology is often considered to be a spiritual spinoff of the ecology movement.
- Bright green environmentalism is a currently popular sub-movement, which emphasizes the idea that through technology, good design and more thoughtful use of energy and resources, people can live responsible, sustainable lives while enjoying prosperity.

[NOTES]

[1] green movements 绿色运动：又称生态运动,是在20世纪60年代末能源危机、生态失控的环境下,70年代初产生的保护人类环境的运动,绿色运动既是维护生态平衡的生态运动,也是反对核武器、确保人类社会和平发展的政治运动。

[2] conservation movement 自然资源保护运动：是19世纪末20世纪初,美国为了应对西部开发中出现的资源浪费与破坏及其引起的环境问题,兴起的一场资源与环境保护运动,是美国环保运动的第一个高峰。它是在联邦政府干预下、自上而下地展开的,其内容十分广泛,包括保护森林资源、建立国家公园、保护土地资源、保护野生动物、保护矿产资源、进行自然资源保护的国内国际合作。这个运动促进了群众性资源环境保护组织的成立,如塞拉俱乐部(1892),世界上第一个自然保护组织,其宗旨是谋取公众和政府的支持与合作,欣赏和保护太平洋沿岸可供进入的山区的森林和荒野；奥杜邦协会(1905),美国最早的鸟类保护组织；全国公园和自然保护协会(1919)。

[3] Hot Springs National Park 热泉国家公园：最早建于1832年,被称为热泉国家保留地(Hot Spring National

UNIT 11　ECOLOGY AND ENVIRONMENT PROTECTION

Reservation),虽然当时主要服务于游憩业,但却是最初的自然保护。

[4]　Henry David Thoreau 亨利·大卫·梭罗:19 世纪美国最具有世界影响力的超验主义作家,关注人类生存状况的生态主义哲学家,当代环境主义运动的先驱。著有散文集《瓦尔登湖》(*Walden*; or, *Life in the Woods*)和论文《论公民的不服从权利》。《瓦尔登湖》中,梭罗记录了他在瓦尔登湖与大自然亲密接触的两年中对自然深刻的思考,该书以其丰富的生态思想内涵,被视为生态文学的经典之作。

[5]　George Perkins Marsh 乔治·帕金斯·马什:美国地理学家、外交家、自然资源保护论者。在科学方面,他主要研究人地关系和自然保护,主张人类应保护自然,改良自然,强调人破坏自然的危险性。著有《人与自然》等,该书首次从伦理学上探讨自然保护问题,是自然保护的第一本教科书。

[6]　Bikini Atoll 比基尼环礁:是属于马绍尔群岛的堡礁,由 23 个小岛环绕着一个面积为 229.4 平方英里的泻湖组成。美国从 1946 年到 1958 年在比基尼环礁共进行了 20 多次原子弹和氢弹爆炸试验。2010 年 7 月 31 日被列入世界文化遗产。

[7]　Rachel Carson 蕾切尔·卡逊:美国海洋生物学家,她的小说《寂静的春天》(*Silent Spring*,1962)详述了滥用滴滴涕等杀虫剂带来的严重的环境危害,对农业科学家的实践活动和政府政策提出挑战,并号召人们迅速改变对自然世界的看法和观点,引起强烈争议;卡尔逊因此被看做现代环境运动的先驱。

[8]　*The Population Bomb*《人口爆炸》:作者美国人口生态学家保罗·埃里奇(Paul. R. Ehrlich)认为世界人口的迅速增长(主要是发展中国家)与过剩已经超过了地球生态环境的承载能力,正威胁着整个人类的生存。

[9]　*The Limits to Growth*《增长的极限》:是罗马俱乐部(Club of Rome)1972 年发表的第一份研究报告,涉及人口问题、粮食问题、资源问题和环境污染问题(生态平衡问题)等当前重大问题。这些问题也早已成为世界各国政府和人民不容忽视,亟待解决的重大问题。

[10]　United Nations Conference on the Human Environment 联合国人类环境会议。

[11]　UN Environment Program 联合国环境规划署:简称 UNEP。

[12]　Clean Water Act 清洁水法。

[13]　Clean Air Act 清洁空气法。

[14]　Endangered Species Act 濒危物种法。

[15]　National Environmental Policy Act 国家环境政策法案。

[16]　Environmental health movement 环境卫生运动。

[17]　Progressive Era 进步时代:是指 1890 年至 1920 年期间,美国历史上一股很有影响力的社会运动和思潮,由于贪婪和腐败的盛行引发,包括政治、经济政策、社会公正和促进道德水准普遍提高等方面的改革。

[18]　Gaia Theory 盖娅理论:英国大气物理学家洛夫洛克(J. Lovelock)于 1970 年代初提出的盖娅假说(Gaia hypothesis),也称为盖娅理论,认为地球上的生命与非生命部分通过相互作用而共同形成一个整体的动态系统,地球是一个巨大而复杂的"活体",它囊括整个生物圈,是一个动态"巨生理系统"。

[GLOSSARY]

acid rain	酸雨	global climate change	全球气候变化
conservation movement	自然资源保护运动	green movements	绿色运动
Deep Ecology	深层生态学	hydrogen bomb	氢弹
environmental health movement	环境卫生运动	nuclear proliferation	核扩散
environmental justice	环境正义	ozone depletion	臭氧耗损
environmental movement	环境运动	preventive medicine	预防医学
genetically modified organism	转基因生物	radioactive fallout	放射性物质泄漏

[NEW WORDS]

affluence	n.	富裕,富足	magnitude	n.	数量(值),大小;巨大,重要性
aground	adv.	搁浅,触礁,在地上	mercury	n.	水银,汞
cabin	n.	小木屋,茅舍,小室,工作间	Minamata	n.	水俣病(汞中毒引起的一种严重神经疾病)
conservationist	n.	自然资源保护论者,自然资源保护学家,自然资源保护主义者	offshore	adj.	离岸的,海面上的
de facto	adj.	事实上(的);实际上(的)	ozone	n.	臭氧;新鲜的空气
debate	n.	讨论,争论,辩论	pacifism	n.	和平主义,不抵抗主义,消极态度
depletion	n.	消耗,损耗	poison	vt.	毒害,使中毒
devotee	n.	热爱者,信徒	preventive	adj.	预防性的
environmentalism	n.	环境保护论,环境论	proliferation	n.	增殖,分芽繁殖
environmentalist	n.	环境保护论者,环保人士	racism	n.	种族主义,种族歧视
exponential	adj.	指数的	radioactive	adj.	放射性的
extremist	n.	过激主义者,极端分子	revive	vt.	(使)苏醒,(使)复兴
fauna	n.	动物群,动物区系,动物志	sewage	n.	下水道,污水
flora	n.	植物群,植物区系,植物志	speculative	adj.	投机的
genetical	adj.	遗传的,起源的	spill	vi.	溢出,泄出,撒出
humanity	n.	(总称)人,人类,人道,人性	spinoff	n.	延续,续集
hypothetical	adj.	假设的,假定的	vessel	n.	船,容器;脉管,导管

[难句]

Due to its large membership, varying and strong beliefs, and occasionally speculative nature, the environmental movement is not always united in its goals. 由于其成员众多,身份各异以及其强大的信仰、偶尔的投机性质,环境运动并不总是目标一致的。

[参考译文]

环境运动

[1] 环境运动,其中包括资源保护和绿色运动,是为解决环境问题,集科学、社会、政治多元为一体的运动。

[2] 环境保护论者倡导资源的可持续管理,以及通过改变公共政策和个人行为来管理环境。运动认识到人类是生态系统的参与者(而不是敌人)的身份,将关注点集中在生态、健康和人权。

[3] 环境运动是由各种组织所倡导,从大型机构到普通民众。由于其成员众多,身份各异以及其强大的信仰、偶尔的投机性,环境运动并不总是目标一致的。在其最广泛的运动中,包括市民个人、专业人士、宗教信徒、政治家和极端分子。环境保护论者也常常与其他社会运动相联系,如人类和动物的权利、和平主义。

导言

[4] 美国的环境运动可以追溯到早期的自然资源保护运动和1832年热泉国家公园的建立。两个早期的自然资源保护主义者(亨利·大卫·梭罗和乔治·帕金斯·马什)作为领导人在运动中脱颖而出。梭罗关注的是

马萨诸塞州的野生动物;他在瓦尔登湖一个小木屋研究野生动物并撰写了《瓦尔登湖》。马什在保护资源的必要性方面有着影响力。

[5]　现代环境运动的根源可以追溯到19世纪欧洲和北美揭露忽视环境保护的代价,尤其是疾病,以及普遍存在的空气污染和水污染的代价。但直到第二次世界大战后更广泛的环保意识才开始出现。

[6]　20世纪50年代、60年代和70年代,一些事件说明了人类造成环境破坏的严重性。1954年,日本渔船福龙5号带着23名船员遭受了暴露在比基尼环礁氢弹试验的放射性沉降物的伤害。蕾切尔·卡逊《寂静的春天》(1962年)的出版引起人们关注化学品对自然环境的危害。1967年,托利峡谷号油轮在英格兰西南海岸搁浅,以及1969年加州圣巴巴拉海峡的海上油井石油泄漏。1971年,一项法律诉讼的审判结果引起了国际上对汞中毒数十年的水俣病患者影响的关注。

[7]　与此同时,新的科学研究关注已有的和可能对环境和人类产生的威胁。其中有保罗·R.埃利希,他在《人口爆炸》(1968年)重新关注呈指数增长的人口数量的影响。生物学家巴里·康芒纳发起了关于人口增长、富裕程度和"有缺陷的技术"的论战。此外,科学家和政治领袖组成的罗马俱乐部在1972年发表了《增长的极限》的报告,并呼吁关注人类活动对自然资源造成的日益增加的压力。

[8]　与此同时,核扩散和太空拍摄的地球照片,凸显了科技成就的同时也强调了地球在宇宙中只不过是沧海一粟。

[9]　1972年,联合国在斯德哥尔摩召开了人类环境会议,并首次团结各国代表就全球环境问题进行讨论。这次会议直接促使政府环境机构和联合国环境规划署的建立。美国还通过了新的立法,如清洁水法,清洁空气法,濒危物种法,国家环境政策法案——目前环境标准的基础。

[10]　自20世纪70年代以来,公众意识、环境科学、生态学及技术都得以发展,包括最新的关注点,例如臭氧损耗,全球气候变化,酸雨及转基因生物存在的潜在危险。

运动的范围

[11]　**生物学研究**
- 环境科学是研究环境中物理、化学和生物成分之间的相互作用;
- 生态学,或生态科学,是关于生物体分布和丰富度,以及这些属性是如何受到生物体与其环境交互作用影响的科学研究。

[12]　**主要关注点**
- 环境运动的范围很广泛,可以包括任何有关环境、环保、生物学,以及各种目的和用途的景观、植物和动物保护。
- 自然保护运动力求保护自然区域可持续消费,以及传统活动(狩猎、捕鱼、捕获)和精神享受。

[13]　**其他重关注点**
- 环境卫生运动至少应当追溯到进步时代,它关注城市标准,如清洁的水、高效的污水处理、稳定的人口增长。环境卫生与营养学、预防医学、老龄化和其他具体关系到人类福祉的问题有关。环境卫生也被视为环境状况的指标,或早期预警系统。
- 环境正义运动始于20世纪80年代的美国,寻求结束环境种族主义和防止低收入人群和少数族裔靠近公路、垃圾场和工厂的不平衡。环境正义运动旨在链接"社会"和"生态"环境问题,同时防止实际的种族主义和阶级主义。
- 生态运动可能涉及盖娅理论,以及地球价值和其他人类、科学和责任之间的交互作用。
- 深层生态学往往被认为是生态运动的一种精神延续。
- 光明的绿色环境保护论是目前流行的子运动,它强调通过技术、优良的设计和更谨慎的能源及资源利用,使人们在享受繁荣的同时,获得可靠、可持续的生活。

Reading Material A
Ecological Principles for Managing Land Use I
Ecological Society of America

土地利用管理的生态原则 I

[1] Key ecological principles for land use and management deal with time, species, place, disturbance, and the landscape. The principles result in several guidelines that serve as practical <u>rules of thumb</u>[1] for incorporating ecological principles into making decisions about the land.

Introduction

[2] Humans are the major force of change around the globe, transforming land to provide food, shelter, and products for use. Land transformation affects many of the planet's physical, chemical, and biological systems and directly impacts the ability of the Earth to continue providing the goods and services upon which humans depend.

[3] Unfortunately, potential ecological consequences are not always considered in making decisions regarding land use. In this brochure, we identify ecological principles that are critical to sustaining ecosystems in the face of landuse change. We also offer guidelines for using these principles in making decisions regarding land-use change.

Challenges of ecologically sustainable land use

[4] A critical challenge for land use and management involves reconciling conflicting goals and uses of the land. The diverse goals for use of the land include:
- resource-extractive activities (e.g., forestry, agriculture, grazing, and mining);

UNIT 11 ECOLOGY AND ENVIRONMENT PROTECTION

- infrastructure for human settlement (housing, transportation, and industrial centers);
- recreational activities;
- services provided by ecological systems (e.g, flood control and water supply and filtration);
- support of aesthetic, cultural, and religious values;
- sustaining the compositional and structural complexity of ecological systems.

[5] These goals often conflict with one another, and difficult land-use decisions may develop as stakeholders pursue different land-use goals. Local versus broad-scale perspectives on the benefits and costs of land management also provide different views. In this brochure, we focus on sustaining ecological systems which also indirectly supports other values, including ecosystem services, cultural and aesthetic values, recreation, and sustainable extractive uses of the land.

[6] To meet the challenge of sustaining ecological systems, an ecological perspective should be incorporated into landuse and land-management decisions. Specifying ecological principles and understanding their implications for landuse and land-management decisions are essential steps on the path toward ecologically based land use.

[7] Key ecological principles deal with time, species, place, disturbance, and the landscape. While they are presented as separate entities, the principles interact in many ways.

Time principle

[8] Ecological processes function at many time scales, some long, some short; and ecosystems change through time. Metabolic processes occur on the scale of seconds to minutes, decomposition occurs over hours to decades, and soil formation occurs at the scale of decades to centuries. Additionally, ecosystems change from season to season and year to year in response to variations in weather as well as showing long-term successional changes. Human activities can alter what makes up an ecosystem or how biological, chemical, and geological materials flow through an ecosystem. These in turn can change the pace or direction of succession and have effects lasting decades to centuries. The time principle has several important implications for land use.

[9] The current composition, structure, and function of an ecological system are, in part, a consequence of historical events or conditions and current land uses may limit land use options that are available in the future.

- The full ecological effects of human activities often are not seen for many years.
- The imprint of a land use may persist on the landscape for a long time, constraining future land use for decades or centuries even after it ceases.
- Long-term effects of land use or management may be difficult to predict due to variation and change in ecosystem structure and process. This problem is exacerbated by the tendency to overlook low-frequency ecological disturbances, such as 100-year flooding, or processes that operate over periods longer than human life spans (e.g., forest succession).

Species principle

[10] Particular species and networks of interacting species have key, broad-scale ecosystem-level effects. These focal species affect ecological systems in diverse ways:

- Indicator species tell us about the status of other species and key habitats or the impacts of a stressor.
- Keystone species have greater effects on ecological processes that would be predicted from their abundance or biomass alone.
- Ecological engineers (e. g. , the gopher tortoise or beaver) alter the habitat and, in doing so, modify the fates and opportunities of other species.
- Umbrella species either have large area requirements or use multiple habitats and thus overlap the habitat requirements of many other species.
- Link species exert critical roles in the transfer of matter and energy across trophic levels (of a food web) or provide critical links for energy transfer within complex food webs.

[11] The impacts of changes in the abundance and distribution of focal species are diverse. For example, keystone species affect ecosystems through such processes as competition, mutualism, dispersal, pollination, and disease and by modifying habitats and abiotic factors. An introduced, nonnative species can assume a focal-species role and produce numerous effects, including altering community composition and ecosystem processes via their roles as predators, competitors, pathogens, or vectors of disease and, through effects on water balance, productivity, and habitat structure. The impacts of land use changes on keystone and invasive species can spread well beyond the boundaries of the land-use unit and are difficult to predict prior to changes in their abundance. Changes in the pattern of land cover can affect, and even promote, the establishment of nonnative species.

[12] Trophic levels refer to the stages in food chains such as producers, herbivores, consumers, and decomposers. Changes in the abundance of a focal species or group of organisms at one trophic level can cascade across other trophic levels and result in dramatic changes in biological diversity, community composition, or total productivity. Changes in species composition and diversity can result from land use through alterations to such ecosystem properties as stream flow or sediment load, nutrient cycling, or productivity. The effects of land use on species composition have implications for the future productivity of ecological systems.

Place principle

[13] Local climatic, hydrologic, soil, and geomorphologic factors as well as biotic interactions strongly affect ecological processes and the abundance and distribution of species at any one place. Local environmental conditions reflect location along gradients of elevation, longitude, and latitude and the multitude of microscale physical, chemical, and edaphic factors that vary within these gradients. These factors constrain the suitability of various land uses, as well as defining resident species and processes.

[14] Rates of key ecosystem processes, such as primary production and decomposition, are limited by soil nutrients, temperature, water availability, and the temporal pattern of these factors controlled by climate and weather. Thus, only certain ranges of ecological-process rates can persist in a locale without continued management inputs (e. g. , irrigation of crops growing in a desert). Chronic human intervention may broaden these ranges but cannot entirely evade the limitations of place without a cost.

[15] Naturally occurring patterns of ecosystem structure and function provide models that can guide sustainable and ecologically sound land use. Only those species adapted to the environmental constraints of an area will thrive there. Precipitation limits which species are appropriate for landscape plantings as well as for managed agricultural, forestry, or grazing systems. Further, some places with unique conditions may be more important than others for conservation of the species and ecosystems they support.

[16] Land uses that cannot be maintained within the constraints of place will be costly when viewed from longterm and broad-scale perspectives. For example, establishing croplands and ornamental lawns in arid areas is possible, but draws down fossil groundwater at a rate unsustainable by natural recharge. Only certain patterns of land use, settlement and development, building construction, or landscape design are compatible with local and regional conditions. In terrestrial systems, land-use and land-management practices that lead to soil loss or degradation reduce the long-term potential productivity of a site and can affect species composition. Land use practices can also influence local climate (e.g., the urban heat island concept). Sustainable settlement is limited to suitable places on the landscape. For instance, houses or communities built on transient lake shore dunes, major flood plains, eroding seashores, or sites prone to fires are highly vulnerable to loss over the long term. Ideally, the land should be used for the purpose to which it is best suited.

[NOTES]

[1] rules of thumb 大拇指法则:也称经验法则,用于求解问题的高级的但常常是不确切的经验法则和直觉推理法。

[GLOSSARY]

biomass	生物量	keystone species	关键物种
community composition	群落组成	link species	链接物种
disturbance	干扰	nutrient cycling	养分循环
ecological principle	生态原则	resident species	居留种
flood plains	冲积平原,河漫滩	sediment load	输沙量,泥沙通量
focal species	焦点物种	species abundance	物种多度/物种丰富度
food chain	食物链	stressor	胁迫
food web	食物网	succession	演替
habitat	生境,栖息地	trophic level	营养级
indicator species	指示物种	urban heat island effect	城市热岛效应

[NEW WORDS]

abiotic	adj.	非生物的,无生命的		invasive	adj.	侵略性的,侵害的,攻击的
alter	vt.	修改,改动;改变,改建		keystone	n.	关键,根本
arid	adj.	干旱的,干燥的,贫瘠的		latitude	n.	纬度
beaver	n.	海狸		locale	n.	(事件发生的)场所或地点
biomass	n.	生物量		metabolic	adj.	代谢作用的,新陈代谢的
biotic	adj.	生命的,生物的		microscale	n.	微小的规模,微尺度
brochure	n.	手册,宣传册,简介材料		multitude	n.	大量,许多
cease	vt.	使……停止;中止		mutualism	n.	互惠共生,互助论
chronic	adj.	长期的;持续的;慢性的		nonnative	adj.	非本土
decomposer	n.	分解者,分解体		overlap	vt.	重叠,交叠,与……交搭
decomposition	n.	分解,溶解,还原		overlook	n.	忽视,未注意到;监督,管理
disturbance	n.	干扰,打扰,扰乱		pathogens	n.	病菌,病原体
dune	n.	(尤指海边被风吹积成的)沙丘		precipitation	n.	降水,沉淀,析出
ecosystem	n.	生态系统		predator	n.	捕食者
edaphic	adj.	土壤的		productivity	n.	生产率,生产力
elevation	n.	高度,海拔		prone	adj.	易于……的,很可能……的
erode	vt.	侵蚀,腐蚀		recharge	n.	补给,充电;休整,恢复体力
evade	vt.	逃避,回避,躲避		reconcile	vt.	调解,使和解,使和谐
exacerbate	vt.	使……加重,恶化,使……加深		sediment	n.	沉淀物,沉积物
exert	vt.	发挥,施加,竭力,努力		species	n.	物种,种
extractive	adj.	消耗资源的,抽取的,萃取的		stakeholder	n.	参与方,利益相关者,股东
filtration	n.	过滤,过滤法		stressor	n.	胁迫,紧张性刺激(物)
focal	adj.	中心的,很重要的,焦点的		succession	n.	演替,连续,接连
fossil	n.	化石		successional	adj.	演替的,连续性的,接连的
geomorphologic	adj.	地形学的		suitability	n.	适合,适宜性
gradient	n.	斜度,坡度,梯度		temporal	adj.	时间的,短暂的
groundwater	n.	地下水		tendency	n.	倾向,趋势,癖好
guideline	n.	指导原则,行动纲领,准则		terrestrial	adj.	地球的,陆地的,人间的
habitat	n.	生境,栖息地		transient	adj.	短暂的,片刻的,转瞬即逝的
herbivore	n.	草食动物,食草类动物		trophic	adj.	营养的,有关营养的
hydrologic	adj.	水文的		vector	n.	传病媒介,带菌者;矢量
indirectly	adv.	间接地		versus	prep.	……对……,对抗,反对,相对
intervention	n.	干涉,干预,干扰		vulnerable	adj.	易受伤的,脆弱的,敏感的

UNIT 11 ECOLOGY AND ENVIRONMENT PROTECTION

[参考译文]

土地利用管理的生态原则 I
美国生态学会

[1] 土地利用和管理的关键性生态原则涉及时间、物种、地点、干扰和景观。根据这些原则所制定出的条例成为土地决策中顺应生态原则的实践性经验法则。

导言

[2] 人类是改变地球的主要力量,他们改造土地来生产食物、建造庇护所、制造可用的物品。土地改造影响了地球上的许多物理、化学和生物系统,并直接影响着地球继续提供人类赖以生存的物品和服务的能力。

[3] 不幸的是,潜在的生态后果在做有关土地利用的决策中并非总是被考虑在内。在本手册中,我们界定了在土地利用改变时为维护生态系统所需遵守的重要生态原则。同时,我们还提供了在做土地利用改变决策时运用这些原则的指导方针。

生态可持续的土地利用所面临的挑战

[4] 对于土地利用和管理最重要的挑战是调和土地利用目标和土地使用之间的冲突。土地利用有多种目标:
- 资源消耗性活动(例如林业、农业、畜牧业和采矿业);
- 人居环境中的基础设施(住宅、交通和工业中心);
- 休闲娱乐活动;
- 生态系统提供的服务(例如防洪、供水和过滤);
- 维持美学、文化和宗教方面的价值;
- 维持生态系统成分和结构的复杂性。

[5] 各方利益相关者追求的土地利用目标不同,往往导致这些目标相互冲突。关于土地管理收益与支出以及基于本地或基于大区域的不同角度的考量也会带来不同的观点。在这个小手册中,我们聚焦于维护生态系统及其间接满足的包括生态系统服务、文化和美学价值、娱乐休闲和土地可持续开发利用等其他需求。

[6] 面对维护生态系统所面临的挑战,应该将土地利用和土地管理决策与生态方面的考量结合起来。将生态原则具体化,并理解它们在土地利用与土地管理决策中的含义,是通向基于生态的土地利用之路的重要步骤。

[7] 关键性生态原则涉及时间、物种、地点、干扰因素和景观。尽管它们作为独立的实体存在,但是在许多方面又是相互作用的。当这些生态原则作为分散的个体出现时,它们会以多种方式相互作用。

时间原则

[8] 生态进程作用在许多不同的时间尺度中,或长或短,并且生态系统无时不在发生改变。代谢过程发生在分秒的时间尺度中,分解过程则需要持续数小时乃至数十年,而土壤的形成则需要经历数十年到几世纪的跨度。此外,生态系统随着天气的不同,季复一季、年复一年的变化,也体现了其长期的演替变化。人类的活动能够改变生态系统的组成或生物、化学和地理物质在生态系统中的循环。这些反过来又能够改变演替的节奏和方向,并将这种影响持续几十年至数百年。时间原则对于土地的利用有几个重要的意义。

[9] 当前生态系统的组成、结构和功能,从某种程度上说,是历史事件或条件所导致的结果,而现阶段的土地利用,可能会限制未来土地可供选择的使用方式。
- 人类活动所产生的全部生态影响持续许多年也未必能完全看到。
- 土地利用的印迹可能会在景观中留存很长一段时间,在未来几十年或几个世纪,甚至在印迹消失以后,都会限制未来的土地使用。
- 生态系统结构和进程的多样性和变化性,使得土地利用和管理所导致的长期影响难以预测。然而忽视低频率生态干扰的趋势,使这个问题更加严重,如百年一遇的洪水或者发生周期长于人类寿命的进程(例如:森

· 195 ·

林演替)。

物种原则

[10] 特殊物种和物种之间相互作用的网络有着关键的、大尺度生态系统水平的影响。这些焦点物种以多种方式影响着生态系统:
- 指示物种能够指示其他物种和关键生境的状态或胁迫对它们的影响。
- 关键物种对生态过程有更重要的作用,仅靠它们的丰富度和生物量就可预测生态过程。
- 生态工程师(例如:地鼠龟和海狸)在改变它们栖息地的同时,并且在这样做的时候,将改变其他物种的命运和机遇。
- 保护伞物种要么有较大的用地需求要么占据多个栖息地,从而与许多其他物种的栖息地交叠。
- 链接物种在营养级(食物网中)的物质和能量传递中起着至关重要的作用,或在复杂食物网的能量传递中提供重要的链接作用。

[11] 焦点物种在数量和分布变化上所产生的影响是多样的。举例来说,关键物种通过竞争、共生、分散、授粉和疾病等过程以及改变生境和非生物因子来影响生态系统。一个被引进的非本土物种可以担当起焦点物种的角色而产生许多影响,包括通过它们作为天敌、竞争对手、病原体或病毒载体的角色及通过对水平衡、生产力和生境结构的影响,来改变群落组成和生态过程。因土地利用的改变而对关键物种和入侵物种的影响可以传播到远超出这块土地的范围,并且在它们的数量发生变化之前难以预测。土地覆盖格局的变化会影响甚至促进外来物种的建立。

[12] 营养级是指食物链中的各个层次,如生产者、草食动物、消费者和分解者。某一个营养级中焦点物种或生物群数量的变动会向下影响其他营养级,并最终导致生物多样性、群落构成或总生产力的显著变化。土地利用通过改变水流、输沙量、养分循环或生产力等生态系统特性,导致物种组成和生物多样性的变化。土地利用对物种组成的影响可以预示出生态系统未来的生产力。

地点原则

[13] 在任何一个地方,气候、水文、土壤、地形因素以及生物的相互作用强烈地影响着生态过程和物种的丰富性及分布。当地的环境条件反映了该地区沿坡度变化的海拔、经度、纬度和随坡度变化的众多微尺度物理、化学和土壤因素。这些因素限制了各类土地用途的适宜性,并且确定居留种及生态过程。

[14] 关键生态系统进程的速率,如初级生产和分解,受限于土地养分、温度、可用水量以及由气候和天气控制的这些因素的时间模式。因此,只有在生态进程特定范围内的速率可以不需要持续的管理投入(例如为生长在沙漠中的作物灌溉)在这个场所中自行维持。长期的人类干扰可能会扩大这些范围,但是不能在没有投入的情况下完全不发生超出土地利用局限的情况。

[15] 在自然状态下形成的生态系统结构和功能模式提供了可供参考的可持续和生态健康的土地利用模式。只有那些适应一个区域环境限制的物种才能健壮生长。降水量限制了适合景观种植的植物种类,以及适合农业、林业或牧业管理系统的种类。此外,一些拥有独特条件的地方因对物种和它们所支持的生态系统的保护而要比其他地方更加重要。

[16] 从长远和宽角度来看,土地利用如果不能在场地所局限的范围内得以维护,代价将是昂贵的。举例来说,在干旱贫瘠的地区建立耕地和观赏草坪是可能的,但抽取原生地下水进行浇灌的速率是自然恢复难以支持的。只有某些土地利用、定居和开发、房屋建设和景观设计模式能够与当地及区域的条件相匹配。在陆地系统,由土地利用和土地管理实践所导致的土壤流失或退化会减少该场地长期的潜在生产力并影响物种组成。土地利用实践也会影响当地气候(例如,城市热岛效应)。可持续的居住场所受制于拥有景观的适宜区域。例如,在短暂的湖滨沙地、冲积平原、被侵蚀的海滨或易发生火灾的地方建立房屋或社区从长远来看是极容易损毁的。因此理想的情况是因地制宜。

Reading Material B
Ecological Principles for Managing Land Use II
Ecological Society of America
土地利用管理的生态原则 II

Disturbance principle

[1] The type, intensity, and duration of disturbance shape the characteristics of populations, communities, and ecosystems. Disturbances are events that disrupt ecological systems; they may occur naturally (e.g., wildfires, storms, or floods) or be induced by human actions, such as clearing for agriculture, clearcutting in forests, building roads, or altering stream channels. The effects of disturbances are controlled in large part by their intensity, duration, frequency, timing and the size and shape of the area affected. Disturbances may affect both above- and below-ground processes and can impact communities and ecosystems by changing the number and kinds of species present; causing inputs or losses of dead organic matter and nutrients that affect productivity and habitat structure; and creating landscape patterns that influence many ecological factors, from movements and densities of organisms to functional attributes of ecosystems.

[2] Land-use changes that alter natural-disturbance regimes or initiate new disturbances are likely to cause changes in species' abundance and distribution, community composition, and ecosystem function. In addition, the susceptibility of an ecosystem to other disturbances may be altered.

[3] Land managers and planners should be aware of the prevalence of disturbance in nature. Disturbances that are both intense and infrequent, such as hurricanes or 100-yr floods, will continue to produce "surprises." Ecosystems change, with or without disturbance; thus, attempts to maintain landscape conditions in a particular state will be futile over the long term. Attempts to control disturbances are generally ineffectual and suppression of a natural disturbance may have the opposite effect of that intended. For example, suppression of fire in fire-adapted systems results in the buildup of fuels and increases the likelihood of severe, uncontrollable fires. Similarly, flood control efforts have facilitated development in areas that are still subject to infrequent large events (e.g., the 1993 floods in the upper Midwest), resulting in tremendous economic and ecological impacts. Land-use policy that is based on the understanding that ecosystems are dynamic in both time and space can often deal with changes induced by disturbances.

[4] Understanding natural disturbances can help guide land-use decisions, but the differences between natural and human-made disturbances must be recognized. Continued expansion of human settlement into disturbance-prone landscapes is likely to result in increased conflicts between human values and the maintenance of natural-disturbance regimes necessary to sustain such landscapes.

Landscape principle

[5] The size, shape, and spatial relationships of landcover types influence the dynamics of populations, communities, and ecosystems. The spatial arrangement of ecosystems comprises the landscape and all ecological processes respond, at least in part, to this landscape template. The kinds of organisms that can exist are limited by the sizes, shapes, and patterns of habitat across a landscape.

[6] Human-settlement patterns and land-use decisions often fragment the landscape or otherwise alter land-cover patterns. Decreases in the size of habitat patches or increases in the distance between habitat patches of the same type can greatly reduce or eliminate populations of organisms, as well as alter ecosystem processes. However, landscape fragmentation is not always necessarily destructive of ecological function or biodiversity because a patchwork of habitat types often maintains more types of organisms and more diversity of ecosystem process than does a large area of uniform habitat. Making a naturally patchy landscape less patchy may also have adverse affects.

[7] Larger patches of habitat generally contain more species (and often a greater number of individuals) than smaller patches of the same habitat. Larger patches also frequently contain more local environmental variability. This variability provides more opportunities for organisms with different requirements and tolerances to find suitable sites within the patch. In addition, the edges and interiors of patches may have quite different conditions, favoring some species over others. The abundance of edge and interior habitat varies with patch size; large patches are likely to contain both edge and interior species, whereas small patches will contain only edge species.

[8] The extent and pattern of habitat connectivity can affect the distribution of species by making some areas accessible and others inaccessible. The amount of connectivity needed varies among species and depends on two factors: the abundance and spatial arrangement of the habitat and the movement capabilities of the organism.

[9] While gradual reduction in habitat may have gradual effects, once a certain threshold is reached, the effects become dramatic. Land-cover changes are most likely to have substantial effects when habitat is low to intermediate in abundance and small changes may cause large impacts.

[10] The ecological importance of a habitat patch may be much greater than is suggested by its size and distribution across the landscape. Some habitats, such as bodies of water or riparian corridors, are small and discontinuous, but nevertheless have ecological impacts that greatly exceed their spatial extent. For example, the presence of riparian vegetation, which may occur as relatively narrow bands along a stream or as small patches of wetland, generally reduces the amount of nutrients being transported to the stream. This filtering by the vegetation is an ecologically important function because excess nutrients that unintentionally end up in lakes, streams, and coastal waters are a major cause of eutrophication, acidification, and other water quality problems. Thus, the presence and location of particular vegetation types can strongly affect the movement of materials across the landscape and can contribute to the maintenance of desirable water quality.

UNIT 11 ECOLOGY AND ENVIRONMENT PROTECTION

[GLOSSARY]

biodiversity	生物多样性	eutrophication	富营养化
clear cutting	皆伐	human settlement environment	人居环境
community	群落	land cover	地表
connectivity	连接度	patch	斑块
ecosystem	生态系统	population	种群

[NEW WORDS]

acidification	n.	酸化	infrequent	adj.	少有的,不经常的,罕见的
be subject to	vt.	易受……的,受支配,服从	intermediate	adj.	中间的,中级的
buildup	n.	逐渐积聚,建造,增强	patchwork	n.	拼凑物,混杂物
destructive	adj.	破坏(或毁灭)性的	patchy	adj.	斑驳的,杂凑的,不完整的
disrupt	vt.	使混乱,扰乱,使中断,瓦解	prevalence	n.	普遍,盛行,流行,传播
dynamics	n.	力学,动力学,动态	prone	adj.	有……倾向的,易于……的
eliminate	vt.	消除,除掉,除去	regime	n.	状况,条件,状态,方式
eutrophication	n.	富营养化	riparian	adj.	河边的,水滨的,河岸的
exceed	vt.	超过,超越	severe	adj.	严厉的,剧烈的,严重的
facilitate	vt.	帮助,促进,使容易	substantial	adj.	重大的,可观的,实质的
favor… over	vt.	偏爱……而不喜欢……	suppression	n.	抑制,制止,镇压
fragment	vt.	(使)碎裂,破裂,分裂	susceptibility	n.	感受性,易感性,敏感性
	n.	片段,碎片,断片	threshold	n.	阈值,极限,临界值;入口
fragmentation	n.	分裂,破碎	tremendous	adj.	巨大的,可怕的,非常的
futile	adj.	徒劳的,无用的,无效的	uniform	adj.	一致的,一样的,均匀的
induce	vt.	引起,导致;引诱,劝导	unintentionally	adv.	非故意地,非存心地
ineffectual	adj.	无效的,不起作用的	variability	n.	可变性,易变性,变化性

[难句]

  Disturbances may affect both above- and below-ground processes and can impact communities and ecosystems by changing the number and kinds of species present; causing inputs or losses of dead organic matter and nutrients that affect productivity and habitat structure; and creating landscape patterns that influence many ecological factors, from movements and densities of organisms to functional attributes of ecosystems. 干扰既影响地上进程,也影响地下进程,并且对群落和生态系统产生冲击,这些影响主要是通过以下作用产生的:改变当前物种数量和种类;引起对生产力和生境结构有影响的有机物和营养物的输入和流失;创造对许多生态因子(从生物运动和密度到生态系统功能属性)有影响的景观格局。[长句分译]

[参考译文]

土地利用管理的生态原则 Ⅱ

干扰原理

[1] 干扰的类型、强度和持续时间形成了种群、群落和生态系统的特征。干扰是扰乱生态系统的事件,它们可能是自然发生(比如野火、暴风雨或洪水)的,也可能是人类活动引起的,比如农业清除、森林皆伐、修建道路或改变河道。干扰的作用效果很大程度上是由它们的强度、持续时间、频率、发生时间和受影响区域的大小和形状所控制。干扰既影响地上进程,也影响地下进程,并且对群落和生态系统产生冲击,这些影响主要通过以下作用产生:改变当前物种数量和种类;引起对生产力和生境结构有影响的有机物和营养物的输入和流失;创造对许多生态因子(从生物体运动和密度到生态系统功能属性)有影响的景观格局。

[2] 土地利用的转变会改变自然干扰方式或引起新的干扰,这有可能引起物种丰度和分布、群落组成和生态功能的改变。此外,生态系统对于其他干扰的敏感性也可能被改变。

[3] 土地管理者和规划者应该注意到在自然界常发生的干扰。强烈而罕见的干扰,譬如飓风或百年一遇的洪水,将不断制造"奇袭"。生态系统的改变,或受或不受干扰影响,因此,试图将景观条件状况维持在一种特定的状态,长期来看是徒劳的。试图控制干扰,总体来说也是无用的,对自然干扰的抑制也许会产生与预期相反的效果。比如,在火灾适应系统中对大火的抑制会导致燃料积累并增加使火势变严重且无法控制的可能性。类似的,控制洪水的努力能促进区域发展,却仍然容易受到罕见大灾难(譬如1993年中西部北部的大洪水)的破坏,导致巨大的经济、生态冲击。基于对生态系统在时间和空间上是动态变化的理解的土地利用政策可以处理由干扰所引起的变化。

[4] 了解自然干扰有助于指导土地利用的决定,但是必须认识自然干扰和人为干扰之间的不同。人类居住地不断扩张到易受干扰的景观中,有可能导致人类价值与对维持这些景观具有必要性的自然干扰状态维护之间产生更大的冲突。

景观原理

[5] 地表类型的尺度、形状和空间关系影响种群、群落、生态系统动态。生态系统的空间布局构成了景观,并且所有的生态过程或多或少受到景观模式的影响。能够生存的各种生物体都受限于景观中生境的尺度、形状和格局。

[6] 人居环境模式和土地利用决策经常破坏景观或是改变地表形式。生境斑块尺度减小或同类型生境斑块距离增加会极大程度上减少或消除生物种群数量,也会改变生态系统进程。然而,景观破碎对于生态功能或生物多样性来说不一定是破坏性的,因为比起一大片均质的生境,拼合的生境类型往往能维持更多生物种类和更多样的生态过程。使一个原本自然斑驳的景观不再斑驳也可能会产生不利影响。

[7] 相似的生境,较大的生境斑块一般比较小的斑块包含更多物种(并且常常是更多数量的个体)。较大斑块也常常包含更多的环境易变性。这种易变性为有着不同需求和耐受性的生物提供更多机会,来寻求斑块中适合的场地。此外,斑块边缘和内部可能有完全不同的条件,适合于一些物种而不适合另外一些。生境丰富的边缘和内部的富饶程度随着分地的尺度大小而改变。生境边缘与内部的丰富程度因斑块尺度不同而不同,大的斑块可能既包含边缘物种,也包含内部物种;然而小斑块将只拥有边缘生物。

[8] 生境连接程度和格局会通过使区域可达或不可达来影响物种的分布。物种所需要的连接量各不相同,并且取决于两个因素:生境的丰富程度和空间分布以及生物的移动能力。

[9] 虽然生境逐渐减少会逐渐产生影响,但是一旦达到某一极限,这种影响将变得极其巨大。当生境丰富度降低到中间值时,地表变化极有可能产生巨大的作用,并且小的改变就可能会引起大的冲击。

[10] 生境斑块的生态性也许比它的尺度和在景观中的分布重要得多。一些生境,比如水体或河岸廊道,尽管小型且不连续,但是它们的生态影响力却大大超越其空间范围。例如,水岸植被往往出现在溪边相对狭窄的带状用地或小型湿地,它们能够大大减少进入水体的营养物数量。植被的过滤作用有着重要的生态功能,因为

排入湖水、溪流和海岸水体中的过剩营养是引起富营养化、酸化和其他水质问题的主要成因。因此,特殊植被类型的存在可以强烈地影响穿越景观的物质运动,并且有助于维持良好的水质。

[思考]

(1)环境运动产生的根源是什么?

(2)环境问题的对策或关注点有哪些?

(3)土地利用管理的生态原则有哪些?

UNIT 12 METHODS OF PLANNING AND DESIGN

TEXT

Site Volume[1]
John Ormsbee Simonds
空　间

[1]　Much of the art and science of land planning is revealed to the planner when it is first realized that one is dealing not with areas but with spaces. As an example, a playground composed of play equipment set about on a dull base plane has little child appeal, while the same apparatus arranged within a grouping of imaginative play spaces can provide endless hours of delight. It is a matter of designing the volumetric enclosure and spatial interconnections to suit the use. The creation of well-organized interior and exterior spaces for uses of any type and scope is our goal as environmental designers.

[2]　**Spatial qualities**

The essence of a volume is its quality of implied containment.

A confined space may be static. It may hold interest, induce repose. It may direct and concentrate interest and vision inward. The whole spatial shell may be made seemingly to contract and bear down, to engender a feeling of intensity or compression.

Alternatively, a space may open out. It may direct attention to its frame and beyond. It may fall away or seem to expand. It may seem fairly to burst outward. It may impel outward motion to its perimeter and to more distant limits.

A space may be flowing and undulating, suggesting directional movement.

A space may be so developed as to have its own sufficient, satisfying qualities. It may appear complete within itself or incomplete, a setting for persons or objects.

A space may be in effect a vacuum.

A space may have expulsive pressure.

A space may be developed as optimum environment for an object or a use.

A space may be so designed as to stimulate a prescribed emotional reaction or to produce a predetermined sequence of such responses.

A space may dominate an object, imbuing the object with its particular spatial qualities. Or it may be dominated by the object, drawing from it something of its nature.

UNIT 12　METHODS OF PLANNING AND DESIGN

A space may have orientation inward, outward, upward, downward, radial, or tangential.

A space may relate to an object, or another space and may gain its very meaning from the relationship. It may relate to vista or view, the rising or setting sun, a sunlight slope, the starlit sky, or the welcome evening breeze. A complex space assumes to a degree the qualities of its component volumes and should relate them into a unified entity.

Spaces may vary from the vast to the minute, from the light and ethereal to the heavy and ponderous, from the dynamic to the calm, from the crude to the refined, from the simple to the elaborate, and from the somber to the dazzling. In their size, shape, and character they may vary endlessly. Clearly in designing a space for any given function, we would do well first to determine the essential qualities desired and then to do our best to provide them.

From its hollowness arises the reality of the vessel; from its empty space arises the reality of the building… (<u>Lao-tse</u>[2])

A creation in space is an interweaving of parts of space… (<u>Laszlo Moholy-Nagy</u>[3]) (Fig. 1)

Architecture… is the beautiful and serious game of space… (<u>Willem Dudok</u>[4]) (Fig. 2)

[3]　Spatial size

It is well known that the size of an interior space in relation to people has a strong effect on their feelings and behavior. This fact may be illustrated graphically in the accompanying diagrams (Fig. 3).

Exterior spaces have similar psychological attributes. On an open plain, timid persons feel overwhelmed, lonesome, and unprotected; left to their own devices, they soon take off in the direction of shelter or kindred spirits. Yet, on this same plain, bolder persons feel challenged and impelled to action; with freedom and room for movement they are prone to dashing, leaping, yahooing. The level base plane not only accommodates but also induces mass action, as on the polo field, the football field, the soccer field, and the racetrack.

Fig. 1　Laszlo Moholy-Nagy　　Fig. 2　Willem Dudok

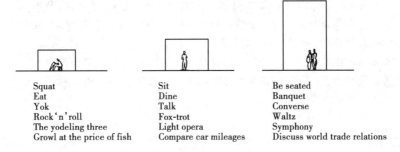

Squat　　　　　　　　Sit　　　　　　　　　　Be seated
Eat　　　　　　　　　Dine　　　　　　　　　Banquet
Yok　　　　　　　　　Talk　　　　　　　　　Converse
Rock'n'roll　　　　　　Fox-trot　　　　　　　Waltz
The yodeling three　　　Light opera　　　　　　Symphony
Growl at the price of fish　Compare car mileages　　Discuss world trade relations

Fig. 3　The size of interior space has a strong effect on people's feelings and behavior.

If, upon this unobstructed surface, we set an upright object, it becomes an element of high interest and a point of orientation for the visible field. We are drawn to it, cluster about it, and come to rest at its base. No small factor in this natural phenomenon is the human atavistic tendency to keep one's flanks protected. A vertical plane or wall gives this protection and suggests shelter. Increased protection is afforded by two intersecting upright planes. They provide a corner into which our subject may back and from which he or she can survey the field for either attacker or quarry. Additional vertical planes define spaces that are further controlled by the introduction of overhead planes. Such spaces derive their size and shape and degree of enclosure from the defining planes, acting together and counteracting.

A volume may be stimulating, or it may be relaxing. It may be immense, suggesting certain uses, or it may be confining, suggesting others. In any event, we are attracted to those spaces suited to our purpose, be it hiking, target shooting, eating grapes, or making love. We are repelled by, or at least have little interest in, those spaces that appear to be unsuited to the use we have in mind.

[4] **The elements of containment**

In a large measure all spaces acquire their being and character from the elements that contain them. Because each, element so used will imbue the space in some degree with its own qualities, it must be well related not only to all other such elements but also to the essential resultant character desired for the space.

Lines, forms, colors, textures, sounds, and odors all have certain predictable impacts on the human intellectual-emotional responses. If, for example, a certain form of color says or does things to the observer, this is reason enough to employ such a form or color in the shaping of those structures, objects, or spaces that are to convey this message. Surely, if the abstract expression of a given line violates the proposed expression of structure, object, or space, it should be used only with studied intent. Every line evident in the form or planes has its own abstract design connotation. This must be in keeping with the intended nature of the space. There follows a graphic demonstration of variations in abstract line expression.

[NOTES]

[1] 本文节选自约翰·奥姆斯比·西蒙兹(John Ormsbee Simonds)的《景观设计学:场地规划设计导则》(*Landscape Architecture:A manual of Site Planning and Design*)。

[2] Lao-tse 老子(公元前571—前471):我国古代伟大的哲学家和思想家,道家学派创始人。

[3] Laszlo Moholy-Nagy 拉斯洛·莫霍伊·纳吉(1895—1946):匈牙利设计师、画家、摄影师、包豪斯学院教授。受构成主义影响,致力于工业技术与艺术的整合。

[4] Willem Dudok 威廉·杜多克(1884—1974):荷兰建筑师。作品具有阿姆斯特丹学派注重个性表现和风格派强调几何形体的双重特点,重要作品有希尔佛赛姆市政厅(Town Hall Hilversum)、冯德尔学校(Vondel School)等。

[GLOSSARY]

spatial interconnection	空间联系	volumetric enclosure	空间围合

[NEW WORDS]

apparatus	n.	仪器	ponderous	adj.	沉重的,冗长的,笨重的
atavistic	adj.	隔代遗传的,返祖现象的	predetermined	adj.	业已决定的,预先决定的
attributes	n.	属性,标志,象征,特质,特性	predictable	adj.	可预料的,可预见的
cluster	n.	丛生,成群,使聚集	prescribed	adj.	规定的,指定的
concentrate	vt.	集中,浓缩,聚集,全神贯注	prone	adj.	倾斜的,陡的
convey	vt.	搬运,转让,传达	radial	adj.	径向的,放射状的,辐射的
dash	vi.	使猛撞,猛掷,冲撞	refine	vt.	精制,提炼,改进
dazzling	adj.	眼花缭乱的;耀眼的	repose	n.	静止,休息
dynamic	n.	动力的,动力学的,动态的	resultant	n.	结果,生成物
elaborate	adj.	精细的,精心的,详尽的	scope	n.	范围,观测设备
engender	vt.	使产生,引起;产生	somber	adj.	昏暗的,忧郁的,阴沉的
ethereal	adj.	轻的,天上的,像空气的	stimulate	vt.	刺激,鼓舞,激励,起促进作用
expulsive	adj.	逐出的,开除的	sufficient	adj.	充足的,充分的
exterior	n.	外部	tangential	adj.	切向的,相切的,切线的
hollowness	n.	凹陷,空虚,空旷,不实	timid	adj.	胆小的,羞怯的
imbue	vt.	使吸入,浸染,充满	undulate	vi.	使起伏,使波动;波动,起伏
imbue	vt.	使感染,鼓吹,使蒙受	unobstructed	adj.	无阻的,不受阻的
interior	n.	内部	vacuum	n.	真空
kindred	adj.	同类的,血缘的,同族的	vertical	n.	垂直线,垂直位置
lonesome	adj.	寂寞的,人迹稀少的	vessel	n.	导管,容器
optimum	adj.	最佳,最适宜	volume	n.	容积,体积,容量
orientation	n.	定向,定位,方位	volumetric	adj.	容积的,体积的
perimeter	n.	周长,周界,周边,边缘			

[参考译文]

空 间

[1] 当规划师第一次觉悟到人们所涉及的不是区域而是空间时,许多土地规划的艺术和科学才会展现在他们面前。例如,一个游乐场,如果游乐设施置于枯燥乏味的平地之上,对孩子们是很少有吸引力的。如果同样的设施安排在富有想象力的一组游乐空间中,却可令人乐不思归。这就涉及如何设计空间围合和空间联系以适应用途的问题。为各种类型和规模的功能创造具有良好组织的内部和外部空间,是我们环境设计者的宗旨。

[2] **空间特性**

容积的本质在于其不言而喻的容纳性。

追求宁静的限定空间可以是静态的。它可向内引导且集中兴趣或视线。整个空间外壳似乎是为了收缩或压倒一切而建造,以产生激动或压抑感。

另外,空间也可以是外向的。它可以把注意力引向边框甚至更远的地方。它可以无限延展或看起来要扩展。它有时似乎要向外迸发。它可以驱使外向运动直至外围边界甚至更远。

空间可以是流动和起伏的,引导定向的运动。

空间可高速发达,以至于自身具有足够的、令人满意的特性。它既是自我完善、自成一体,也可成为人或事物的背景。

空间,其实可以是一种真空状态。

空间,可以有一种排斥力。

空间可成为物体或特定用途的最佳背景。

空间可以设计用来激发既定的情感反应或产生一系列预期的反应。

空间可支配物体,使事物融入它独特的空间特性中。它也可为事物所主宰,由物体获得某些自身的本质。

空间可以是内向的、外向的、上升的、下降的、辐射的或切向的。

空间可以与物体或其他空间相联系,并从这种联系之中获得最真实的意义。它可以与远景近景、日出日落、阳光明媚的山坡、星光灿烂的夜空或者撩人的习习晚风相呼应。一个复合空间在一定程度上限定了局部空间的特性,并且将其联成统一的整体。

空间的变化可从大到小,从轻盈飘渺到凝重沉闷,从动态到平静,从粗犷到精致,从简单到精巧,从阴郁到灿烂。它们的尺寸、形状、特征可以无止境地变化。很明显,在为任何特定功能设计空间时,我们首先要很好地确定那些最需要的特征,并竭尽全力展现它们。

"埏埴以为器,当其无,有器之用;凿户以为室,当其无,有室之用,故有之为利,无之为用"——老子

"空间内的创造就是局部空间的交织……"——拉斯洛·莫霍伊·纳吉

"建筑学,……是美丽而庄严的空间游戏"——威廉·杜多克

[3] 空间尺度

众所周知,与人相关的内部空间的尺度极大地影响着人的情感和行为。这点可用下面的附图加以说明。

外部空间具有相似的心理影响力,站在广阔的草原上,懦弱的人会觉得压抑、孤独且缺乏保护。如果听任其自行其是,他们会很快奔向遮蔽所或志同道合的人群。然而在同一平原上,坚强的人会觉得受到了挑战且跃跃欲试;因为那里有自由和活动的空间,他们想飞跑、想跳跃、想欢呼。水平地面不只容纳,同时也诱导了大规模活动的开展,例如马球场、橄榄球场、足球场和田径场上的活动。

如果在这片一览无余的土地上设置一个笔直的突出物,它就会成为引起高度兴趣的要素和可视区域的方向标识点。我们会被吸引过去,围绕着它聚集起来,并在其基面上休息。此微小的自然现象反映了人类固有的自我防卫的倾向。一个竖向平面或墙则提供这种保护且意味着庇护。两个相交叉的竖直平面增强了防御感。它们提供一个人们可以退入的角落,从那里可以瞭望整个区域,看是否有侵袭者或猎物。垂直面的增加可界定出更多的角落和空间,它们进一步通过增加顶面覆盖量来强化空间的控制。这类空间的尺度、形状和围合度是通过相互作用、相互制约的、界定空间的平面体现的。

空间可令人激动或使人放松。它可以很大,提示某些特定用途;也可以是有限的,提示另外一些用途。无论我们做什么,远足、射击、吃葡萄还是谈情说爱,我们都会被那些适合我们需要的空间所吸引。对于那些与我们想象中用途不相符的空间,我们会排斥它们,或者至少我们对其不感兴趣。

[4] 空间要素

广而言之,所有空间都从其组成要素中获得生命与个性。因为在某种程度上,每一个这种要素的自身性质都浸染于空间中,它不仅要很好地与所有其他这类要素相联系,也要与为空间设计所预期的本质性特征相呼应。

线条、形体、颜色、质地、声音和气味都对人的理智-情感反应产生某些可预知的影响。例如,如果某种色彩的形式能对观察者传达信息或产生一定影响,这就有充分的理由采用这样的形式或色彩去塑造那些要传达这种信息的构筑物、对象或空间。当然,如果给定线条的抽象表达背离了构筑物,物体或空间的预定表达形式,它

就只能在深思熟虑后加以运用。在形体或平面中,每一条明显的线条都有其自身的含义且必须与所在空间的预期本质保持一致。

Reading Material A

The Conceptual Plan

John Ormsbee Simonds

概念规划

[1] A seed of use—a cell of function—wisely applied to a receptive site will be allowed to develop organically, in harmonious adaptation to the natural and the planned environmental.

[2] We have by now developed a comprehensive program defining the proposed nature of our project. We have become fully aware of all features of the total environs. Up to this point, the planning effort has been one of research and analysis. It has been painstaking and perhaps tedious, but this phase is of vital importance because it is the only means by which we can achieve full command of the data on which our design will be based. From this point on, the planning process becomes one of integration of proposed uses, structures, and site.

Plan concepts

[3] If structure and landscape development are completed, it is impossible to conceive one without the other, for it is the relationship of structure to site and site to structure that gives meaning to each and to both.

[4] This point perhaps raises the question of who on the planning team—architect, landscape architect, engineer, or others—is to do the "conceiving". Strangely, this problem, which might seemingly lead to warm debate, seldom arises, for an effective collaboration brings together experts in various fields of knowledge who, in a free interchange of ideas, develop a climate of perceptive awareness and know—how. In such a climate, plan concepts usually evolve more or less spontaneously. Since the collaboration is arranged and administered by one of the principals (who presumably holds the commission), it is usually this team leader who coordinates the planning in all its aspects and gives it expressive unity. It is the work of the collaborators to advance their assigned planning tasks and to aid in the articulation of the main design idea in all ways possible (Fig. 1).

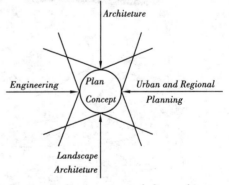

Fig. 1 A planning team including architect, landscape architect, engineer and urban planner

The site-structure diagram

[5] When planning a project or a structure in relation to a land area, we first consider all the various uses to be fitted together and accommodated. For a high school, for instance, we would determine the approximate architectural plan areas and their shapes—the general plan areas required for service, parking, outdoor classrooms, gardens, game courts, football fields, track, bleachers, and perhaps fu-

ture school expansion. Over a print of the topographic survey (or site analysis map) (Fig. 2) we should then indicate, in freehand line, use areas of logical size and shape in studied relation to each other and to the natural and built landscape features. Having thus roughed in the site use areas, we may at last block in the architectural elements of the project. The result is the site-structure diagram (Fig. 3).

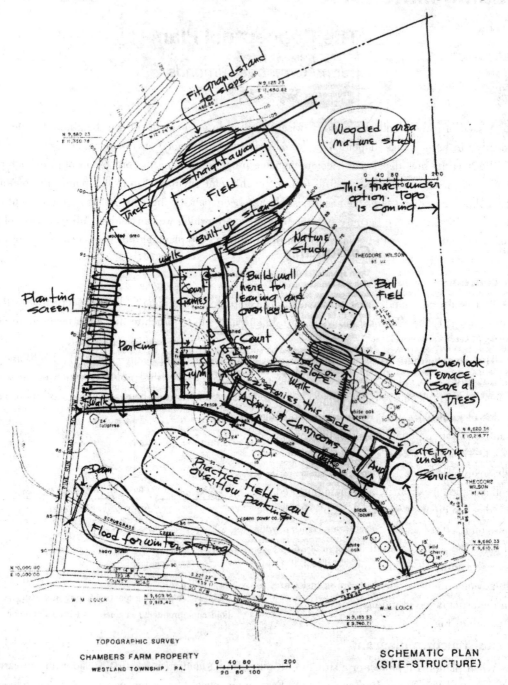

Fig. 2　Schematic plan (site-structure)

UNIT 12　METHODS OF PLANNING AND DESIGN

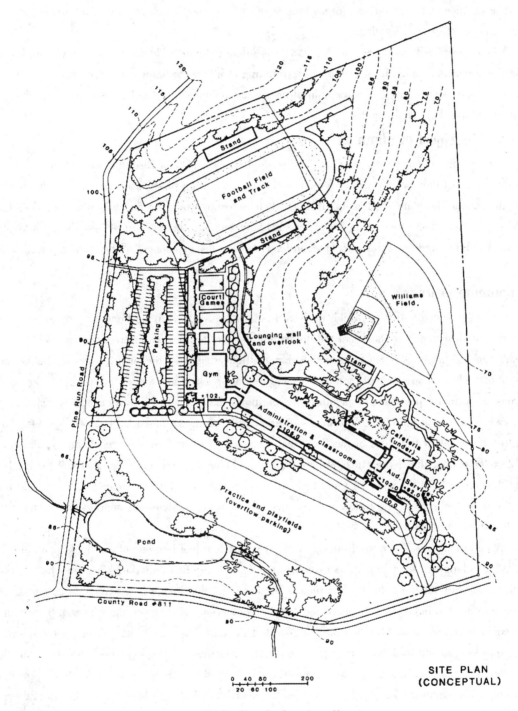

Fig. 3　Site plan (conceptual)

[6] The balance of the planning process is a matter of comparative analysis and refinement of details—a process of creative synthesis.

[7] A good plan, reduced to essentials, is no more than a record of logical thought. A dull plan is a record of ineffectual thinking or of very little thinking at all. A brilliant plan gives evidence of response to all site factors, a clear perception of needs and relationships, and a sensitive expression of all components working well together.

The creative aspect of planning

[8] Planners may create in the material, forms, and symbols of their disciplines an object, space, or construction that they believe will engender in the users a certain predictable experience. In effect, users will recreate the planned elements through their perception of them and will thus by led to the desired experience. For when we perceive, we actually recreate through our senses the form-giving process. An understanding of this phenomenon leads us to a clearer concept of the creative function of design.

The planning attitude

[9] We can only create that for which we have first developed empathy and understanding. A shopping mall? As designers, we must feel the quickening tempo, the pull and attraction, the bustle, the excitement of the place. We must sense the chic boutique displays, the mouth-watering sights and smells of the bakery shop; we must see in our mind the jam-packed counters of the hardware store and the drugstore with its pyramids of mouthwash, perfume, nail polish, hot-water bottles, and jelly beans. We must see in the market the heaps of grapefruit, oranges, rhubarb, Brussels sprouts, bananas; whiff the heady fragrance of the floral stalls; picture the shelf on shelf of bargain books, the bolts of cotton prints, the sloping trays of peppermints and chocolate creams. We must feel the brightness of the sunshine on the sidewalks and the coolness and protection of the shaded doorways and arcades We must feel crowds and traffic and benches and trees and perhaps the sparkle and splash of a fountain or two. And then we can start planning.

[10] A children's zoo? If we would design one, we must first feel like one of the flocking children, the gawking, clapping, squealing kids; we must appreciate the delight, the laughter, the chatter, the confusion, and the rollicking thrill of the place. We must feel the diminutive, squeaky "cuteness" of the mouse town, the bulk and immensity and cavelike hollowness of the spouting whale with its dimly illumined interior. We must know the preening strut of the elegantly wandering peacocks, the quack, quack, quacking of the waddling ducks, the soft furry whiteness of the lop-eared rabbits, and the clop, clop, clopping and creaking harness and the awed delight of the pony ride. We must, in our minds, be at the children's zoo, and we must see it, hear it, feel it, and love it as a child would love it as we make our plans.

[11] Are we to design a parkway, hotel plaza, terminal, or bathing beach? If we would create them, we must first have a feeling for their nature. This self-induced sensitivity we might call the planning attitude. Before we mature as planners, it will be intuitive.

UNIT 12　METHODS OF PLANNING AND DESIGN

[NEW WORDS]

arcade	n.	连拱廊	heady	adj.	易使人醉(发晕)的,令人陶醉的
articulation	n.	清晰度,咬合,关节	heap	n.	(一)堆,大量,许多
awe	vt.	使敬畏,使惊惧,使惊叹	hollowness	n.	穴,孔,洞坑,洼地
bargain	n.	特价商品,减价品,便宜货	illumine	vt.	照亮,照明
bleacher	n.	(球场的)露天看台	immensity	n.	无限,广大,巨大
bolt	n.	一卷布,一卷纸,螺栓,插销	intuitive	adj.	直觉的,本能的,天生的
boutique	n.	精品店,时装精品屋	jelly	n.	果冻,果酱,胶状物
brussels sprout		球芽甘蓝	lop-eared	adj.	垂耳的
bulk	n.	(巨大)物体,(大)块,(大)体积	peppermint	n.	薄荷糖,薄荷
bustle	n.	忙乱,热闹,嘈杂,喧闹	perceptive	adj.	洞察力强的,敏锐的,理解力强的
chatter	n.	喋喋不休,唠叨	perfume	n.	香水,香气,芳香;香味
chic	adj.	漂亮的,时髦的,雅致的	preen	vt.	把自己打扮得漂亮,自满,自负
clap	vi.	拍手,鼓掌,欢呼	receptive	adj.	善于接受的,易于接受的
clop	n.	马蹄声,得得响,脚步声	rhubarb	n.	食用大黄,大黄
creak	vi.	嘎吱作响	rollick	vi.	欢闹,耍闹
diminutive	adj.	小得出奇的,特小的,微小的	sparkle	n.	闪烁的光,光亮,活力,烟花
dimly	adv.	暗淡地,昏暗地,朦胧地	splash	n.	溅泼声,喷溅,飞溅
drugstore	n.	杂货店,(兼售化妆品等的)药房	squeaky	adj.	吱吱作响的,轧轧响的
empathy	n.	移情作用,同感,共鸣	squeal	vi.	长声尖叫,号叫
engender	vt.	使产生,造成	strut	n.	趾高气扬的步态,高视阔步的样子
environs	n.	环境,周边,包围	tempo	n.	速度,拍子,节奏
floral	adj.	用花做的,用花装饰的	terminal	n.	(火车、公共汽车或船的)终点站
furry	adj.	毛皮的,覆上毛皮的	track	n.	田径场
gawk	vi.	做笨拙的动作,呆视	tray	n.	盘子,托盘,碟
gawk	vi.	呆呆地看	waddle	vi.	摇摇摆摆地走
hardware	n.	五金器具,硬件	whiff	n.	一阵气味,一阵
harness	n.	马具,挽具			

[难句]

A brilliant plan gives evidence of response to all site factors, a clear perception of needs and relationships, and a sensitive expression of all components working well together. 杰出的规划对所有场地因素都给予充分考虑,明确认识各种需要和关系,精心处理所有相互作用的局部。[词性转换]

· 211 ·

[参考译文]

概念规划

[1] 用途的种子——功能的细胞——一旦明智地应用于一块乐于接受的土地,将和谐的适应于自然的与规划的环境,并得以有机的茁壮生长。

[2] 迄今为止,我们已经完成了一项综合的工作程序:确定了目标的性质。我们已经开始完全了解整体环境的所有属性。一直到现在为止,我们的努力都属于研究分析的范畴。这一阶段是辛苦的,也许还有些沉闷,但是它确实至关重要,因为这是我们完全掌握设计所立足的数据的唯一途径。从现在开始,规划过程成为将建议用途、构筑物与场地的统一和整合过程。

规划构思

[3] 如果同时考虑建筑与景观的建设,那么脱离一方去构思另一方都是不可能的。因为正是建筑——场地之间的这种联系才赋予双方各自和共同的意义。

[4] 在这一点上,也许会产生一个问题:在规划队伍中由谁——建筑师、景观设计师、工程师抑或其他人来作"构思"工作?奇怪的是,这个问题,看起来会引起激烈的争论,却很少发生,因为一种卓有成效的合作把不同知识领域的专家集合到一起,大家在自由的思想交流中,创造一种感悟和启迪的气氛。在这样一种气氛中,通常或多或少的,规划概念会自然产生。既然这种合作是由某一位负责人(假定他控制规划队伍)组织和管理的,那么通常也正是他来把规划的方方面面进行协调,并取得统一的表达。合作者们的工作是充分完成各自指派的规划任务,并在提出的主体设计思想中尽可能予以帮助。

场地——构筑物图

[5] 当规划一个与一定场地相关的工程或建筑时,我们首先考虑场地需要提供的、将被组织在一起的各种功能。以一所中学为例,我们会确定大概的建筑规划的区域以及它们的形状——对需要的服务设施、停车场、户外教室、花园、体育场、橄榄球场、田径场、露天看台,兴许还有将来的学校扩展用地进行总体用地规划。在地形测量图(或场地分析图)的复制地图上,我们就可以用徒手线条表示出合理尺度和形态的用地范围。它们彼此之间的关系以及它们与自然和人工景观特征的关系都是经过精心研究的。经过这样一番对场地用地范围的粗加工后,我们终于可以加入项目的建筑要素。最终成果就是场地——构筑物图。

[6] 规划过程的平衡是比较分析和对细节的改进——一个创造性的综合过程。

[7] 一个优秀的规划本质上不过是一个逻辑思辨过程的记录;一个蹩脚的规划则是考虑失当或者毫无思考的产物。杰出的规划对所有场地因素都给予充分考虑,明确认识各种需求和关系,精心处理所有相互作用的局部。

规划的创新

[8] 规划师们能利用材料、形式、专业符号创作出他们相信会给使用者带来某种可预见体验的物体、空间,以及构筑物。实际上,使用者会根据他们自己对规划元素的感受来进行"再创新",从而产生令人满意的体验。因为我们感知的过程其实是通过我们的感官进行形体再造的过程。对这种现象的理解使我们清楚地认识到设计的创新功能。

规划态度

[9] 我们只有首先投入感情去理解才能创造作品。就说一座购物广场吧,作为设计师,我们得感受场所中加速的节拍、吸引和诱惑、喧嚣以及热闹的场面。我们得注意到小商店别致的陈设,烤饼店那令人馋涎欲滴的视觉和嗅觉刺激;我们脑海中得浮现出五金店拥挤的柜台,还有杂货店里一堆堆的漱口水、香水、指甲刀、热水杯以及果冻豆等。我们得看见市场中成堆的葡萄、柑橘、食用大黄、球芽甘蓝以及香蕉;呼吸得出花棚浓郁的香气;描画得出一架架的折价书,一卷卷的印花棉布,一盘盘的薄荷糖和巧克力奶油。我们得感受到阳光照射到人行道上的刺目明亮,感到入口及拱廊处的阴凉和庇护。我们得感受到拥挤的人群和交通,感受到长椅、树

木,甚至一两处喷泉的闪烁和泼溅。然后,我们就可以开始进行规划了。

[10] 再比如儿童动物园。如果我们想要进行设计,首先应当觉得自己就像是这群小家伙中的一员,跌跌撞撞着,拍着手尖叫着;我们必须理解在这块土地上的欢乐、笑声、喋喋不休、迷惑以及欢闹。我们必须能体会到老鼠城迷你的、吱吱叫的娇小可爱;喷水鲸鱼笨重巨大,空空的肚子像山洞一般,里面还有朦胧的照明。还得想象出:优雅漫步的孔雀高视阔步,仪态万千;摇摇摆摆的鸭子"嘎嘎嘎,嘎嘎嘎"地叫个不停;垂耳的兔子软绵绵、毛绒绒的,就像一团白雪;骑着小马,蹄儿得得,鞍儿悠悠,心中慌慌又陶陶。在做规划时,我们的心一定要像在儿童动物园里,能够看到它,听到它,感觉到它,像一个孩子那样爱恋着自己的天地。

[11] 还有景观路、宾馆广场、车站,以及海滨浴场,如果要对它们进行设计,首先要对它们的特性有所感受。这种自发式的感受我们可以称之为规划态度。成为一个成熟的规划师之前,规划态度就是一种直觉。

Reading Material B
The Planning-Design Process
John Ormsbee Simonds
规划设计过程

[1] **Commission**

Most planning-design interviews and commission awards are based upon experience and reputation.

Experience is gained by education and training, acceptance of increasingly demanding assignments, engaging consultants as required, and doing all necessary research.

Reputation is gained by full, prompt, and excellent performance. The successful professional has a widening coterie of pleased and vocal clients.

Effective professional service agreements embody a clear, simple statement of intent—who does what, how, when, and for how much compensation.

Remuneration for services is usually arranged in one of the following forms:

Lump sum, with phased payments[1]

Time, plus reimbursement[2] for travel, material and related expenses

A percentage fee based on construction costs[3]

Consultation is normally provided on a per diem plus reimbursement basis.

Form of agreement:

A verbal agreement is enough.

A letter of confirmation is better.

A standard professional agreement is better yet.

With large, complex, or long-term projects, as with many public agencies, a detailed legal instrument of agreement is prescribed.

[2] **Research**

(an exercise in gaining awareness)

The basic tool in land planning is a topographic survey, meeting a specification to provide all, and only, the information needed.

规划—设计过程
建筑设计、景观设计、工程

委托	调查	分析	综合	施工	运行
客户需求的说明 服务内容的确定 协议的执行	测量 数据收集 访问面谈 观察 拍照	场地分析 政府条例的分析 限制条件 可能性 策划	草案研究 比较分析 影响评价 调整 充实 实施方法	施工文件拟定 中标合同 施工监理 检查清单	定期访问 调整；改进 运行观察 学习
初次会晤	专业服务合同	底图： 补充文件、数据	综合计划	产生初步规划 并作经费估算	项目完成

The planning design process
Architecture, landscape architecture, engineering

Commission	Research	Analysis	Synthesis	Construction	Operation
Client statement of need Definition of services Execution of agreement	Survey Data collection Interviews Observation Photography	Site analysis Review of governing regulations Constraints Possibilities Program development	Schematic studies Comparative analysis Impact assessment Accommodation Consolidation Method or methods of implementation	Preparation of construction documents Contract award Supervision of construction Punch list checkout	Periodic visits Adjustment;improvement Performance observation Learning
Initial meeting	Professional service agreement	Base maps supporting file data	Comprehensive program	Developed preliminaries and estimate of cost	Completed project

UNIT 12 METHODS OF PLANNING AND DESIGN

Data collection begins with a listing of all materials required together with a notation of the most likely source.

Maps, reports, and other useful data are available in public agencies and planning offices, often without charge.

Interviews with potential users, agency staff members, and public officials not only yield helpful information but also build in an understanding of the project and a sense of contribution.

Research includes the study of past and present examples and a knowledge of innovative trends. It is a continuing process of travel, observation, reading, and experimentation.

Visits to the site are essential. A photographic record keyed to a location map is always beneficial.

Surveys, base maps, and all related information are to be organized into a convenient project reference file kept complete and updated.

[3] Analysis

Supplementary planning information, observations, and notes can be recorded on prints of the topographical survey and overlays.

Constraints such as land use and density limitations, easements, areas of ecological sensitivity, hazards, and difficult terrain or subsurface conditions are noted.

Favorable site aspects and features are also described.

Governmental regulations[4], standards, and requirements are reviewed and underlined.

Finally, in the analysis phase, a comprehensive development program is formulated.

This will respond to the stated intent as modified in the light of the survey information and data obtained. It will include:

A statement of goals and objectives

A summary of preliminary findings

A description of the project components and their interrelationship

Proposals as to conceptual alternatives

An outline of performance standards

[4] Synthesis

Schematic studies are prepared to explore the plan alternatives[5]. These are kept simple and diagrammatic to explain as directly as possible the conceptual idea as it relates to the givens of the site.

As the schematics evolve, they are subjected to a comparative analysis of their positive and negative values and net yields.

Unsuitable schemes are rejected or modified, promising concepts are improved, and other schematic approaches suggested by the reviews are added to the array of contenders.

Insofar as feasible, all constructive ideas and recommendations are accommodated, negative environmental impacts ameliorated, and benefits increased.

When the most likely plan approaches have been delineated and compared, the best is selected for conversion into a developed preliminary plan and an estimate of cost.

[5] Construction

Upon approval of the developed preliminary plan and estimate, detailed construction documents are prepared. These comprise plans, details, specification, and bidding forms[6] to be issued as a package. A final estimate of cost and a cash-flow analysis are also in order.

In form, bids are invited on the basis of either a cost-plus proposal (when top quality is a prerequisite and when conditions are uncertain or changes anticipated) or the lowest responsible bid[7] (when economy and budget limitations are the decisive factors).

Professional services normally include supervision, "observation," or consultation during the bidding, contract award, and construction process.

Supervision is to be firm, fair, and expeditious. Field adjustments[8] are to be welcomed if the project is thereby improved and if all parties have a clear understanding of the nature of the change and its cost implications.

During construction it is well to have the ongoing maintenance superintendent present to gain an understanding of the project installation and conditions.

In advance of construction completion a punch list[9] is provided by the supervisor as the basis for final inspection and acceptance[10].

[6] Operation

Prior to project completion the thoughtful planner will provide the owner with a sheet of instructions or, on large projects, a concise manual to govern continuing operation and maintenance.

Many professional service agreements provide for continuing consultation. In any event, the conscientious planner will return for periodic visits to observe, learn, and advise as to suggested improvement.

There is no better lead to future commissions than a demonstration of continued interest in the project's success and the client's satisfaction.

Excel, and exceed expectations.

[NOTES]

[1] lump sum, with phased payments 总额承包,分期兑现。
[2] time, plus reimbursement 按时计费,外加凭据报销。
[3] construction cost 工程造价。
[4] governmental regulations 政府法规。
[5] plan alternatives 可选方案。
[6] bidding forms 标书。
[7] cost-plus proposal or the lowest responsible bid 附加费用标和最低责任标。
[8] field adjustment 现场调整。
[9] punch list 竣工查核事项表。
[10] inspection and acceptance 检查与验收。

UNIT 12　METHODS OF PLANNING AND DESIGN

[NEW WORDS]

agreement	n.	协议,协定,契约,合同	ongoing	adj.	继续进行的,不断前进(发展)中的
ameliorate	vt.	改善,改进,改良	overlay	n.	叠加图,覆盖图,重叠,覆盖
bid	vt.	投标,出价	per diem		每日,每天,按日
compensation	n.	补偿,报酬	preliminary	adj.	初步的,预备的,开端的
conscientious	adj.	有责任心的,谨慎的,尽责的	prompt	adj.	立刻的,迅速的,准时的
contract	n.	契约,合同	regulation	n.	法规,规则,规章,规章制度
coterie	n.	小圈子,小集团	reimbursement	n.	报销,退款,酬劳,偿付,赔偿
delineate	vt.	描……的外形,画……的轮廓,勾画	reputation	n.	声望,名望,名声,名誉
diagrammatic	adj.	图解的,图表的,概略的	schematic	adj.	要领的,纲要的,示意的,概略的
	n.	示意图,简图,原理图,图解	survey	n.	测量图,测量,勘察
easement	n.	在他人土地上的通行权,地役权	verbal	adj.	口头的(而非书面),言语的
expeditious	adj.	迅速而有效率的,迅速的,敏捷的	vocal	adj.	口头的,口述的,有声的
instrument	n.	正式的文件,文书	without charge		免费,不计价
net	adj.	净的,纯的	yield	n.	收益量,投资利益,利润,税收

[难句]

The successful professional has a widening coterie of pleased and vocal clients. 成功的专家总有一个由满意的并为之宣传的客户组成的大圈子。[引申译法]

[参考译文]

规划设计过程

[1] 委托

多数规划设计的委托和中标都基于规划师的经验和声望。

经验的获得需要通过教育及训练,承担难度不断提高的任务,聘请必要的顾问,以及进行所有的必要的研究。

声望靠充分、快捷、优异的工作表现获得。成功的专家总有一个由满意的并为之宣传的客户组成的大圈子。

高效的专业服务协议包含一个简明扼要的要求陈述:谁做什么,怎么做,什么时间做,以及有多少报酬。

服务报酬通常以下面几种形式中的一种来安排:

总额承包,分期兑现;

按时计费,外加凭据报销,包括旅差费、原料费及相关开支;

以工程费为基数的百分比式付费。

咨询则一般按日付费外加凭据报销。

协议的形式:

通常口头协议已足够；
如有确认文件更好；
标准的专业协议则更好；
对于大型、复杂或长期的项目，如同与许多公共代理人签约一样，一个详细合法的协议样本是必备的。

[2] **调查**

（获得感知的一种实践活动）

土地规划中，基础的工具是符合特定要求的地形测量图，它提供而且只提供需要的信息。

数据收集开始于一份所需资料清单以及可能来源的注释。

地图、报告和其他有用数据一般在一些政府和规划部门常可免费获得。

与一些潜在用户、机构成员及公务员交流，不仅可以获得有用信息，而且可建立对这一项目的理解和支持的氛围。

研究包括过去和现在的实例分析，掌握创新动向。它是外出旅行、观察、阅读及实践的一个连续过程。

实地调查是必不可少的。拍摄照片并使之与位置图相对应是有益的。

测量图、底图，以及所有相关资料应当汇编成一个方便的项目档案，并需保持完整和不断更新。

[3] **分析**

补充的规划资料，观察数据和标注可记录在地形测量图及叠加图上。

记载下一些限制因素，比如：土地利用及密度限制、公用通行区、生态敏感区、危险区、不良地形及地下情况。

场地的优越方面及特征也应描述。

政府法规、标准和要求也要考虑和强调。

最后，在分析阶段，一个深入的发展计划形成了。

它与预定目的相适应并在调查资料和获得数据基础上修改。它应包括：

目的及目标的陈述；
初步成果概述；
项目组成部分及其相互作用关系的描述；
提出可选方案的构思；
项目操作的指标纲要。

[4] **综合**

草案研究是为研究可选方案所准备的。它们要保持简明和图解性，以便尽可能直接解释与特定场地的特性相关的规划构思。

随着规划草案的进展，进一步可以对它们的优缺点及纯收益作比较分析。

不合适的方案将被放弃或要加以修正。好的构思应当采纳并改进。其他讨论中新提出的方案应加入到方案列表中以供比较。

只要有可能，所有建设性的思想和建议都要包括在内，减少负面的环境影响，增进有益之处。

当最有可能的方案已初具轮廓并相互比较过后，选出一个最好的，并转化成初步规划和费用估算。

[5] **施工**

初步规划和概算获批后，详尽的施工文件开始准备。这些文件包括：规划图、详细设计图、明细说明书和准备发送的招标书，最终估算及现金流动分析也应在内。

就形式而言，招标分为两种：附加费用标和最低责任标。前者以工程质量为最高标准，而条件不确定或会有变化；后者则在经 s 费有限时，以经济为决定因素。

专业服务一般包括：监理、"观察"，以及招标、签订合同及建设过程中的咨询。

监理应当严格、公平和迅速有效。关于现场调整,如果这一项目因此得以改进,如果所有人对这一改进的性质和费用都有清楚认识,那么这样的调整是受欢迎的。

在施工中,让工程进度的主管人员理解项目建设情况与条件是很有益处的。

工程完工之前,监理人应当提供一个检查目标清单作为最后检查与验收的基础。

[6] 运行

在项目完成之前,考虑周到的规划师会给业主提供一份说明,或者对大一些的项目提供一个简明的手册,以指导长期运作和维护。

许多专业服务协议提供长期的咨询。无论如何,尽责的规划师会定期回来巡查,进行观察、研究并提供改进意见。

没有什么方法比持续地关注项目成败及委托人的满意度的行动更能吸引未来地委托任务了。

卓越的工作,不断地超越期望。

[思考]

(1) 空间有哪些主要特征?
(2) 概念规划需要考虑哪些内容?
(3) 规划设计过程包括哪几个主要阶段?每个阶段有哪些工作内容?

UNIT 13 LANDSCAPE DETAIL DESIGN PRINCIPLES

TEXT

Site Systems

John Ormsbee Simonds[1]

场地系统

[1] As a logical extension of the principles of site-project unification the concept of site systems deserves special attention. The term implies simply that all site improvements are conceived to be constructed and function in a systematic way.

[2] Natural systems

As a starter with such an approach, the natural systems are preserved insofar as feasible. It has been previously noted that every parcel of land is directly related to the surrounding landscape and may help to provide protection from the winds, storms, and erosion. It may contribute its share of surface and subsurface water. It may help to ensure the continuity of vegetative growth and wildlife habitat. Its ground forms and cover may also give a visual continuity to the landscape. The evident continuation of these relationships helps to assure for the planned development a strong and satisfying tie to the surrounding environs.

[3] Drainage

With few exceptions the natural site provides for storm runoff across its surface without causing erosion. The ground-stabilizing roots and tendrils or living plants knit the soils and absorb precipitation. The fallen twigs and leaves also form an absorptive mat to keep the soil moist and cool the air. The natural swales, streambeds, and river gorges of the undisturbed landscape provide for the most efficient storm-water flow, while marshes, ponds, and lakes provide the ultimate storage and recharge basins. Any alteration to this established network is both disruptive and costly. The movement of materials is required, new storm drainageways must be shaped, and often extensive artificial storm-sewer systems must be constructed. Usually, with the installation of roofs, paved areas, and sewer pipe, the amount and rate of runoff is increased, to the detriment of the project site and downstream landowners.

Experience would suggest that artificial drainage devices be minimized and that they and the natural drainageways be planned together as a balanced system.

UNIT 13 LANDSCAPE DETAIL DESIGN PRINCIPLES

[4] Movement

Planned paths of pedestrian and vehicular movement that oppose the existing ground forms generate the problems and costs of earthwork, slope retention, interception gutters, storm-sewer connections, and the establishment of new ground covers. When such routes are aligned instead to rise and fall with the natural grades, to follow the ridge lines and ravines, or to trace a cross-slope gradient that requires no heavy cuts or fills, they not only are more economical to build but are also better to look at and more pleasant to use.

Well-designed walks, bicycle trails, and roadway also provide interconnecting network of movement that assure regional continuity, are particularly suite to the type of traffic to be accommodated, and take into account all such factors as safety, efficiency, and landscape integrity. Materials, sections, profiles, lighting, signing, and planting are coordinated and designed as an integrated system.

[5] Lighting

Site illumination does many good things. It provides safety in traffic movement and crossings, it gives warning of hazards, and it serves to increase security and reduce vandalism. It interprets the plan arrangement by giving emphasis to focal points, gathering places, and building entrances. It demarcates and illumines paths of interconnection, serving as a guide-on. With accent lighting, fine architecture or site areas of exceptional significance or beauty can be brought into visual prominence.

Well-conceived lighting gives clarity and unity to the overall site and to each subarea within it, especially if planned form the start as a coordinated systems.

[6] Signs

Graphic informational systems are closely allied with site illumination, since the two are usually interdependent and complementary. Street and route lighting obviously must be planned together with the positioning of related directional signs. Often light standards provide support for signs and informational symbols.

Signs, like lighting, are best developed as a hierarchy, each sign being designed in terms of its size, color, and placement to best serve its particular purpose and all existing together as a related family. When the system is kept simple and standardized, the signing gives its own sense of order and clarity to the trafficway pattern and landscape plan.

[7] Planting

Planting of excellence is also systematic. It articulates and strengthens the site layout. It develops an interrelated pattern of open, closed, or semienclosed spaces, each shaped to suit its planned function. Planting extends topographical forms, enframes views and vistas, anchors freestanding buildings, and provides visual transitions from object to object and place to place. It severs as backdrop, windscreen, and sunshield. It checks winter winds. It catches and channels the summer breeze. It casts shadow and shade. It absorbs precipitation, freshens the air, and modifies climatic extremes.

Aside from serving these "practical" functions, plants in their many forms and varieties are also decorative. But even their decorative quality is more pleasing if there is an evident reason behind their selection and use.

Fine plantings, like any other fine work of design, have a fundamental simplicity and a discernible order. Many experienced landscape designers limit their plant lists to a primary tree, shrub, and ground cover and one to three secondary trees, shrubs, and supplementary ground cover—grasses, herbs, or vines, with all other supporting and accent plants comprising no more than a small fraction of the total.

Except in urban settings, the large majority of all plants used will be native to the region and will therefore fit and thrive without special care.

Essentially, each plant used should serve a purpose, and all together should contribute to the function and expressiveness of the plan.

[8] **Materials**

Just as the palette of plant materials is limited in the main to those which are indigenous, so it is also with the materials of construction. Wall stone from local quarries seems most appropriate. Crushed stone and gravels exposed as aggregate, bricks made of local clays, lumber from trees that grow in the vicinity, and mulches made of their chipped or shredded bark all seem right in the local scene. Even the architectural adaptation of the natural earth, foliage, and sky colors relates the constructions to the regional setting.

From small-home grounds to campus, to park, to large industrial complex, site installation and maintenance costs can be reduced and performance improved by the standardization of all possible components, materials, and equipment. Use only the affordable best, therein lies quality and economy.

The reduction of the number of materials used to a small and selective list lends simplicity and unity to the planned development.

[9] **Operations**

All projects must be planned to work, and work efficiently. Each building and each use area of the site must operate well as an entity, and all, together, as a well-organized complex. This can be achieved only if all components are planned together as an integrated system.

[10] **Maintenance**

To be effective maintenance must be a consideration from the earliest planning stages. This presupposes that all maintenance operations have been programmed. It also assumes that storage for the required materials and equipment is provided, that access points and ways are strategically located, that convenient hydrants and electrical outlets are installed, and that maintenance needs are reduced insofar as practical.

It also means that the number of construction materials and components and thus the replacement inventory of items that must be kept stocked are reduced to a workable minimum. This requires standardization of light globes, bench slats, anchor bolts, sign blanks, curb templates, paint colors, and everything else. Usually a reduction in the quantity of items stocked can result in improved quality at significant savings. This is possible only if the maintenance operation is planned from the start as, or converted to, an efficient system.

Unity with diversity is the key to identification signs. Shapes, sizes, and letter forms may vary with the information to be conveyed. Materials, mountings, and colors are usually standardized.

[NOTES]

* 本文节选自约翰·奥姆斯比·西蒙兹(John Ormsbee Simonds, 1913—2005)《景观设计学:场地规划设计导则》(*Landscape Architecture:A Manual of Site Planning and Design*):西蒙兹是20世纪举世公认的景观设计和环境规划的带头人和思想家。他曾在卡内基梅隆大学任教15年,著写了一批脍炙人口的书籍及大量文章。他的匹兹堡环境规划设计事务所(EPD)完成了大批优秀设计作品,其中包括芝加哥植物园、佛罗里达的佩利肯湾,以及弗吉尼亚州立法会的州域环境行动计划。西蒙兹是美国风景园林协会的研究员和前主席,并获该会颁发的最高荣誉奖:ASLA奖章。他还是总统环境特别工作组、佛罗里达州政府自然资源特别工作组的成员,英国皇家设计研究院的研究员。西蒙兹还拥有超凡的个人魅力,曾有一位杰出的建筑史学家这样评价他:"西蒙兹是美国最受人尊敬的风景园林师,不仅因为他以多种方式服务于他的职业,而且因为他具有聪明、诚实和意志力坚强的品质,正是这种品质赋予他与丘吉尔的作风相当的崇高地位。"

[GLOSSARY]

drainage	排水	lighting	照明	signs	标志
graphic informational system	标示系统	planting	种植	site illumination	场地照明

[NEW WORDS]

absorptive	adj.	吸收性的,有吸收力的	enframe	vt.	配上框子,把……装在框内
accent	n.	强调,重点	environs	n.	周围的事物,附近,近郊
aggregate	n.	骨料,集料	foliage	n.	树叶(总称),枝叶
ally	vt.	结盟,结合,联姻	freestanding	adj.	独立的,不依靠支撑物的
anchor	vt.	(把……)系住,(使)固定	gorge	n.	山峡,峡谷
articulate	vt.	清楚地表达,明确表达	gutter	n.	排水管,排水沟,引水槽
backdrop	n.	背景,周围的景物	hierarchy	n.	分级,分层,层次体系
bolt	n.	螺栓,插销,闩	illumination	n.	照明,光照强度,照度
clarity	n.	清楚,明晰,清澈	indigenous	adj.	土生土长的,生来的,固有的
continuity	n.	连续(性),持续(性),衔接	insofar	adv.	在这个范围,到这种程度
curb	n.	侧石,边石,路缘	interception	n.	拦截
demarcate	vt.	定……的界线,区分	interconnection	n.	相互连接,互连
detriment	n.	损害,伤害,造成损害的事物	knit	vt.	(使)紧密地结合,紧凑,编织
discernible	adj.	可看出的,可识别的	lumber	n.	木材,木料
disruptive	adj.	破坏的,破裂的,扰乱的,中断	marsh	n.	湿地,沼地,沼泽地
downstream	adv.	在下游,顺流地	mat	n.	垫子,席子,小地毯
earthwork	n.	土方(工程),挖土,填土,做土方	mounting	n.	底板,座架,装配,安装,裱装

continued

mulch	n.	覆盖层,护根	streambed	n.	河床
outlet	n.	出口,出路	subarea	n.	分区,子区
precipitation	n.	降水,沉淀,沉积	swale	n.	沼泽地,洼地,湿地
quarry	n.	采石场,露天矿场	systematic	adj.	系统的,规划的,有计划的
ravine	n.	峡谷,深谷,沟壑,沟谷	tendril	n.	(植物的)卷须
retention	n.	保持,保留,滞留	therein	adv.	那就是,此即,缘此,其中
runoff	n.	径流,流量,溢流	twig	n.	细枝,嫩枝,小枝
sewer	n.	阴沟,污水管,下水道	unification	n.	统一,一致,联合,标准化
shred	vt.	切成条状,撕碎	vandalism	n.	故意破坏,破坏行为
slate	n.	板条,狭板	vicinity	n.	附近,邻近
stabilize	vt.	(使)稳定,(使)稳固,(使)平衡	vine	n.	藤本植物,攀缘植物
storm-sewer	n.	雨水管,暴雨排水沟	windscreen	n.	风挡,挡风玻璃

[难句]

1. Just as the palette of plant materials is limited in the main to those which are indigenous, so it is also with the materials of construction. 正如品种繁多的植物材料的选用总体上限于那些本地物种,建筑材料也是这样。[一词多义]

2. Use only the affordable best, therein lies quality and economy. 在承受范围内使用最好的,就能做到既保证质量,又经济合算。[加词]

[参考译文]

场地系统
约翰·奥姆斯比·西蒙兹

[1] 作为"场地—项目统一化"原则的合理延伸,场地系统的概念值得特别注意,该术语简明地暗示了任何场地的改进都应系统地进行建设和发挥其功能。

[2] **自然系统**

作为规划途径的第一步,自然系统应尽可能地被予以保护。前文已经谈及每一块土地都与周边景观直接相关,并有可能使场地免受大风、暴雨及侵蚀的危害,能贡献地表水、地下水,有助于保证植被生长和野生动物生境的连续性。地形和地被物还可提供景观上的视觉连续性。这些关系间明显的连续性有助于确保规划地开发与环境保持紧密且令人满意的联系。

[3] **排水**

雨水流过自然地表,一般不会引发侵蚀,极少数情况除外。植物通过密布的根系固定土壤,吸收降水。枯枝落叶也会形成一种能保持土壤水分、冷却空气的吸收性垫层。自然洼地、河床、河谷等未受破坏的景观可产生最有效的暴雨径流,沼泽、池塘、湖泊则充当基本的蓄洪和补水凹地。对这种成型网络的任何改变都是有破坏性、高代价的。因为这时需要运输所需物料、筑建新的排暴雨通道和大型的暴雨人工下水系统。一般情况下,随着屋顶、铺装地面和下水管道的增加,排水的数量与速度会大大增加,以致对项目基地和下游土地所有者造成损害。

经验表明了人工排水设施应尽可能少用,人工与自然排水通道应同时规划,形成一个平衡的系统。

[4] 运动

不适合现状地形的人行和车行规划线路会导致填挖土方、保持坡度、跨越沟渠、连接下水道以及新建地被等诸多问题和花费。相反,要是这些线路依据自然坡度、沿着山脊线和沟谷,或是采用不需太多填挖的沿等高线方式来安排起伏的话,那么不但在修筑上更为经济,而且更美观和适用。

精心设计的人行步道、自行车道和车行道同样构成保障地域连续性的运动连通网络,它们尤其要与需承担的交通类型想匹配,还要考虑诸如安全、效率及景观统一性的多方因素。材质、剖面、外形、照明、标识以及种植应作为一个整合的系统进行协调和设计。

[5] 照明

场地照明设施可以带来很多好处:在交通行进及穿越时保证安全,发出危险警示,增加安全性,减少破坏行为;通过对设计重点、人流聚集地和建筑入口的强调来体现设计意图;它划清并照亮了相互连接的路线,起着导向作用;通过重点照明,美观的建筑、有特别意义或美感的场地区域可形成视觉的聚焦点。

精心构想过的照明使整体场地及其内部的各个分区更具清晰性和统一性,尤其是在规划之初就将其作为一个协调的系统来考虑。

[6] 标志

标志系统与照明系统密不可分,因为这通常是相互依存、互为补充的。显然,街道照明应与相关的方位标志的定位一起规划,灯柱常常充当标志及消息牌的基座。

标志像照明一样最好组织成一个等级序列:每一标志的尺寸、色调和布置的设计都应服务于各自的特定目的,整体则统一表现为一个相互关联的群体。只有整个系统保持简明、规范,标志才能为交通格局和景观布局提供秩序和明晰的信息。

[7] 种植

优秀的种植同样是系统化的。种植能够表达和强调场地的布局,构成开放空间、封闭空间或半封闭空间相互联系的格局,使每一空间都与其规划功能相适。通过种植,可以拓展地形,可以构成框景,加固构筑物,提供单体与单体、地方与地方之间的视觉过渡带;充当背景、屏风、阳棚;既能阻挡冬季寒风,又能疏导夏季微风,还能洒下阴影,带来绿荫;吸收降水,清新空气,调节气候。

除了这些实用功能之外,多姿多彩、种类众多的植物还可带来视觉上的愉悦。如果选用某种植物有某种理由的话,这种植物的美感还会大大增强。

好的种植,如同其他任何优秀的设计成果:简洁明快,秩序井然。许多有经验的景观设计师往往遵循以下模式配置植物:一种基调树、灌丛、地被植物,1~3种调配树种、灌丛和辅助性的地被;禾本草、阔叶草或藤蔓,以及其他占很小一部分的辅助性植物。

除非在城市环境中,大部分选用的植物应是本土植物,无需特别护理就可繁茂生长。

基本上,每种植物应服务于一种目的,所有植物的整体应有利于规划的功能和表达。

[8] 材料

正如品种繁多的植物材料的选用总体上限于那些本地物种,建筑材料也是这样。墙石从当地采石场选用,碎石和砾石作为骨料,砖用当地黏土制成,木料采自周边地区的树木,削下的树皮切碎还可作为覆盖料,这些在当地景观中看来很合适。进而,把自然中的土地、枝叶和天色作为建筑形、质、色之源也能把建筑物与区域背景联系起来。

从小小的家庭院落到校园、公园到大型产业综合体,尽可能采用的标准化部件、材料、设备可以减少场地施工和维修费用,提高工作性能。在承受范围内使用最好的,就能做到既保证质量,又经济合算。

使用材料的数量缩减到一个精确的小范围,有助于规划地开发的简洁性和统一性。

[9] 运作

所有项目经过规划都应能运作，并且能高效运作。每一建筑，每一用地不仅应能作为良好的单体，还应能一起构成一个有机的整体。这只有当所有部分作为一个整体的有机组成加以规划时才能得以实现。

[10] 维护

有效地维护在规划初期就应加以考虑。这需要假设已经做好了所有维护操作的计划，所需材料、设备的存放场所已经有了，出入口和通路在策划上已做了定址，便利的水电接口已安装好了，维护的需要已降至可实际运作的程度。

这也意味着建材、部件、存货的数量也减少到可操作的最小量。这需要灯罩、长椅板条、固定栓、标志板、道牙模铸、涂料及其他东西的标准化。通常，存储物品数量的减少既可以节约，又能提高质量。当然，这只有当维护操作在一开始就被作为或转化为一个高效的系统来规划时，才得以实现。

变化中的统一是识别性标牌的关键。形状、尺寸、字符样式可随所传达信息的变化而变化，材料、框架和颜色通常应标准化。

Reading Material A

Significance of Landform[1]

Norman K. Booth[2]

地形的重要性

[1] Landscape architects utilize a variety of physical design elements to meet their objectives in creating and managing outdoor spaces for human use and enjoyment. Among these elements, landform is one of the most important and ever present. Landform serves as the base for all outdoor activity and may be thought of as both an artistic and utilitarianelement in its design applications.

[2] "Landform" is synonymous with "topography" and refers to the three-dimensional relief of the earth's surface. In simple terms, landform is the "lay of the land". At the regional scale, landform may include such diverse types as valley, mountains, rolling hills, prairies and plains. These landform types are typically referred to as "macrolandforms". At the site scale, landform may encompass mounds, berms, slopes, level areas, or elevation changes, via steps and ramps, all of which may be generally categorized as "microlandforms". At the smallest scale, "minilandforms" might include the subtle undulations or ripples of a sand dune or the textural variation of stones and rocks in a walk. In all situations, landform is the surficial ground element of the exterior environment.

[3] Landform has great significance in the landscape because of its direct association with so many other elements and aspects of the outdoor environment. Topography affects, among other things, the aesthetic character of an area, the definition and perception of space, views, drainage, microclimate, land use, and the organization of functions on a particular site. Landform also has an impact on the role and prominence of other physical design elements in the landscape including plant material, pavement, water, and buildings. All these other physical design elements plus additional components utilized in the landscape must at some point come to rest on, and relate to, the ground's surface. Few items seen or manipulated as design elements in the outdoors float in space consequently, an alteration

in the landform at a particular point also means a change in the spatial delineation, appearance, and sometimes the function of the other physical elements also located at this point. The shape, slope, and orientation of the ground's surface influences everything on it and above it. While landform has a rather direct impact on all the other physic design elements, it is not necessarily the most important of all. This of course varies greatly with the particular situation and scope of consideration.

[4] Because all other design elements must at some point relate to the ground plane, landform is the one common component in exterior environment. It can be considered a thread that ties all the elements and spaces of the landscape together into a continuum that ends along the horizon or at water's edge. In regions and sites of level topography, this commonality can function as a unifying factor, visually and functionally connecting other components in the landscape (see Fig. 1). Conversely, this unifying capability is lost in hilly and mountainous areas where ridges and high points tend to segment the land into separate spaces and use areas.

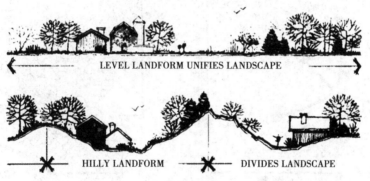

Fig. 1 Landform

[5] Landform has other noteworthy influences on the outdoor environment as well. Landform can be thought of as establishing the underlying structure of any given portion of the landscape. It acts like the framework of a building or the bones of an animal; it formulates the overall order and form of the environment. Other elements are seen as being a covering or facade on top of this frame. Thus in evaluating a given site during the site a-

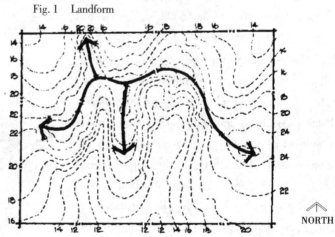

Fig. 2 Development of site should follow centeral ridge and be linear

nalysis phase of the design process, it is often wise to study the topography early, particularly if it is not flat or uniform. The sites landform can suggest to the designer the overall organization and orientation that various uses, spaces and other elements should take to be compatible with the inherent composition of the site, as shown in Figures 2 and 3. In both figures, the size and shape of the site are the same. Yet the landform configuration in Figure 2 suggests a linear layout of elements to follow the ridge line while in Figure 3 the landform permits a more sprawling and multidirectional arrangement. An experienced designer is able to skillfully "read" the topography of a site or region and interpret its implications for design or management of that area.

[6] Similarly, landform can be considered as a setting or stage for the placement of other design elements and functions. It is the foundation for all exterior spaces and land uses. This is why the ground surface is often referred to as the "base plane" (i.e., the starting point for the evolution of a design solution). As such, one of the first tasks in the design process is typically to obtain a "base sheet" or topographic plan of the site. As in Figure 4, this base sheet usually shows contour lines, property lines, existing structures, roads, and sometimes vegetation. The base sheet can be obtained from one-site property and topographic surveys or by aerial surveys of the site.

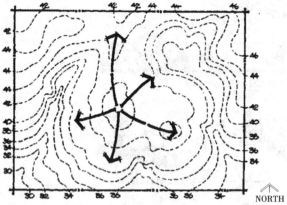

Fig. 3　Development of site should be placed on high point and can be multidirectional

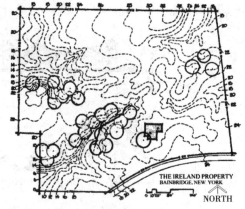

Fig. 4　Example of a base plan

[7] Having obtained a base sheet of the site's landform, the designer is then able to utilize it as the foundation for developing a design solution. All design concepts and proposals are prepared and studied as tracing paper overlays on top of the base sheet. One of the early steps in this process is to develop a functional diagram of the proposed uses on the base plane as illustrated in Figure 5. In doing this, the designer studies the relationship of the proposed uses to each other as well as to the existing landform. This diagrammatic organization of the base plane or stage of the site is critical because its layout affects the order, scale and proportion, character or theme, and functional quality of the outdoor environment. A well-established arrangement of the base plane provides a sound footing for the integration of other design elements including vertical and overhead planes. On the other hand, a poorly organized base plane contributes to problems throughout an environment that usually cannot be easily compensated for by

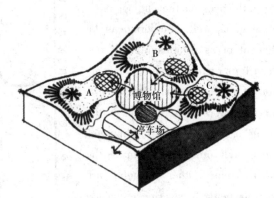

Fig. 5　Functional diagram on base plane

skillful design during subsequent phases. It must be pointed out here, however, that the landscape architect's attention should not be limited only to the base plane even though its arrangement is crucial.

Ultimately, the designer must be concerned with the three-dimensional experience of a of a design and how it will feel to be in it.

[8] The significance of landform to the landscape architectural profession is further emphasized by the name itself: landscape architecture. *Webster's New Collegiate Dictionary*[3] describes "landscape", among other definitions, as "a picture representing a view of natural inland scenery". It further defines "landscape" as "the landforms of a region in the aggregate". The word "land" by itself is defined by Webster's as "the surface of the earth and all its natural resources". It can be clearly seen from these particular definitions that earth and soil along with their three-dimensional from are inherent in the concept of landscape. "Landform" and "landscape" are mutually supportive in terms of definition. If "architecture" is then defined as "the art or science of building," it is possible to interpret "landscape architecture" as the art or science of building on and with the earth's surface.

[9] One of the landscape architect's unique and distinguishing characteristics is the ability to work sensitively with and manipulate landform. While several other professions also deal with various aspects of landform, none of them does so with the same depth of knowledge and skill as the landscape architectural profession includes the modification and stewardship of the earth's surface for our use and enjoyment.

[**NOTES**]

[1] 本文节选自诺曼·K.布思的《风景园林设计要素》(*Basic Elements of Landscape Architectural Design*)。

[2] Norman K. Booth 诺曼·K.布思,美国著名景观设计师、美国风景园林师协会(ASLA)会员、俄亥俄州大学教授。

[3] *Webster's New Collegiate Dictionary* 即《韦伯大学字典》,韦式字典是指以"Merriam-Webster"为商标的一系列字典,通常简写为(M-W)。韦式字典是美国人自己出的字典,在权威性方面相当于中国的新华字典。韦式字典对单词的英文解释可以帮助考生理解单词的精确含义,同时避免受一些中文释义的误导。

[**GLOSSARY**]

berm	坡台,阶地	landform	地形,地貌	pavement	人行道,铺过的路面
exterior	外部	layout	布局,安排,规划图	ramp	斜坡,坡道
flat	平地	macrolandform	大地形	ridge line	脊线
high point	顶点,极限	microlandform	微地形	slope	倾斜,斜坡
hilly	丘陵的,多小山的	minilandform	小地形		

[NEW WORDS]

aesthetic	adj.	美的,美学的,具有审美意味的	overlay	v.	覆盖,覆盖物,轮廓纸
base sheet	n.	底图,基础图	prairie	n.	(尤指北美的)大草原,大牧场
component	n.	成分,组件	prominence	n.	突出,显著,突出物,卓越
continuum	n.	连续体,统一体	ripple	n.	涟漪,涟波
definition	n.	定义,清晰度,解说	rolling	adj.	山丘的,起伏的,绵延的
delineation	n.	描绘	scope	n.	范围,余地,视野,眼界
drainage	n.	排水系统,排水,排水区域	segment	vt. &vi.	分割,划分
dune	n.	(由风吹积成的)沙丘	sprawling	adj.	蔓延的,杂乱无序伸展的
encompass	vt.	围绕,包围,包含或包括某事物	stewardship	n.	管理工作,管事人之职位及职责
facade	n.	外表,建筑物的正面,假象	subtle	adj.	微妙的,精细的,敏感的
framework	n.	框架,骨架,结构,构架	synonymous	adj.	同义词的,同义的,类义的
horizon	n.	地平线,视野,眼界,范围	textural	adj.	组织上的,肌理的
inherent	adj.	天生,固有的,内在的	thread	n.	线,线状物,路线,思路
manipulate	vt.	操纵,操作,处理,巧妙地控制	three-dimensional	adj.	三维的,立体的,空间的
mound	n.	土堆,土丘,一大堆	topography	n.	地貌,地形
multidirectional	adj.	多方面的,多向的	undulation	n.	波动,波荡
noteworthy	adj.	值得注意的,显著的,重要的	utilitarian	adj.	功利的,实用的,有效用的
orientation	n.	方向,定向,适应,向东方	visually	adv.	形象化地,外表上,看得见地

[参考译文]

地形的重要性
诺曼·K.布思

[1] 风景园林师通常利用种种自然设计要素来创造和安排室外空间以满足人们的需要和享受。在运用这些要素进行设计时,地形是最重要也是最常用的因素之一。地形是所有室外活动的基础。同时也可以认为它在设计的运用中既是一个美学要素,又是一个实用要素。

[2] "地形"是"地貌"的近义词,意思是地球表面三度空间的起伏变化。简而言之,地形就是地表的外观。就风景区范围而言,地形包括如下复杂多样的类型,如山谷、高山、丘陵、草原以及平原。这些地表类型一般被称为"大地形"。从园林范围来讲,地形包含土丘、台地、斜坡、平地,或因台阶和坡道所引起的水平面变化的地形。这类地形统称为"小地形"。起伏最小的地形称"微地形",它包括沙丘上的微弱起伏或波纹,或是道路上石头和石块的质地变化。总之,地形是外部环境的地表要素。

[3] 在景观中,地形有很重要的意义,因为地形直接联系着众多的环境因素和环境外貌。此外,地形也能影响某一区域的美学特征,影响空间的构成和空间感受,也影响景观、排水、小气候、土地的使用,以及影响特定场地中的功能作用。地形还对景观中其他自然设计要素的作用和重要性起支配作用。这些要素包括植物、铺地

材料、水体和建筑。所以，所有涉及要素和外加在景观中的其他因素都在某种程度上依赖地形，并相联系。可以说，几乎任何设计要素都与地面相接触。因此，某一特定环境的地形变化，就意味着该地区的空间轮廓、外部形态，以及其他处于该区域中的自然要素的功能的变化。地面的形状、坡度和方位都会对在依附其上的一切因素产生影响。不过，虽然地形对其他设计要素有着直接的影响，但它尚不足以称为所有因素中最重要的因素。当然，这一切要取决于特定的场所和对重要性的看法。

[4] 由于其他设计要素必须在不同程度上与地面相接触，因而地形便成为室外环境中的基础成分。它是连接景观中所有因素和空间的主线，从而使它们一直延续到地平线的尽头或水体的边缘。在平坦的地方，地形的这一普遍作用便是统一和协调，它可以从视觉和功能方面将景观中的其他成分交织在一起（图1）。相反，在丘陵或山区，地形的统一作用便失去了效果。因为在这些地区，山脊和高地常常将整个区域分割成各个独立的空间和用地。

[5] 地形对室外环境还有其他显著的影响。地形被认为是构成景观任何部分的基础结构因素。它的作用如同建筑物的框架，或者说是动物的骨架。地形能系统地制定出环境的总顺序和形态。而其他因素则被看做是叠加在这构架表面的覆盖物。因此在设计过程中的基址分析阶段，正确估价某一已知场地时，最明智的做法是首先对地形进行分析研究，尤其是该地形既不平坦，又不均匀时。基址地形的分析，能指导设计师掌握其结构和方位。同时也暗示风景园林师对各不同的用地、空间以及其他因素与场地地形的内在结构保持一致，如图2和图3两图所表示的地形的面积大小都相同，形状都相似，但图2所表示的地形结构，表明其设计要素应以线型排列布局，顺应瘠地的走向，而图3表明该地形容许各要素放射性和多向性布局。有经验的设计师完全能熟练地"读懂"一场地或一区域的地形图，并能理解那一地区对设计或布局的意义。

[6] 另外，地形还可作为其他设计因素布局和使用功能布局的基础或场所，它是所有室外空间和用地的基础。这也就是为什么常称地面为"基础平面"的缘故。正因为如此，设计程序中首要任务之一，通常是要绘制"基础图"或场地的地形图。如图4所示，这种原地形图通常绘有等高线、地界线、原有构筑物、道路及现存的植物。原始地形图可通过现场勘测、地图测绘或航测等方式绘制而成。

[7] 有了原地形图，设计师方能将其作为进行规划设计的基础。所有的设计思想和方案，都可以在覆盖于原地形图上的透明纸上进行研究推敲和绘制。在该过程中的第一步，是在原地形图上大致画出用地的功能分区图，如图5所示。在此基础上，设计师才好研究各用途之间的相互关系，以及它们与原有地形的关系。这种功能分区布局图是很重要的，因为它的布局会影响室外环境的序列、比例尺度、特征或主题，以及环境质量。一张结构合理，布局完美的平面图，可以为其他设计要素，包括垂直面和顶平面的统一奠定基础。另一方面，不好的平面布局，会给整个环境带来许多问题，而这些问题在以后的设计阶段，即使使用巧妙的设计也难以弥补。不过，在此还应指出，尽管平面布局图关系重大，但风景园林师的注意力也不能完全被限制在这上面，而应考虑到设计的三维空间，以及实际的感觉效果如何。

[8] 对于风景园林职业来说，地形的含意被其职业的名称进一步突出强调了。风景园林一词在《韦伯大学字典》中对其含义是这样描述的："地球的表面以及它所有的自然资源"。从以上特殊定义可以清楚地看到，陆地和土壤它们的三维形式都是风景园林概念的固有性质。从定义而言，"地形"和"风景园林"一词互为联系。如果说"建筑学"的定义为"建造房屋的科学或艺术"的话，那么也就有理由将"风景园林"解释为在地球表面和使用地表进行营造的艺术或科学。

[9] 由此看来，风景园林师独特而显著的特点之一，就是具有敏锐地利用地形和熟练地使用地形的能力。虽然其他一些职业也与地形的各个方面有关联，但没有一个能像风景园林师那样，具有透彻的知识和技能去使用地形。此外，风景园林业的职业特点就是为公众的使用和享受而改变和管理地球表面。

Reading Material B

Spatial Characteristics of Plants[1]

Nick Robinson

植物的空间特性

[1] The spatial characteristics of plants are those that contribute to the space structure of the landscape. They include habit, crown shape, foliage density and speed of growth, and en masse they determine the spatial composition of the planted environment.

Spatial Functions of Plants in the Human Landscape

[2] When we are designing spaces for people, the size of plants relative to the dimensions of the human figure is critical. Simply to distinguish areas on a plan by canopy height amounts to an important design stage, because it is height that determines much of the spatial framework and controls vision, movement and physical experience.

[3] Danish landscape architect Preben Jakobsen[2] identified the most useful size categories for the designer as ground level, up to knee height, knee to waist height and below or above eye level (Jakobsen, 1977). The kinds of plants that fall within these ranges are as follows:

Canopy height	Plant Type
Ground level	Mown grasses and other turf plants, ground-hugging and carpeting herbaceous plants and shrubs
Below knee height	Prostrate and dwarf shrubs, sub-shrubs, low-growing herbaceous plants
Knee-waist height	Small shrubs and medium growing herbaceous plants
Waist-eye level	Medium shrubs and tall growing herbaceous plants
Above eye level	Tall shrubs and trees

[4] When it comes to actual dimensions, these heights will, of course, vary for different people. For adults, this variation will be only marginal and would rarely affect our choice of species. For children of different ages and for people in wheelchairs, however, the height difference would be significant, and we must take this into account and allow for their different spatial experience.

Ground-level Planting (Carpeting Plants)

[5] This lowest growing vegetation forms a foliage canopy very close to ground level, and often not more than a few centimeters thick. Plants include grasses and other turf species when mown or grazed, absolutely prostrate shrubs and creeping herbaceous plants. Its primary spatial role is as a "floor" that allows both free vision and movement. This enables it to perform a number of roles:

- On even, firm ground carpeting plants can provide a pedestrian circulation surface, although less hard-wearing than a pavement. The most wear tolerant species include many of the turf grasses that, when grazed or mown regularly, form surfaces suitable for relaxing, walking, play, sport, cycling and occasional vehicles. This durability accounts for much of the value and popularity of lawns, meadows and other grasslands in both public and private landscapes.

• A uniform carpet of mown grass or ground-hugging, smooth-textured groundcover can be used to enhance the visual effect of ground modeling by closely follow the contours. Species include prostrate chamomile (Chamaemelum nobile "Treneague") or piripiri (Acaena sp.). Breaks of slope can be emphasized by a change to a groundcover of contrasting foliage.

• Ground level vegetation can be used to make two-dimensional patterns. Carpets of foliage, used alone or combined with boulders, gravel and paving materials, form a tapestry of colour, texture and pattern across the ground surface.

Shrubs and Herbaceous Plants below Knee Height (Low Planting)

[6] Shrubs and herbaceous plants that form a higher canopy but still below knee height have further possibilities in spatial design. Many of them come within the category of "groundcover", that is, species that are well adapted to the local conditions and competitive enough to exclude most of the unwanted, self-colonizing "weed" plants. In addition to this labour-saving benefit, low planting has the spatial role of allowing freedom of vision while defining an edge and deterring (though not preventing) movement.

• Low planting can, when used by itself, form a visual platform or ground plane like carpeting plants.

• It can be combined with taller herbaceous species, shrubs or trees growing up through it. This situation is like a foundation or wash in painting, or a "ground" against which the "figure" is to be seen. In this way, low planting can give a common ground or platform that unifies other planting and elements in a composition.

• Many prostrate species that form low groundcover will trail down walls and banks and form hanging curtains (prostrate rosemaries—Rosemarinus officinalis—are a classic example). Trailers and climbers can be planted to form a continuous mantle of foliage over vertical and horizontal surfaces. Foliage will cascade down banks and walls and flow over flatter ground, masking the angles between vertical, horizontal and inclined planes. By clothing new and old alike, climbers can give a sense of belonging and maturity to new structures or earthworks that have been inserted in an established landscape.

• Low planting has an essential role at the edges between hard and soft landscape and between soft landscape areas of differing uses. Tall shrubs need room to spread laterally without encroaching on circulation space. Low planting can provide a groundcover over which the taller species extend freely without the need for frequent cutting back or shaping. If this groundcover spreads over pavement or grass some incidental or natural "pruning" will result from tramping. Where traffic is light, occasional trimming is needed.

Knee to Eye Level Planting (Medium Height Planting)

[7] Planting that grows to between knee height and eye level can have a similar design role to a low wall, fence or rail. It becomes a barrier to movement and can be used to limit access but it leaves views open and makes little difference to sunlight. This opens up a number of spatial uses for medium-

height planting.

- It can separate areas for safety reasons: for example, keeping people or vehicles away from steep slopes, water or from each other.
- It can be used to acknowledge and emphasize desire lines or pathways where visual enclosure is not wanted.
- It can be used to maintain a distance between people and buildings and other private areas, in this way giving privacy while not growing above windowsill level and reducing light.
- It can define a building cartilage fringing a building or other structure can visually anchor it to the ground and link it to the surrounding landscape. This is particularly important when a building or other structure is introduced into a landscape characterized by generous existing vegetation.

Planting Above Eye Level (Tall Shrubs/Small Tree Planting)

[8] Shrubs and small trees with a canopy extending above eye level form a visual and physical barrier. So tall planting with a close knit canopy can, in a similar way to a wall or fence, separate, enclose, screen and shelter on a smaller scale than is possible with larger tree planting.

- In the human scale landscape of parks, gardens, courtyards, streets and playgrounds tall planting gives privacy and shelter and screen intrusions like car parking, service areas and refuse bins.
- Like a wall or fence, tall planting can make a backcloth to ornamental planting such as herbaceous borders and display beds. Clipped "formal" hedges have traditionally played this role in gardens, but looser shrub planting can also be effective. Classic hedging plants include yew, Taxus baccata, beech, Fagus sylvatica and hornbeam, Capinus betulus for tall hedges in northern Europe. Monterey cypress, Cupressus macrocarpa, and totara, Podocarpus totara, make fine clipped hedges in warmer climates.
- Because of its size, tall planting can play an accompanying role to buildings. Its visual mass is similar to small buildings so it can be used to balance areas of their masonry or cladding.
- An isolated pair of tall shrubs or a gap in mass planting creates a frame. It can frame a whole vista or attract attention to a focus or landmark. This kind of arrangement not only focuses attention, but also invites exploration. Like an arch or gateway, it suggests a different place to be discovered.
- When planted as individuals or small groups, choice tall shrubs have the size and presence to act as specimens and a feature or visual focus within a human scale landscape.

Tree Planting

[9] The sizes of trees are of the same order of magnitude as buildings, roads, bridges and smaller industrial developments. Tree planting can therefore be used for screening, separating, sheltering, enclosing, accompanying and complementing these larger structures. When tree species grow freely to produce a clear main stem or bole with their canopies above head height they leave the space above the ground open except for the vertical pillars of their boles. This offers a quite different type of spatial element.

[10] Mature heights of trees range from about 5 meters in species such as weeping pear (Pyrus sal-

UNIT 13   LANDSCAPE DETAIL DESIGN PRINCIPLES

icifolia "Pendula") and akeake (Dodonaea viscosa), to over 40 meters in European ash (Fraxinus excelsior), New Zealand kahikatea (Dacrycarpus dacrydioides), and some conifers from the west coast of North America and many Australian eucalypts (Eucalyptus species, especially E. regnans the mountain ash). For design purpose it is helpful to divide trees into small: mature height 5-10 meters; medium: 10-20 meters; and tall: 20 meters.

● Small trees are of similar height or lower than the majority of buildings of two storeys, so their influence in the urban environment is mainly local to the spaces between buildings.

● Medium trees can create spaces that contain smaller buildings and therefore have a greater effect on the spatial structure of urban landscape.

● Tall trees are less common in urban areas because of the space they demand, although naturally tall growing species are often planted in streets and gardens only to be lopped or pruned once they begin to shade or dominate nearby buildings. The size of trees over about 20 meters enables them to form the part of the primary spatial structure of streets, squares and parks. In the rural landscape large trees create a large-scale framework.

● Medium and tall tree planting can play a crucial role in integrating massive industrial buildings, like power stations, into the surrounding landscape. Tree belts and plantations enveloping and extending outwards from such sites provide screening of near distance views. From greater distances, although they cannot hide structures on the scale of cooling towers or turbine houses they can visually anchor them to their supporting landscape and screen the lower level ancillary development, temporary buildings and car parks. This is a vital landscape role because the low-level clutter is often the most disturbing part of large-scale industry.

● The ability of trees to screen and obscure views from further away than shrub planting can be made use of to manipulate views as the observer moves through the landscape. Carefully located gaps in planting open up vistas or frame a focus at just the right moment. Like a window or an archway, a frame of branches or foliage directs attention and focuses the mind on what is beyond it.

● A single specimen or small group of trees, on the other hand, itself acts as a focus. Being an isolated object, it occupies a small area in our field of vision and our eye trends to rest on it. A tree with a distinctive feature such as autumn colour or picturesque habit will make a particularly notable focus. Large tree specimens or groups have this effect at some distance and so provide focus and landmarks in the largest-scale rural landscape.

● When single specimens or small groups of trees accompany buildings the relationship between the form of the tree and of the building can be interesting. Humphrey Repton[3] formulated a rule, in the picturesque tradition, prescribing which tree forms best accompanied different architectural styles. He recommended that buildings in the classical style with broad, stable proportions and shallow roof angles, be accompanied by the resting lines and upright forms of fastigiated trees such as spruces or firs. Conversely, the rising pinnacles and steep pitched roofs of the Victorian Gothic Revival would be complemented by stable rounded or horizontal spreading trees such as cedar of Lebanon, English oak or chestnuts.

● A further architectural role of tree planting might be the linking of varied building styles. A

· 235 ·

simple, regular line of one species can provide uniform frontage or a free-standing counterpoint to an architectural facade. Its continuity can bind together different building styles so that the architectural variety adds interest within a unifying green framework.

[**NOTES**]

[1]　本文节选自尼克·罗宾逊(Nick Robinson)的 *The Planting Design Handbook*。
[2]　Preben Jakobsen 普雷本·雅各布森:丹麦景观设计师,其作品有布罗德沃特公园、法恩伯勒技术学院、塔街、阿登·特提尔瑞学院花园等。
[3]　Humphry Repton 亨弗利·雷普顿(1752—1818):英国近代风景园林设计师,最先提出了在园林中应用折中风格。

[**GLOSSARY**]

architectural facade	建筑立面	hedge	绿篱,树篱
archway	拱门,拱道	landmark	地标,界标,里程碑
boulder	卵石	masonry	石工工程,砖石建筑
cladding	覆盖层	pavement	铺装,路面,铺砌层
durability	耐久性,持久性	Victorian Gothic Revival	维多利亚哥特式复兴风格
gravel	砾石	vista	视景线,透视线,街景

[**NEW WORDS**]

akeake	n.	阿克树	cedar	n.	雪松
allow for	v.	考虑到,体谅,留出,给予	chamomile	n.	黄春菊
amount to	v.	意味着,发展成,共计	chestnuts	n.	栗树
anchor	vt.	抛锚,(把……)系住,(使)固定	cladding	n.	覆层,包层,骨架外墙,覆盖层
ancillary	adj.	辅助的	composition	n.	结构,构造,组成
archway	n.	拱门,拱道	conifer	n.	针叶树,松柏类
ash	n.	白蜡树	counterpoint	n.	对应物,对位法,旋律配合
beech	n.	山毛榉	crown	n.	王冠,花冠,树冠
bin	n.	箱子,仓,容器	crucial	adj.	决定性的,至关重要的,关键性的
bole	n.	树干	cypress	n.	柏树
canopy	n.	(树)冠,冠幅	deter	vt.	阻止,制止,威慑,使不敢
carpeting	n.	毛毯,地毯	distinctive	adj.	有特色的,出众的
cartilage	n.	软骨	dwarf	adj.	矮小的
category	n.	种类,范畴,类别	en masse	adv.	在一起,一同

UNIT 13　LANDSCAPE DETAIL DESIGN PRINCIPLES

continued

encroach	vi.	侵犯,蚕食,侵蚀	mow	vt.	刈,割,修剪
envelop	vt.	包围,笼罩,包住	obscure	adj.	含糊的,难解的,模糊不清的
eucalypts	n.	桉树	ornamental	n.	装饰品
facade	n.	(房屋的)正面,屋前空地	pillar	n.	柱,支柱,栋梁
fastigiate	adj.	锥形的,倾斜的,圆束状的	pinnacle	n.	顶峰,顶点,极点,尖顶,尖塔
firs	n.	冷杉	piripiri		新西兰刺果薇
foliage	n.	植物的叶子(总称),叶子及梗和枝	pitched	adj.	(屋顶)有坡度的
formulate	vt.	用公式表示,确切地阐述	proportion	n.	比例,均衡,相称,协调
free-standing	adj.	独立的,不需依靠支撑物的	prostrate	adj.	卧倒的,俯卧的,匍匐的
frontage	n.	(建筑的)正面,前面	prune	vt.	修剪(树木等),整枝,减少,删除
herbaceous	adj.	草本的,绿色的,叶状的	rosemarie	n.	迷迭香
hornbeam	n.	角树	shallow	adj.	浅的
kahikatea	n.	罗汉松	specimen	n.	范例,样品,标本
knit	vt	编织,结合,弄紧	spruces	n.	云杉
labour-saving	n.	省工,节省劳力	tapestry	n.	挂毯,织锦,绣帷
laterally	adv.	横向地,在侧面	totara	n.	桃拓罗汉松
lop	vt.	砍掉,剪去,削去(树枝等)	tramp	vi.	踩,践踏,步行,徒步
magnitude	n.	大小,数量,等级	turbine	n.	涡轮机,汽轮机
mantle	n.	覆盖物,披风,外罩	turf	n.	草皮,草皮块
marginal	adj.	不重要的,微小的,边缘的	weed	n.	杂草,水草,烟草
masonry	n.	石工工程,砖瓦工工程,砖石建筑	weeping	adj.	垂枝的,树枝低垂的
massive	adj.	整块的,宽而大的,结实的	yew	n.	紫杉
mature	adj.	成熟的,壮年的,成年的			

[参考译文]

植物的空间特性

[1]　植物的空间特征是指那些对景观空间结构有影响的因素。它们包括生长习性、冠形、枝叶茂密程度、生长速度,所有这些决定了种植环境的空间构成。

人工景观中植物的空间功能

[2]　当我们在为人们设计空间时,植物的大小尺寸与人体尺度的对比、关联是至关重要的。在规划设计中,通过植物冠幅的高度来界定规划区域是一个重要的设计阶段,因为这些高度决定着区域的空间结构,控制视线、行为活动和物理运动。

[3]　丹麦景观设计师普雷本·雅各布森将最常用的植物大小的种类分为地表高度、地表至膝高、膝盖至腰部高度、腰部到视线以及视线以上高度(1977)。在这些范围内的植物种类如下:

· 237 ·

冠幅高度	植物类型
地表高度	草坪和其他草皮植物,地被植物,地表草本植物和灌木
地表至膝高	匍匐类植物和低矮的灌木,小灌木,低矮的草本植物
膝盖至腰部高度	小灌木和中等高度的草本植物
腰部到视线高度	中等高度的灌木和较高的草本植物
视线以上高度	高灌木和乔木

[4] 当然,在实际的设计尺寸中,这些高度会随着人的不同而有所改变。对于成年人而言,这些变化极其微小,因此几乎不会影响我们对植物种类的选择。然而对于不同年龄段的儿童还有坐轮椅的残疾人而言,这些高度的变化将是显著的,因此我们必须要考虑到他们对不同空间的感受。

地被植物种植(地毯式植物)

[5] 这些最低矮的植被形成了一个非常贴近地表的叶冠,通常厚度不超过几厘米。植物包括草类和其他修剪或者放牧的草坪(场),匍匐类灌木和爬藤类草本植物。它们主要的空间作用就像"地板"一样,允许视线通过和人的自由活动。这样的作用让其承担了几种不同的角色:

- 虽然平整、牢固的地毯式地被植物不如硬质铺地耐磨,但它仍旧能够作为人行通道的面层。最耐磨的植物种类包括草坪,它们在定期修剪或放牧后可以形成适合休闲、行走、玩耍、运动、骑自行车和偶尔机动车通行的表面。这种持久耐用的特性使草坪、草地和草原牧场在私人和公共景观中体现出巨大价值且备受喜爱。
- 统一的地毯式地被植物或紧贴地面,质感平滑的地被植物顺应地形种植可以增强地面起伏的视觉效果。可使用的植物种类包括黄春菊和新西兰刺果薇。斜坡的交界处可以通过使用叶形对比强烈的地被植物来加以强调。
- 地表植物可以组成二维的图案,单独使用地毯植物的叶丛或者与卵石、沙砾及铺装材料结合使用,可以在地表形成由色彩、质感和图案所构成的织锦效果。

低于膝高的灌木及草本植物(低矮植物)

[6] 可以形成更高冠幅但高度仍然在膝盖以下的灌木及草本植物可以为空间设计提供更多的可能性。它们当中的许多属于"地被植物",能够很好地适应本土环境并且有足够的竞争力去排除大部分不受欢迎的、自播繁衍能力强的"杂草"植物。除了可以大大节省养护所需的人工劳动外,低矮植物还有在界定边界并为人的活动形成障碍(虽然不是阻止)的同时,保证视线自由的空间作用。

- 低矮植物单独使用时,可以像地毯植物一样形成一个视觉的平台或地平面。
- 也可以与生长较高的草本植物、灌木及乔木结合在一起。这如同画面中的底色,或者是为了突出"图"所用的那个对比强烈的"底"。通过这种方式,低矮植物可以提供一个共有的背景或平台,以此来整合构图中其他植物和元素。
- 许多可形成低矮地被的匍匐植物会顺着墙或堤岸生长并形成悬垂的植物维幔(匍匐迷迭香就是一个典型的例子)。栽植悬垂和攀援植物可在垂直面和水平面上形成连续的植物叶面。植物叶片如瀑布般从堤岸和墙上倾泻而下并流过地面,将垂直面、水平面及斜面之间的夹角掩盖。如同给新的和旧的物体穿上相同的外衣,攀援植物可以使建成景观中的新结构及土方工程给人一种隶属于原有环境的从属及成熟感受。
- 低矮植物在硬质与软质景观交界处及不同用途软质景观的衔接处有着必不可少的作用。高大的灌木需要横向延伸的空间以避免在后期的生长中挤占通道空间。低矮植物可以提供一个地被层,使在其上生长的较高植物可以自由延伸而不需要经常修剪和造型。当这个地被层蔓延到了铺装或草地上时,偶然或自然的踩踏就能达到修剪的作用。当通行较少时,偶尔的修剪也是必要的。

膝盖到视线高度的植物(中等高度的植物)

[7] 膝盖到视线高度的植物在设计上与矮墙、篱笆和围栏有相似的作用。它可以作为行为活动的障碍,用来

限制人们接近,但其同样可以保持视线的开敞,且对阳光所造成的影响微乎其微。这样中等高度的植物提供了许多空间上的作用:

- 从安全意义上分隔空间:例如,使人或车辆远离陡坡、水,或使人车分离。
- 在视线不希望被封闭的地方,可以用来识别和强调期望路线或通道。
- 使行人与建筑以及其他私人领域之间保持一定的距离,从而在植物高度不超过窗台的情况下避免光照的减少且保护隐私。
- 帮助界定建筑或其他构筑物的边界,在视觉上使其扎根于地面并与周围景观环境融为一体。当建筑或其他构筑物是建造在一个以大量现有植物为特征的景观环境中时,这种情况显得尤为重要。

高于视平线的植物(高灌木或小型乔木种植)

[8] 高于视平线的灌木和带有树冠的小型乔木可以形成视觉和物理屏障。所以,具有浓密树冠的高大植物在某种程度上类似于墙体和栅栏,以较小规模的种植实现比种植更高大树木更可能产生的分隔、围合、屏蔽和遮挡效果。

- 在人体尺度的公园、花园、庭院、街道以及游乐场景观中,高大的植物提供了私密空间和庇护区域以屏蔽外界的干扰,如停车场、服务区以及垃圾箱。
- 就像墙体和栅栏一样,高大的植物能够为具有装饰性的植物形成背景,如由草本植物所构成的边界线和展示性植床。修剪"规整"的绿篱在花园中扮演了这一传统的角色,但更加松散的灌木种植也能达到同样的效果。典型的绿篱植物主要包括紫杉属的紫杉、山毛榉属的山毛榉和角树属的角树,常用作北欧地区的高大绿篱。柏木属的大果柏木和罗汉松属的桃拓罗汉松,可以在温暖的气候条件下成为良好的修剪树篱。
- 高大的植物因其较大的尺度能够扮演建筑物配景的角色。由于它们在视觉上的体量与小型建筑相近,所以可以用来平衡建筑的砖石和贴面。
- 一对独立在外的高大灌木或在大规模种植中的一个间隙可以构成一个框架,它可以将全部的景色聚焦于框架内,或将人们的注意力集中到主景观或地标物上。这种设计不仅可以吸引注意力,同时也可以引导人们进行新的探索。如同拱门和通路一样,它们暗示了对于另一场所的探寻。
- 当树木被单独或形成小型组合种植时,精选的高大灌木以其尺度和仪态作为标志、景点或视觉焦点存在于人体尺度的景观中。

乔木种植

[9] 乔木的尺寸正如建筑、道路、桥梁和小型机械部件一样,具有一定的大小序列。所以可以通过树木种植来屏蔽、分隔、遮庇、围合、搭配以及完善那些较大的构筑物。当树木自由生长到形成一根清晰的主干或是带有树冠的树干高过头顶时,除了垂直的树干外,在树冠和地面间留下了开阔的空间。由此形成的是一种与众不同的空间元素。

[10] 成年乔木高度从5米到40米不等,5米的树种如垂枝洋梨树和阿克树,超过40米的如欧洲白蜡树、新西兰罗汉松、生长于北美西海岸的一些针叶类树种乃至不少澳大利亚桉树等。为了不同的设计意图,我们将树木分为三类:小型,成年高度在5~10米;中型,10~20米;大型,20米以上。

- 小型乔木的高度一般等于或略低于大多数建筑两层楼的高度,所以它们对城市环境的影响主要表现在建筑间的局部空间。
- 中型乔木可以营造出包容较小建筑的空间。因此,它对城市景观的空间构造会有更大的影响。
- 由于高大的乔木需要更大的空间,因此它们在城市中的使用并不那么普遍,然而自然生长的高大树种经常被种植在街道两旁和花园里,当它们长到能提供较大的遮阴效果或对附近的建筑体形成主导地位时就会被修剪。当树木的尺寸高于20米时,它们将形成街道、广场和公园主要空间结构中的一部分。在乡村景观中,大树可以构成一个大尺度的框架。
- 中型和高大的乔木种植在将大规模工业建筑群融入到周围景观中可以起到关键性作用,比如将发电站与环境相融合。树木种植带和种植区包围这些场地,并向外延伸,能对近距离视线形成屏障。而对于更远的距离,虽然它们不能隐藏诸如冷却塔或涡轮机房这样的大尺度建筑,但是可以从视觉上把这些建筑融合于其周边的环境中,并隐藏低层的辅助配套设施、临时建筑和停车场等。因为低层建筑的零乱常常是大型工业最为恼人的部分,所以种植中型或高大的乔木在这里扮演着一个极其重要的角色。

- 在较远距离屏蔽或模糊视线方面,乔木种植优于灌木种植,正因为如此,这种特征常常被用来控制随观察者在景观中移动的视线。通过仔细考量,在种植中营造出间隙能形成开阔视景或在适当的时刻框住一个景观焦点。如同一扇窗户或拱门,由树枝或树叶形成的景框将注意力和关注点转移到了"这之后有什么"的问题上。

- 一棵孤植树木或一小片丛植树木,本身亦可成为景观焦点。作为一个独立在外的物体,它在我们的视线范围内占据了一小块位置,因此我们的视线倾向于停留在上面。一棵有着与众不同特征的树,比如可在秋天变色的或者具有如画般特质的,将成为特别引人注目的焦点。在一定的距离范围内,大型标志性树木种植或丛植都可以产生这种效果,因此在大型乡村景观中能够成为景观焦点或地标。

- 当孤植的标志性树木或一小片丛植树木配衬着建筑物种植时,树木形态和建筑形态间的关系是非常有趣的。亨弗利·雷普顿为了创造如画般美妙的景致,制定了一个规则,规定了以何种树木的形式来配合不同类型的建筑风格可以达到最好的效果。他建议,对于宽大的有稳定比例和浅屋顶角的古典建筑,应该配合有静止线条的、向上挺直的圆锥形乔木,比如云杉和冷杉。反之,对于尖塔形和陡峭斜面屋顶的维多利亚哥特式复兴建筑,应该用稳重的圆形树型或者水平向扩张的树型来搭配。比如黎巴嫩雪松、英国橡树或栗树。

- 乔木种植在建筑上的另一个更深远的作用应该是连接多种建筑风格。同类树木形成的简单而整齐的线条能为建筑提供统一的正立面或在建筑正面塑造一个与其相衬的独立对应物。其连续性可以将不同的建筑风格相连而在一个统一的绿色框架中,为不同建筑增加趣味。

[思考]

(1)如何在场地的规划设计和维护管理方面体现系统性?

(2)地形的重要性从哪几个方面体现?

(3)从构成空间的角度来看,植物可以分为哪些类型?它们各有哪些空间特性?

PART VI

Engineering and Technology
第六部分 工程与技术

UNIT 14 LANDSCAPE ENGINEERING

TEXT

Site Development Guidelines
John Ormsbee Simonds
场地开发导则

A checklist of helpful considerations, these may vary in some instance with site or climatic conditions and material finishes.

[1] **Excavation and grading**

Keep to an absolute minimum.

Balance the on-site cut and fill[1]. Off-site borrow or disposition is expensive.

Protect trees and established ground covers.

Remove and stockpile the topsoil.

Avoid working the soil when it is wet, powder-dry, or frozen.

Provide positive surface drainage away from building to swales, gutters, drain inlets[2], or outfalls.

Reestablish ground covers without delay.

Unprotected soils cause erosion and siltation.

Most slopes are best blended into the natural landform.

[2] **Slopes(earth cut or fill)**

Do not exceed the angle of repose of the soils[3] being cut or placed or a slope of 1 on $1^{1/2}$ maximum (1 foot of vertical rise of $1^{1/2}$ feet of horizontal distance).

A slope of 1 on 2 maximum is recommended for mulched or planted embankments.

A slope of 1 on 3 maximum is preferred for lawn areas to facilitate mowing.

Place fill material in uniform layers of 6 on 8 inches of loose material.

Allow for soil shrinkage (or swelling in some instances); 3 to 5 percent shrinkage is normal in compacted fills.

Provide mechanical compaction. Natural compaction by the eventual settlement of loosely placed soils is seldom uniform or complete.

All fills should be compacted fills, placed on prepared benches cut through topsoil and overburden.

Thrust benches and positive drainage must be provided at the base of major fills.

[3] **Steps**

Avoid steps, if possible, except when they are used as a landscape feature.

Always consider the handicapped and provide alternate access ramps.

Avoid use of single steps. They are hazardous.

The risers in architectural flights of steps should be of uniform height. In free-form or naturalized flights of steps (where consistency is not anticipated), riser and tread[4] dimensions may vary widely within a given flight.

Good footing is essential. On concrete steps a wood float[5] or light broom finish[6] or the use of abrasive fines[7] is suggested.

In rough terrain particularly, perrons (as in a stepped ramp) may be desirable on slopes ranging in grade from 16 to 25 percent.

[4] **Lawn and seeded areas**

Provide a 1 percent minimum gradient for lawn areas (a fall of 1 foot each 100 feet).

A $1^{1/2}$ percent to 2 percent slope is preferred to ensure more positive surface drainage.

Swales should have a gradient of 1 percent minimum, 4 percent maximum. (On steeper grades a loose stone or paved gutter is required.)

Provide a 6-inch fall from buildings in the first 20 feet.

A 4-inch compacted topsoil section is considered the minimum for new lawn construction.

A 6-to 8-inch section (or deeper) is recommended when soils are impervious or overly porous or when topsoil is abundant.

[5] **Walk paving**

Provide a 1 percent minimum longitudinal or cross slope[8].

A slope of $1^{1/2}$ percent is recommended for terraces.

A pitch of 8 percent is considered maximum for walks if no handrail is provided.

With a handrail, the walk pitch can be steepened to 15 percent (for a short ramp distance only).

Width: a confined walkway requires a minimum width of 2 feet per person for comfortable passing. In the open, where people's shoulders can overhang the walk edge, the outer pedestrian lanes can be reduced by up to 6 inches. A width of 5 feet 0 inches for a typical low-volume community walk will allow three persons or one person and a baby carriage to pass. If bicycle use is anticipated (such a joint use is not generally recommended), a walk width of 6 feet 0 inches is required for a bicycle and two persons or for two bicycles to pass.

Capacity: each 2 feet of width will accommodate between 50 and 60 persons per minute, or an average of 3 300 persons per hour. This holds as well for shuffling crowds, strolling window-shoppers, or students walking briskly across a campus, since as the rate of movement increases, the person-to-person spacing increases accordingly. Rates and capacities vary climatic conditions, surface textures, and gradients. They are to be adjusted for intermittent movement as at crossings, constrictions, and counter pedestrian flow[9], which can reduce capacities by up to 50 percent.

[6] **Roads and driveways**

In planning the approach drive[10] or roadway consider:

Sight distance: provide sufficient horizontal and vertical sight distance to give 10-second minimum observation time at permitted approach speeds.

Adequate sight distances at intersections

An attractive introduction and portal

Sequential revealment of views, site features, and buildings

All-weather drivability and safety

Recognition of topography, sun angles, and storms

Minimum length and minimum landscape disruption

A pleasurable driving experience

Align roads and drives (and walkways) so that adjacent swales, gutters, and/or sewers will have continuous gravity flow, with minimum grading or depth of trenching.

A longitudinal gradient of $1^{1/2}$ percent is preferred; 1 percent is considered minimum.

When flatter grades are necessary, the road must be crowned or cross-sloped to drain, as:

Concrete, 1/4 inch per foot

Bituminous, 3/8 inch per foot

Gravel, 1/2 inch per foot

Use a dishes (concave) section only in narrow lanes or minor service drives.

All road and drive intersections should be approximately perpendicular (90 degrees).

Horizontal and vertical curvature is subject to design speed and topography.

Horizontal curves are normally true arcs.

Vertical curves are parabolic. Radii vary from 30 feet at the entrance to public roads to 600 feet or more on private drives.

On private drives and in natural areas both the horizontal and the vertical curvature may follow the topography freely without need for geometric computation.

In such cases the grading equipment is guided by prelocated field stakes or flags and responds with a light touch to the existing landforms.

[7] **Recommended private local street widths (on-street parking prohibited)**

Dwellings served	Paving width
1-5	12 feet, single lane, optional to 500 feet
1-20	16 feet, two lanes
21-50	18 feet, two lanes
50 plus	20 feet, two lanes

For public roads, see local requirements:

18 feet minimum for two lanes

10 to 12 feet per lane normal

[8] **Parking**

Allow a normal stall width of 8 feet 6 inches minimum to 12 feet maximum; 10 feet 0 inches is a comfortable average.

Stall marking: while a single divider stripe will suffice, a double 3-inch line, 12 to 16 inches on center with a half circle at the aisle end, is recommended.

Parking compound (two or more courts): for area-capacity calculation (approximate) allow 300 square feet of paved parking area per standard car, plus approach ramps, distributor loops, planting medians, turnabouts, collector walks, and buffer areas.

[9] **Site drainage**

Preserve the natural drainageways insofar as feasible.

Preclude concentrated surface runoff to downgrade properties.

Avoid trapped water pockets.

Provide underdrains at road edge and low points.

Conduct surface water by swale, gutter, or buried pipe to storm-sewer mains or outfall.

If storm inlets and lateral sewers are needed, compute the required capacity and then use the next larger size.

Keep the site drainage system unobtrusive.

[10] **Site furnishings**

In the selection and placement of lighting standards and fixtures, recreational equipment, informational signs, benches, movable tables, seating, etc., consider:

Functional suitability

Compatibility of form, material, and finish

Durability

Long-term cost: a higher initial expense that yields longer life with less required maintenance is usually good economics.

Durability is to be stressed. Site equipment and furniture must be designed to withstand the effects of the elements, including sun, expansion-contraction[11], wind stress, moisture, and sometimes salt spray[12], frost, or ice.

Plan a coordinated family of shapes, materials, and finishes.

Generally, use strong, simple shapes, native materials, and natural finishes.

Black, grays, and earth tones are basic, with bright colors reserved for accent.

Standardize components such as lighting globes, signposts and blanks, bench slats, bolts, and stains.

Invest in the best.

[11] **Landscape planting**

Strive always for utmost simplicity.

Stress quality, not quantity. One well-selected, well-placed plant can be more effective than 100 plants scattered at random.

When budgets are limited, economize on the extent of the lawn and plant area, but invest in soil quality and depth, larger plant pits, soil preparation, and provision for irrigation.

Lawn area are best given a well-defined and pleasant shape and (in an architectural context particularly) edges with paving, curb, or mowing strip.

Install no lawn area or plant without a predetermined purpose.

Select each plant to best serve the purpose intended.

In the use of plant materials consider:

Need

Suitability

Appearance in all seasons

Appearance in all stages of growth

Compatibility of form, texture, color, and association in the total building and site composition

Hardiness, cultural requirements, and degree of maintenance needed

As a result, use only indigenous or naturalized materials except for bedding plants and container-grown exotics.

Plants used for backdrop, screening, shade, or space definition are generally selected for strength and cleanliness of form, richness of texture, and subtlety of color.

Plants to be featured are selected for their sculptural qualities and for ornamental twigging, budding, foliage, flowers, and fruit.

They are to be place strategically for optimum display.

Ground covers and mulches do much to enrich a fine planting.

In landscape planting of excellence restraint is the key.

[NOTES]

[1] cut and fill 填方与挖方
[2] drain inlet 排水口
[3] repose of the soil 土壤安息角
[4] riser and tread 阶高和阶宽
[5] wood float 木抹子
[6] broom finish 混凝土面扫处理
[7] abrasive fines 细磨面
[8] longitudinal or cross slope 纵坡或横披
[9] counter pedestrian flow 逆行人流
[10] approach drive 入口车道
[11] expansion-contraction 热胀冷缩
[12] salt spray 盐雾

UNIT 14　LANDSCAPE ENGINEERING

[NEW WORDS]

aisle	n.	过道,通道	mowing	n.	割草,修剪草坪
align	vt.	排成直线,排成一行	mulch	vt.	用护根覆盖,护根
bench	n.	挡板,工作台	off-site	adj.	异地,工地外
bituminous	adj.	沥青的,含沥青的	on-site	adj.	现场的,在工地上,就地
briskly	adv.	迅速地,轻快地,活泼地	optimum	adj.	最适宜的,最有利的,最佳的
buffer	n.	缓冲	outfall	n.	河口,排水口
checklist	n.	清单,一览表	parabolic	adj.	抛物线的
climatic	adj.	气候的,风土的,水土的	perpendicular	adj.	垂直的,成直角的,正交的
compaction	n.	压紧,夯实	pit	n.	深坑
concave	adj.	凹的,凹面的	pitch	n.	斜度,坡度,程度,强度
concrete	n.	混凝土	porous	adj.	多孔的,能穿透的,能渗透的
constriction	n.	阻塞物	portal	n.	入口,门面,壮观的大门
curvature	n.	弯曲,曲率,曲度	preclude	vt.	阻止,排除,预防
economize	vt.	节省,节约,减少开支	predetermined	adj.	预定的,预设的
embankment	n.	堤坝,路堤,岸堤,护坡	sequential	adj.	按次序的,相继的,顺序的
excavation	n.	发掘,挖掘,开凿	sewer	n.	阴沟,污水管,下水道
fixture	n.	固定装置	shrinkage	n.	收缩,缩水
grading	n.	土工整理,减小坡度,造坡	shuffle	vi.	逐渐移动,拖着脚步走
gravel	n.	砾石,碎石,砂砾,粗砂	siltation	n.	沉积作用,淤积
guideline	n.	指导方针,行动纲领,准则	stake	n.	桩,柱,标桩,篱笆桩
gutter	n.	排水管,排水沟,阴沟	stall	n.	车位
handicapped	adj.	残疾的,弱智的	stockpile	vt.	贮存,堆放
handrail	n.	栏杆,扶手	swale	n.	洼地,沼泽地,汇水区,排水沟渠
impervious	adj.	不可渗透的,透不过的	swelling	n.	膨胀,润胀
indigenous	adj.	土生土长的,生来的,固有的	topsoil	n.	表土,表土层,土壤表层
intermittent	adj.	间歇的,断断续续的	tread	n.	踏步板,梯面,踏面
intersection	n.	交叉点,十字路口	trench	n.	深沟,地沟,沟渠
lateral	adj.	侧面的,从旁边的,横的	turnabout	n.	回车,旋转,回旋
longitudinal	adj.	经度的,纵向的	unobtrusive	adj.	不引人注目的,不显眼的
median	adj.	在中间的,通过中点的,中央的	utmost	adj.	极度的,最大的,最远的

[参考译文]

场地开发导则

下面是场地开发中应考虑的条目,在某些实例中可能会因场地或气候条件而变化。

[1] 开挖与造坡
- 保持绝对最小值。
- 平衡现场填挖量。异地取土及弃土成本昂贵。

· 247 ·

- 保护树木及已植地被。
- 挖去表土并予保存。
- 避免在土壤潮湿、过干、冻结时施工。
- 排水应从建筑物向四周导入汇水区、水沟或排水口。
- 及时重建地被。
- 没有保护的土壤会引起侵蚀和淤积。
- 大多数的坡地最好改造成自然地形。

[2] 坡地(土方填挖)

在坡地改造时,不要超过土壤的安息角或是1:1.5的坡地(1英尺的垂直距离对1.5英尺的水平距离)。

不大于1:2的坡地建议加覆盖物,或是做成有植物的护坡。

不大于1:3的坡地可以作为草坪以便于修剪。

填疏松土料时,以6~8英寸为一层,均匀覆加。

考虑土壤收缩(有时是膨胀);3%~5%的收缩量在紧压填充中是正常的。

实行机械夯实。依靠松散填土的最终稳定所形成的自然压实很难达到均匀或完全紧实。

所有的填充都应是紧压填充,将土置于挖去表土的土槽中加压夯实。

在大规模填土地段,基部必须有挡板墙和排水槽。

[3] 踏步

尽可能避免踏步,除非它们作为一种景观特征存在。

通常要考虑残疾人,为他们提供其他通行坡道。

避免使用单个的踏步,这很危险。

建筑踏步的阶高应保持一致。对于自由形式或自然化的踏步段并不强求连续性,阶高和阶宽的尺寸在给定范围内可以变化较大。

有良好的落脚面亦很重要。在混凝土踏步上,建议使用木抹面、浅扫面或细磨面处理。

在困难地段,坡度16%~25%的斜坡上宜使用石台座(形成多级坡道)

[4] 草坪和播草地区

草坪地区坡度不小于1%(每100英尺下降1英尺)。

常采用1.5%~2%的坡度,能确保地表排水的通畅。

洼地坡度保持在1%~4%(较陡的坡度须有松散石块或铺砌排水沟以防侵蚀)。

距离建筑物20英尺的范围内应有6英寸高度落差。

新的草坪在铺种时应至少有4英尺厚的夯实表层。

当土壤渗透性差,过于多孔,或表土充足时,建议采取6~8英寸的表土(或更厚)。

[5] 步道铺设

需要不小于1%的纵坡或横披。

阶地建议采用1.5%的坡度。

无扶手时步行道的最大倾斜度为8%,有扶手时倾斜度可达15%(仅适于较短的坡道距离)。

宽度:受限的步道至少应为每个行人提供2英尺的宽度以便于舒适地通过。在开敞地段,行人的肩部可超出步道边缘,因此外侧人行道宽可减少6英寸。典型的低容量社区步道通常宽5英尺,可以容许三个行人或一个行人带一辆婴儿车通过。如果允许自行车通行(这样的混合使用通常不推荐),步道宽度需要6英尺,可以容许一辆自行车和两个人或两辆自行车通过。

容量:每2英尺的宽度上可容纳50~60人/分钟,即平均3 300人/小时。由于运动速度增加,行人间距也相应增加,所以以上承载力也适于运动的人群,漫步的逛街者,或快速穿过校园的学生。速率和容量随气候条件、路面质地和坡度的变化而变化。对于受干扰的运动路段,如交叉路口、拥塞段、逆行人流等能使容量降至50%,速率与容量需要调整。

[6] **道路与车道**

在规划入口车道或道路时,应考虑:

视距:提供足够的水平、垂直的视线距离以保证在允许速度范围内有10秒的观察时间。

- 交叉口有足够的视距
- 醒目的标示和入口门面
- 景色、场地特征和建筑的展示序列
- 在各种天气均可安全驾车
- 对地形、太阳角以及暴雨的考虑
- 经济上合理的长度和最小的景观破坏
- 令人愉快的驾驶体验

使道路和车道(以及步道)的路线尽量不干扰,邻近洼地、排水沟和/或下水道的重力排泄系统使填挖方达到最小、排水沟的深度达到最小。

最好保证至少1.5%的纵向坡度,1%为最低限度。

当需要平缓一些的坡度时,道路必须做成拱形或横坡以利排水,比如:

- 混凝土,每英尺1/4英寸
- 沥青,每英尺3/8英寸
- 砾石,每英尺1/2英寸

只有狭窄的巷道或小型服务车道才用凹形路面。

所有的道路和车道交叉处应近于垂直(90°)。

水平的、垂直的弧线曲率应符合设计速度和地形。

水平弯道一般是正圆弧线。

垂直弧线是抛物线型的。半径变化很大,从公共道路入口的30英尺到私人车道的600英尺以上。

对于私人车道和自然区域,水平和垂直的曲度可随地形自由变化,不必拘于计算要求。

在这种情况下,推土机械根据事先树于场地中的标杆或旗帜施工,只对现状地形做轻度改造。

[7] **居住区街道的推荐宽度(禁止沿街停车)**

住户服务规模	铺装面宽度
1~5	12英尺,单车道,路长在500英尺之内
1~20	16英尺,两车道
21~50	18英尺,两车道
大于50	20英尺,两车道

对于公共道路,依当地情况而定:

通常双行车道至少18英尺

每条正常车道10~12英尺

[8] **停车场**

标准车位宽度为8英尺6英寸~12英尺;平均10英尺较为舒适。

车位标示:尽量单一的分隔线已足够,但最好有两条3英寸宽的直线,通道端点处有一半圆,两线中心距离为12~16英寸。

就停车场的容量来说,一辆普通汽车约需300平方英尺的铺装面积,外加入口坡道、场内疏散回路、植物分隔带、回车空间、步行通道和缓冲区。

[9] **场地排水**

尽可能保持自然排水道。

避免集中的地表汇流,以免场地受损。

避免洼陷的水坑。

在路边和地势低处采取地下排水。

通过洼地、排水沟或埋设的管道把地表水引导至主雨水管道或出水口。

如果需要雨水进口和侧方排水道,应计算所需容量,然后采用大一号的排水管。

保持场地排水系统不引人注目。

[10] **场地装饰**

在选择并摆设灯光设备、娱乐器材、信息标示、长椅、活动桌凳等时,应考虑:

• 功能上的适宜性

• 形式、材料、加工的兼容性

• 耐久性

长期费用:初期投入较高,然而寿命较长,维护较少,常常是经济合理的。

要着重强调耐久性。场地设施和装置应设计得能承受各种因素的影响,包括光照、热胀冷缩、风力、潮湿及偶尔的盐雾、霜冻或结冰。

形式、材料和加工应有一个一体化的计划。

通常采用坚固简洁的形状、本地材料和自然加工。

以黑色、灰色和土色为基调,用亮色加以强调。

灯罩、标志柱和标志牌、座椅板条、螺栓和颜料等部件应标准化。

要投资于最好的。

[11] **景观种植**

尽量简单化。

强调质量而非数量。一棵精心挑选、精心布置的植物比随意分散栽种的一百棵植物更为有效。

当预算有限时,在草坪和种植面积上应节约,应将投资用在土壤质量和厚度、足够大的种植坑、土壤配制和灌溉方面。

最好通过铺装、路牙石或修剪的嵌边(在以建筑为背景时尤其强调)为草坪界定一个精心设计的形状。

没有既定目标时,不要先布置草坪和植物。

植物选择应能最好地服务于目的。

在使用植物材料时应考虑:

• 需求

• 适宜性

• 四季的状态

• 不同生长期的形态

• 形体、质地、色彩及与整体建筑和场地之间的兼容性

• 耐寒性、文化需求,以及需要的维护程度

作为常规,除了温床植物和容器里的珍奇品种,一般采用本土物种或归化物种。

用作背景、屏蔽、遮阳或构成空间时,这类植物应从形体的力度和简洁性、质地的丰富性和色彩的微妙性等方面来挑选。

欲突出显示的植物应从造型、枝、叶、花及果实的装饰性方面来挑选。

巧妙地放置它们以达到最优的展示效果。

地被植物和覆盖物对于丰富一个优秀的种植设计很有帮助。

优秀的景观种植中,节制是关键。

Reading Material A

Landscape Engineering[1] I

Michael Laurie

风景园林工程 I

[1] Landscape engineering or grading is a fundamental technical aspect of landscape architecture. It involves the remodeling of existing land form to facilitate the functions and circulation of the site plan and to ensure adequate drainage. Thus a knowledge of grading technology is useful in the site planning process. Detailed leveling is needed to make connections between architecture and landscape, between indoors and out. Psychologically there is perhaps more satisfaction in a house or building being directly on and related to solid land. Being on the earth, dug into it, is probably a positive psychological urge and need.

[2] The site-to-structure relationship is a visual as well as a functional matter. The floor levels of buildings should be higher than the surrounding ground. Surfaces adjacent to and outside buildings should also slope away from the building so that rain water will not easily enter the structures or undermine foundations.

[3] At Foothill College, the original landscape was extensively reshaped to fit the buildings and open space into buildable and usable land. The sensitivity with which this was done resulted in a landscape that did not look as though it had been remodeled. Buildings were set well back from the edge of the hill, permitting them to sit comfortably on level ground[2].

[4] There are two principal relationships between buildings and land. The land may be graded or adjusted to suit the architectural or engineering requirements, or the architecture may be adapted to meet variations in ground level so that the original surface is disturbed less (Fig. 1). Buildings that stand on piloti and do not touch the ground at all require little grading except to provide access. The land will be disturbed to some extent during construction and will result in changed light and moisture conditions under the structure. On the other hand, construction that assumes conventional foundations results in a building-to-ground relationship all around the structure. Single-level houses on steep hillsides require extensive cut and fill grading, eliminating the original soil relationship which may lead to unstable conditions resulting in erosion, landslides, floods, and a complete destruction of the ecosystem. The appropriate way to site buildings depends on a careful analysis of the land, its slope, soil, geology, and so on. In addition, the initial decision to build in any area should result from land suitability studies[3], so that truly destructive process can be prevented or special construction techniques instituted to match the conditions.

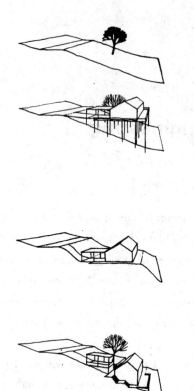

(a) A sloping site

(b) A single-story house set on stilts leaves the ground relatively undisturbed. Tree can be saved.

(c) A single-story house set on conventional foundations requires considerable cut and fill resulting in steeper slopes at each side of lot.

(d) A split-level house with retaining walls maintains shallow slopes at each side of the lot and the tree can be saved. The house is closely related to the form of the landscape.

Fig. 1

The Principles and Technology of Grading

[5] Let us now proceed to the more conventional problems in which buildings and use areas are sited on suitable land in which the site plan essentially creates a new and working landscape to replace the existing "natural" condition. The technology of grading represents a unique skill of landscape architecture and is connected with the process of placing the diagram of the site plan onto a given topographical landscape. Our concern is not only with fitting or connecting buildings to the land but also with sitting use areas such as playing fields, parking lots, and circulation routes. All of these have specific criteria for slope, foundation depth, and drainage.

[6] Landscape engineering is concerned with economical development and with sensitivity to the existing conditions. The basic principles and goals of grading may be summarized as follows.

- The ground surface must be suitable for the intended purpose or use.
- The visual result should be pleasing; indeed the purpose of the grading may be purely aesthetic, to screen views or create symbolic land form.
- The resulting ground surface must have positive drainage.
- The grading plan should attempt to keep the new levels as close as possible to the original state of the land. In nonurban areas especially, existing landscape represents an ecological balance, a natural drainage system, and a developed soil profile.
- When ground is reshaped it should be done positively and at the scale of the machinery. Grading machinery is by definition gross in nature and subtle details are difficult to achieve except by hand

labor.

- Topsoil should be conserved wherever possible. It may be stripped, stockpliled, and reused after heavy grading.
- In the grading operation, the quantity of cut should approximately equal the amount of fill. This eliminates the need to import soil or to find a place to dump unwanted material.

Grading Plans

[7] Grading plans are technical documents and are instruments by which we show and calculate changes to the three-dimensional surface of the land (Fig. 2). Contour lines are used to indicate the extent of that change. Depending on the scale, contours show relative elevation at intervals of 1 foot, 5 feet or more. Existing contours are shown as dashed lines. The proposed new land form is shown by solid lines drawn where this varies from the existing form. The difference between these lines shows where land is to be cut and where filled and, in general, the extent and nature of the change. Such drawings, showing two sets of contour lines, express the difference between the existing condition and the design intent. From these drawings quantities of cut and fill can be calculated. The plans must be accurate if calculations and therefore cost estimates are to be reliable.

[8] Site planning and grading take care of the adjustment necessary between fixed levels, structures, and use areas within the boundaries of the site. Fixed levels or control elevations[4] include the levels of existing trees or vegetation to be retained, existing and proposed buildings and roads, the levels at the boundary of the site, existing land forms to be included in the design, lakes and natural swales, existing and proposed underground utilities. Such levels constrain the reshaping possibilities.

[9] The terminology of the grading plan is simple. Contour lines are abstractions which join all points of the same elevation above a fixed datum. Spot elevations provide additional information beyond that given by the contour lines. They indicate micrograding; that is, specific levels that lie between the intervals of the contour lines. They show level differences needed to ensure drainage on "flat" surfaces and to indicate specific levels at important points in the plan. Typical locations at which spot elevations are shown are at tops and bottoms of steps, tops of retaining walls, outside entrances to buildings and their inside floor levels.

[10] Another term frequently used is grade or gradient. This refers to the rate of slope between two points, expressed as a percentage, or as a ratio of horizontal distance to vertical change in elevation, or as an angle. Thus a 1 percent slope is the same as 100-1. A 10 percent slope is the same as 10-1 or a 6° angle. A 50 percent slope is as the same as 2-1 or a 26°30′ angle, and a 100 percent slope is 1-1 or 45° angle. Key variables in figuring gradients are the horizontal distance between two points and the vertical change in elevation.

[11] Frequently the surface has to be changed to provide some specific gradient or to fit into some maximum or minimum slope criteria. The process of developing a grading plan involves the manipulation of the three factors: gradient (G), horizontal distance[5] (L), and difference in elevation[6] between two points (D). If, for example, the existing G between two points 100 feet apart is too great for an entrance road (say 15 percent), then either the vertical difference must be made less (10 feet) or the horizontal distance must be increased (to 150 feet). The variables interact as follows:

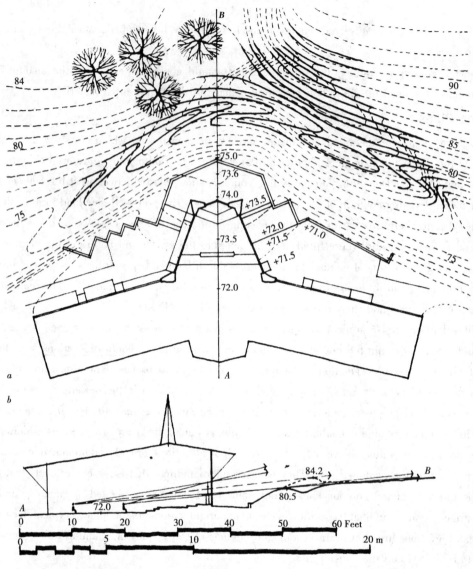

(a) Grading plan deals with drainage on the hillside and on the paved surfaces. Usable space is provided around the church and the steepness of the original slope is reduced. (b) Section.

Fig. 2

$$G = D/L; \quad L = D/G; \quad D = L \times G$$

Altering these variables produces solutions with different economic and aesthetic implications (Fig. 3).

[12] In landscape design there are generally accepted maximum grades. These figures have been developed by engineers, architects, and landscape architects from experience and practice. They become determinats of form in site planning, and they are also economic factors. For example, walks and paths usually have a crowned cross section and if possible should not have a longitudinal gradient of more than 6 percent in areas of cold winters, or 8 percent in milder climates where frost is not per-

A road from A to B in plan 1 would have a gradient of 15 percent. By cutting into the hillside and extending the road length beyond the plan area, the gradient between A and B can be reduced to 10 percent(plan 2). Alternatively, a gradient of 10 percent can be achieved within the plan area and without extreme changes to the land as shown in plan 3. The length of road is increased, however.

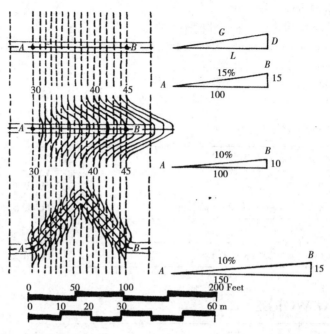

Fig. 3

sistent. These are preferred slopes for long distances of path. For shorter distance, ramps with a gradient of up to 12 percent can be used; over 12 percent, steps become the most reasonable way to overcome changes in level, but they should be avoided wherever possible. For wheelchairs and bicycles, steps are an obstacle and a hazard. Flights of three or fewer steps become an irritation for movement, although they can serve as informal seating, especially on school grounds and college campuses.

[13] Steps are purely functional in changing elevation in the shortest possible space and subject to preferred dimensions derived from the physical act of walking developed over centuries of building in both vernacular and Renaissance modes. For example, one "rule" states that the riser multiplied by the tread should equal approximately 74 inches. This allows alternative dimensions, for example, 6-inch riser and 12-inch tread, 5-inch riser and 15-inch tread, 4-inch riser and 18-inch tread. As the riser is reduced, the tread becomes wider. However, when the tread becomes excessively wide, the rhythm of walking and the dimensions of a stride must enter into the calculations. Designers should measure and record steps they find comfortable.

[14] Roads and driveways also have preferred maximum gradients. Although 6 percent is considered desirable, 8 to 10 percent is permissible for short distances. In exceptional cases (in San Francisco, for example, where fortunately the weather is mild) streets are found at 15 percent and more. These standards and rules of thumb must be viewed in terms of the specific problem to solved and its context of geography, custom, and use.

[NOTES]

[1] 本文节选自迈克·劳瑞(Michael Larie, 1932—2002)的 An Introduction to Landscape Architecture。

[2] level ground 水平地面
[3] land suitability studies 土地适宜性研究
[4] fixed levels or control elevations 固定标高或控制高程
[5] horizontal distance 水平距离
[6] difference in elevation 高差

[GLOSSARY]

contour line	等高线	land form	地形
cost estimation	造价预算	land suitability	土地适宜性,用地适宜性
cross section	横断面,横截面	longitudinal gradient	纵坡
grading	土方整理	retaining walls	挡土墙
grading plan	竖向设计	spot elevations	高程点

[NEW WORDS]

buildable	adj.	可建造的	permissible	adj.	可允许的,许可的,容许的
contour	n.	轮廓,等高线	piloti	n.	(多层建筑的)底层架空柱
criteria	n.	标准,准则	positively	adv.	断然地,肯定地,确实地
crown	vt.	形成……顶部,给……加拱顶	psychologically	adv.	心理上地,心理学地
datum	n.	数据,已知数,基准点	reshape	vt.	改变……的形状,重塑,改造
flatness	n.	平坦,平滑;平直度,平面度	stilt	n.	支柱,支撑物,桩子
flight	n.	阶梯,阶梯步级	stride	n.	一跨(的宽度),一步
frost	n.	霜冻,严寒天气	subtle	adj.	精细的,微妙的,敏锐的
gradient	n.	坡度	suitability	n.	适宜性,适合性
gross	adj.	总的,毛的,显而易见的	terminology	n.	术语,专门用语,专门名词
irritation	n.	激怒,恼怒,令人恼火的事	undermine	vt.	逐渐损坏,侵蚀……基础
landslide	n.	滑坡,崩塌,塌方,山崩	urge	n.	驱策力,推动力;驱策,激励
leveling	n.	水准测量,水准,平整	variables	n.	变量,参数,变数
mild	adj.	温暖的,暖和的,温和的	vernacular	adj.	本国的,本地的
obstacle	n.	障碍物,绊脚石,妨碍	wheelchair	n.	轮椅

[参考译文]

风景园林工程 I

[1] 风景园林工程或土方整理是风景园林的基本技术。它包括重塑现有地形以满足场地规划的功能和交通要求,并确保排水通畅。因此,在场地规划过程中了解土方整理技术很有帮助。建筑与景观之间、室内与室外之间的连接面需要精细平整。住宅或建筑与土地直接相连可能会让人在心理上感觉更加舒适。进入位于地面

上的建筑或许是一种积极的心理冲动和需求。

［2］ 场地-结构关系既是视觉关系也是功能关系。建筑的地面层应该高于周围地面。建筑周边土地还应从建筑开始降坡，以使雨水不易进入建筑内部或破坏建筑基础。

［3］ 在福德希尔学院，原有景观曾进行大规模改造，使其变为能满足建筑和开放空间建造需要的可用土地。通过敏感性设计，这里的景观看起来好像并没有经过改造一样。建筑从山体边缘退后，舒缓地伫立在水平地面上。

［4］ 建筑与土地之间有两个主要的关系，土地可以被重塑或整理以适应建筑或工程需要，建筑也可以被调整以满足不同的地面高程变化以减少对原有地面的干扰。如图1，底层架空的建筑不直接接触地面，除了设置出入口几乎不需要地形整理。施工过程多少会影响土地环境，导致构筑物下光线和湿度条件的改变。但是，假如采用传统基础，会导致所有的建筑-地面关系的改变。建在陡坡上的单层建筑需要大面积的填挖方，会破坏原有土壤状况，导致不稳定地面条件，并最终导致侵蚀、滑坡、水灾和生态学彻底破坏。合适的场地建筑建造方法应依据详细的土地分析，如坡度、土壤、地质情况等。此外，在哪些区域布置建筑的最初决定应来自于土地适宜性研究，这样才能避免破坏，或者采用特殊建造技术来应对特殊场地条件。

土方整理的原则和技术

［5］ 现在让我们着手更传统的问题，即在场地规划中创造新的有效的景观来替代现状"自然"状态的景观，以使这片土地适合建筑和使用。土方整理代表着风景园林专业最独特的技术，与场地设计图纸落实在现状地形景观的过程相关。我们关注的不仅仅是使建筑和土地相连相适，还布置像运动场、停车场、交通路线这样的使用空间。所有这些都有特定的坡度、基础深度和排水要求。

［6］ 景观工程涉及经济性和现状条件敏感性。土方整理的基本原则和目标可以总结如下：
- 地表必须适合预期的目标或使用。
- 视景必须优美宜人；实际上土方整理的目标可能单纯为了美观，屏蔽视线或创造具有象征意义的地形。
- 建成的地形必须排水良好。
- 地形设计应当尽可能使新的地平与土地原来的状态接近。尤其是在非城市地段，现有景观代表着生态平衡，自然排水系统和成熟的土壤成分。
- 在重塑地面时，务必考虑周全，并在机械工作的允许范围内。土方机械基本上只能粗整理，很难做到精细处理，细致整理需要手工才能达到。
- 表土应尽可能地保留。表土可以被剥离、贮存，并在大量土方整理后回用。
- 在土方处理时，挖方量应与填方量基本持平。这样可以避免异处取土或异处弃土。

［7］ 地形设计是关键的技术文件，通过地形设计我们展示和计算三维地表空间的变化。地形等高线用来表示变化的程度。根据比例，等高线能够显示每一英尺或每五英尺的相对高程。现状等高线由虚线表示。不同于现状的新的地形方案由实线表示。实线与虚线错位的部分就是将要挖方和填方的地方，通常也表示着地形变化的程度和性质。这些由两组地形线组成的图纸，用来表达现状地形条件与设计意图的不同之处。挖方和填方量可以通过这些图纸计算出来。如果是用来计算的图纸，务必精确，方能保证造价预算的可信度。

［8］ 场地规划和土方整理要对场地边界内固定标高、构筑物和使用区域之间的高差进行必要的调整。固定标高或控制高程包括需要保留的现状树木或植被的标高，现状或规划建筑和道路的标高，场地边界的标高，需要纳入设计的现状地形，湖面和自然洼地，现状和规划地下设施的标高。这些高程限制着地形重塑的可能性。

［9］ 地形设计的术语很简单。等高线是一个固定基准面上高程相等的所有点的集合。高程点提供等高线以外的高程信息。它们用于标注微小地形变化，即位于等高线之间的特殊点。它们显示出各点之间的高差以确保水平表面上的排水，以及标注规划中重要部位的特定高程。标注高程点的典型位置包括台阶的顶标高和底标高，挡土墙的顶标高，建筑入口的内外地面标高。

［10］ 另外一个常常用到的术语是坡度。坡度是指两点之间的坡率，用百分比表示，或水平距离和高程上垂直变化的比例，或用角度表示。因此，1%的坡就等于1∶100。10%的坡等于1∶10或6°。50%的坡等于1∶2或

26°30′,100%的坡就是1∶1,或45°。计算坡度的关键变量是两点之间水平距离和高程上的垂直变化。

[11] 地面常常需要改变坡度以满足某些特定的最大或最小坡度要求。地形设计的过程包括三个要素的处理:坡度(G),水平距离(L)和两点之间的高差(D)。例如,如果距离100英尺的两点之间的现状坡度对于入口道路来说过大(比如15%),那么必须缩小竖向高差,或增加水平距离(到150英尺)。变量关系如下:

$$G = D/L; \quad L = D/G; \quad D = L \times G$$

改变这些变量会产生具有不同经济性和审美性的解决方案。

[12] 在景观设计中有普遍适用的最大坡度。这些数值由工程师、建筑师和风景园林师通过经验和实践获得。它们是场地设计中决定形式的因素,同时也是影响经济的因素。例如,步道和小径通常在横切面是一个拱顶,如果可能,在有寒冷冬季的地区,纵坡不应超过6%,在霜冻持续不久的温和地区,纵坡不宜超过8%。这些是长距离道路的推荐坡度。对于较短距离,可以采用坡度不大于12%的坡道;如果坡度超过12%,最合理的处理高差的办法是设置台阶,但是台阶应尽量避免。对于轮椅和自行车,台阶是通行的障碍和危险因素。三级或三级以下台阶会使人不悦,然而它们可用作非正式的座椅,尤其用在学校和大学校园中。

[13] 台阶是在尽可能短的距离内变换高程的纯功能因素,几个世纪以来,在本土模式和文艺复兴模式中都从步行行动发展出偏爱的尺寸。例如,有一条"法则"是阶高与阶宽的乘积应大约等于74英寸。这条法则允许几个尺寸组合,例如6英寸阶高12英寸阶宽,5英寸阶高15英寸阶宽,4英寸阶高18英寸阶宽。随着阶高减少,阶宽变宽。但是,当阶宽变得过宽时,行走的节奏和跨步的尺度必须进行计算。设计师应当测量和记录令人舒服的台阶。

[14] 道路和车道也有首选的最大坡度。尽管6%是令人满意的,8%~10%在短距离也是允许的。在特殊情况下(例如旧金山,幸好天气比较温和),街道也有15%甚至更陡的坡度。这些标准和经验法则必须根据具体需要解决的问题和地理、习俗和使用等情况来确定。

Reading Material B

Landscape Engineering Ⅱ

Michael Laurie

风景园林工程Ⅱ

Surface Drainage

[1] Rainfall is the usual source of surface water. When it falls, a proportion of it percolates into the soil, the amount depending on the soil type and vegetative cover. Another proportion of it drains or runs over the surface of the land to some low point on the site. Another proportion may flow off the site, and some will evaporate. That portion of the rain which does not enter the soil or evaporate is called runoff.

[2] Grading and construction provision must be made for this runoff, so that flooding will be avoided and valuable topsoil will not be lost by erosion. One of the functions of the grading plan is to shape the ground in such a way that rain water will flow through the site to collecting points without causing washouts. It is therefore economical and usual to grade the land in such a way that water is collected in channels, grass swales, or gutters (depending on the nature of the project) and directed around buildings and away from major use area into drain inlets connected to a storm sewer system.

[3] Predictions of water runoff quantities are necessary in order that the size of the pipes or swale

dimensions can be calculated and provided adequately so that they can cope with the worst possible conditions. Historical climate data are used to provide 100-, 50-, 25-, or 10-year storm expectancies[1]. The frequency selected depends on whether occasional flooding would be acceptable, as on playing fields and recreation areas. An agricultural engineering formula is frequently used in which the quantity of water (Q) arriving at any point in a watershed is derived from a combination of several variables.

$$Q = ACi$$

[4]   The variables are the area of the watershed in acres (A), a coefficient of runoff (C), and a quantity derived from the amount of rain that can be expected for a selected storm frequency combined with the farthest distance that a drop of water will have to run before reaching the collecting point (i).

[5]   The coefficient of runoff is the most interesting variable. It varies according to two site factors: the condition of the surface and the topography or slope. The coefficient represents the percentage of rain which is not absorbed or delayed before reaching a specified point or drain inlet. Tables have been prepared to give values for a variety of conditions. For example, in urban areas where 30 percent of the surface is estimated to be impervious (roofs, roads, and so on) in a generally flat landscape, 40 percent of falling rain will become runoff for drainage purposes. With a similar percentage of impervious surface in a situation of rolling terrain[2], the runoff may rise to 50 percent. As the amount of impervious surface increases, the quantity of runoff increases. With building roofs and roads that have runoff coefficients between 85 and 100 percent, it can be seen that urban conditions in areas of heavy rainfall either throughout the year or at selected times within the year pose very serious drainage problems in the design of open space and paved plazas. By contrast, in more natural situations the runoff is considerably less.

[6]   To the variables of surface cover and topography is added soil, which can have a considerable impact on the quantity of water that becomes runoff. Thus in a wooded landscape on flat ground with open sandy loam soil the runoff will be about 10 percent. In the same situation but with a heavy clay soil the runoff may be as much as 40 percent. As the topography becomes steeper, the rate of runoff increases. The coefficient of runoff on hilly land with slopes of between 10 and 30 percent with sandy soil will be about 30 percent, which will double if the soil has a high clay content.

[7]   These figures are significant not only for drainage purposes but also for water conservation. Paul Sears argues that the longer water can be kept on or in the land, the better for mankind. Delay of the water cycle in its land phase is desirable, and in certain circumstances spreading water over the land and allowing it to percolate into the soil, replenishing groundwater, is preferable to immediate removal by storm sewers. The zero runoff design at Village Homes, Davis, California, illustrates how this may be done in a favorable situation and climate.

[8]   For drainage purposes there are some minimum gradients for surface. Drainage, especially around buildings or use areas such as playing fields, must be positive. Water must flow if ponds and puddles are to be avoided. It is therefore undesirable to have dead level[3] surfaces. The minimum gradients vary according to the nature of the surface and/or its permeability. For example, the mini-

mum gradients for asphalt is between 1.5 and 2 percent. For smooth finished concrete[4] the minimum is 1 percent, whereas on coarser, exposed aggregate the minimum would be 2 percent since water will not flow as easily over the rougher surface. Bricks laid in sand need a minimum of 1 percent, whereas brick paving laid with grouted joints[5] needs 2 percent. Lawns and grass areas should have a minimum of 1 percent in the open and 2 percent adjacent to buildings and in grass swales.

[9]　The design of a storm sewer system is properly the work of an engineer. However, drain inlets are surface manifestations of the hidden underground system.

[10]　Their position and elevation are related to economics and the constraints of water flow technology. They are the places where landscape design meets engineering, requiring collaboration between the two. They must be designed as part of the surface treatment or pattern rather than incorporated as afterthoughts, as so often seems to happen.

[11]　Just as there are minimum gradients to facilitate positive drainage, so there are maximums. These are important in erosion control. Groundcover or grass planting on banks of 25 percent or more is absolutely necessary to prevent erosion and excessive runoff. In solving drainage problems without a storm sewer system, that is, in natural swales or streams, it is important to know what the effect of the runoff will be downstream. Silting up of streams and channels may occur, thus reducing their capacity to remove water from the area at a subsequent time.

[12]　Roof gardens require very rapid drainage since the weight of water added to soil, paving, and plants places stresses on the structure.

[13]　The slopes on paved surfaces in such situations should be greater than slopes normally required. Special lightweight and porous soil at least 12 inches in depth should be used in planting beds with subsurface drains to remove excess water.

Visual Considerations

[14]　In addition to all the functional considerations discussed so far, grading also has visual implications. The created land form can have an aesthetic appearance in its own right, as at the Stockholm Cemetery (Fig. 1) or at Foothill College. In addition, land form can help to screen out undesirable views such as parking lots and freeways (Fig. 2). With the addition of planting, earth mounds can provide a quick and effective visual barrier and also wind shelter. The sunken fence or Ha Ha[6], product of the eighteenth-century landscape garden, is a grading technique devised to conceal the boundary fence, the division between pastureland and garden. Changes in level may also be devised to separate circulation and social uses in site planning and detailed design (Fig. 3). Geometric land shaping as opposed to natural form can produce interesting and exciting results, as the wind protection earthworks at Sea Ranch show with bold angular forms. Land forms that are not imitations of nature are especially appropriate where new landscapes are being made in the restoration of derelict land[7] or in the disposal of excess fill[8]. Open-air theaters[9] provide other opportunities for creative land shaping.

UNIT 14　LANDSCAPE ENGINEERING

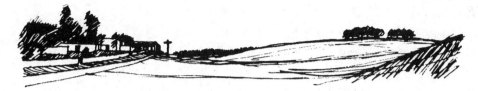

Fig. 1

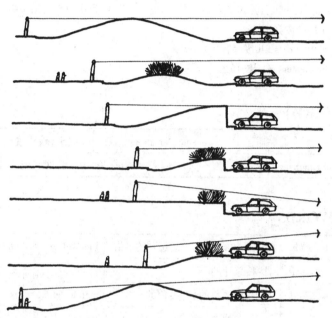

Land form combined with retaining walls and/or planting,and differences in relative levels can be used in design to conceal and separate.

Fig. 2

Articulation of levels to separate circulation and use areas.

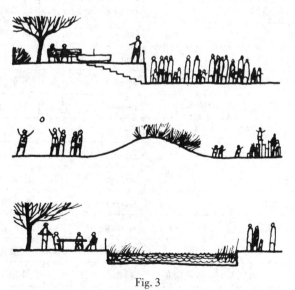

Fig. 3

· 261 ·

[NOTES]

[1] 10-year storm expectancies 十年一遇暴雨
[2] rolling terrain 起伏地区,丘陵区
[3] dead level 绝对水平
[4] smooth finished concrete 光面混凝土
[5] grouted joints 灌浆接缝
[6] Ha Ha 隐垣:英国自然风景园常用的干沟式边界,作为花园与周围田野的分隔。
[7] derelict land 废弃地
[8] disposal of excess fill 废弃物填埋场
[9] open-air theaters 露天剧场

[GLOSSARY]

clay soil	黏质土	sandy loam soil	砂壤土,砂质壤土	surface runoff	地表径流
coefficient of runoff	径流系数	storm sewer system	雨水管网系统	zero runoff design	零径流设计

[NEW WORDS]

afterthoughts	n.	后来添加的东西,事后想到的事物	lightweight	adj.	轻量的,薄型的
aggregate	n.	骨料,总计的	loam	n.	壤土,沃土
angular	adj.	有角的,成角的,有尖角的	manifestation	n.	表现,显现
asphalt	n.	沥青,柏油	mound	n.	土山,土丘,小丘
clay	n.	黏土,泥土	percolate	vi.	渗透,渗入
coarse	adj.	粗的,粗糙的	permeability	n.	渗透性
coefficient	n.	系数	porous	adj.	多孔的
collaboration	n.	合作,协作	puddle	n.	雨水坑,水坑,小塘
cope	vi.	成功地应付,对付	rainfall	n.	下雨,降雨,降雨量
derelict	adj.	被抛弃的,废弃的,被遗弃的	removal	n.	除去,清除
downstream	adv.	在下游,顺流地	replenish	vt.	补充,重新装满,添加,添足
drain	vi.	流干,排空,排出,排水	runoff	n.	径流,流量,溢流
evaporate	vi.	蒸发掉	silt up	vi.	淤塞,淤泥充塞
expectancy	n.	期待,期望	steeper	adj.	险峻的,陡峭的
formula	n.	准则,原则	sunken	adj.	凹陷的,下陷的
grout	vt.	给……灌灰浆	valuable	adj.	宝贵的,有价值的,极有用的
hilly	adj.	多小山的,多斜坡的,多丘陵的	washout	n.	冲蚀
impervious	adj.	不可渗透的,不透水的	watershed	n.	流域分,水岭,分水线

[参考译文]

风景园林工程 Ⅱ

地表排水

[1] 一般降雨是地表水的来源。当雨水降落时,一部分渗透到土壤,渗透量取决于土壤类型和植被覆盖情况。另一部分被管道排走或流经地表到达场地较低的点。还有一部分可能流出场地,还有一些会蒸发。不进入土壤或蒸腾的雨水被称为径流。

[2] 土方整理和建设必须考虑径流,以免发生洪灾和宝贵的表土层侵蚀流失。地形设计的一个功能就是塑造雨水能够流经场地到达汇水点而不引起冲刷的地表面。因此,地形整理经济且常用的做法是,以沟渠、草沟或暗沟(根据项目的性质决定)收集雨水并从建筑周边和主要使用区域排向接入雨水管网系统的排水口。

[3] 为了计算管道或草沟尺寸并提供充分的排水以应对最坏的状况,需要预测雨水径流量。历史气象资料可以提供100年、50年、25年或10年一遇的洪水可能性,在游戏场和游憩区域选择哪一种洪水可能性取决于可接受的淹水频率。一个经常采用的农业工程算式就是根据几个变量计算一个流域到达某一点的水量。

$$Q = ACi$$

[4] 变量分别是流域面积(A),径流系数(C)和在选定频率下从最远处到达集水点之前流经的水量。

[5] 径流系数是最有意思的变量。它根据两个场地因素变化:地表状况和地形或坡度。这个系数代表了在到达某个集水点或排水口之前未被吸收或滞留的雨水百分比。不同地表状况有不同数值,例如,约30%为不透水表面(屋顶、道路等)的总体较平坦的城市区域,40%的降雨会变为排水径流。有相似不透水地表面百分比的丘陵地区,径流会提高到50%。随着不透水地表比例的增加,径流量会相应增加。由于建筑屋顶和道路的径流系数在85%~100%,可以看到在有强降雨的地区,不管是全年还是一年内某一次,在开放空间和铺装广场设计中,这样的城市条件都会造成非常严重的排水问题。相比而言,更加自然的城市表面径流会小得多。

[6] 对于地面覆盖物的变量而言,地形若增加土壤,对变成地表径流的水量有很大影响。因此,在有树木景观的平坦砂壤土地面,地表径流大约为10%。同样的情况下,黏重土壤的地表径流会达到40%。随着地形变陡,径流率也随之增大。坡度10%~30%的砂质土壤的山地,径流系数大约是30%,如果土壤是重黏土,径流系数则会翻倍。

[7] 这些数字对于排水和水体保护都很重要。保罗·西尔斯认为水在陆地中保存的时间越长,对于人类越有利。延缓陆地阶段的水循环是有益处的,在某些情况下,将水分布于陆地,并下渗到土壤,补给地下水,比直接以雨水管道排除更可取。加州戴维斯乡村之家的零径流设计,说明了在有利的地表状况和气候条件下如何实现的。

[8] 对于排水,地表有一些最小坡度要求。建筑周边或像运动场这样的使用区域排水必须良好。如果要避免池塘或水坑,水就必须流动。因此,在完全水平的地表面并不理想。地面最小坡度因地表性质和/或可渗透性而不同,例如,沥青地面的最小坡度为1.5%~2%。对于光面混凝土最小的坡度为1%,然而粗质的露骨混凝土的最小坡度是2%,因为水在粗糙表面不易流动。铺在沙土中的铺砖地面需要最小1%的坡度,然而有灌浆接缝的铺砖地面需要最小2%坡度。开阔区域的草坪和草地的最小坡度应该为1%,靠近建筑以及草沟内的最小坡度为2%。

[9] 雨水管网系统的设计是工程师的工作。但是,排水口是这个系统的地面呈现。

[10] 他们的位置和高程与经济性和水流技术限制有关。他们是景观设计与工程设计的交汇点,需要双方的通力合作。他们必须设计成地表面处理或地表形式的一部分,而不是后来添加的内容,事实上这种情况却经常发生。

[11] 正如设置最小坡度以促进排水一样,坡度也有最大限制。他们对于控制侵蚀非常重要。在坡度大于或等于25%的堤岸,为了防止侵蚀和过大的径流,种植地被或草坪植物是绝对必要的。在没有雨水管网系统的情况下解决排水问题,即自然水沟或溪流,了解径流顺流而下的影响很重要。溪流淤积和渠化可能会发生,进而减少这个区域的排水能力。

[12] 屋顶花园要求非常快速的排水,因为加载在土壤、铺装和植物上的雨水重量会对建筑结构施压。

[13] 这种情况下,铺装表面的坡度应当比普通情况要大。种植坛中应采用至少12英寸深的特殊轻型多孔土壤,并铺设地下排水管以排除多余雨水。

[14] 除了目前讨论的所有功能考虑之外,土方整理还具有视觉意义。创造的地形本身就有一种美学外观,正如斯德哥尔摩墓地或福德希尔学院。此外,地形还有助于屏蔽不佳景观,如停车场和高速公路。再辅以种植和土丘,能提供快速有效的视觉屏障以及风障。产生于18世纪风景园林下沉式围墙或隐垣,就是用来隐藏花园围墙边界的造园技术,以划分牧场与园林。场地规划和细部设计的台级式地形也是用来分隔交通和不同功能的方法。与自然地形不同的几何式地形能创造有趣的和令人兴奋的效果,就像海滨牧场的防风地形采用大胆的有角度地形,这种不模拟自然的土地形式,尤其适合于废弃地修复或废弃物填埋场的新景观营造,露天剧场还为创造性的土地形式提供其他机遇。

[思考]

(1)风景园林工程项目建设过程中的土方工程有哪些需注意的问题?

(2)风景园林工程中地表水的收集与排放有哪些措施?

(3)景观工程中步行道和车行道的工程设计应注意哪些问题?

PART VII

Studies and Development of Landscape Architecture
第七部分 研究与动态

UNIT 15 LANDSCAPE STUDIES AND DEVELOPMENTS

TEXT

Creating Character through Sustainable and Successful Landscape Design[1]

David Ellis(UK)[2]

以可持续方式营造成功景观设计

Abstract: The concept of Eco-towns has aroused great interest but has also generated considerable skepticism and opposition. However, what is indisputable is that the physical context of successful eco-towns—how these are designed, the landscape features and the impressions the scheme makes in terms of "look and feel"—is critical to building a scheme's character, contributing to its success as a sustainable, attractive and enjoyable place to live. This article will identify best practice in landscape design and architecture from PRP's recent European research project and include 1-2 of PRP's current landscape projects, using these to illustrate how best practice is being applied successfully today.

Key words: Landscape Architecture; Eco-town; Planning and Design; Sustainable Development

PRP is very much at the sharp end influencing the character and development of new settlements urban extensions and urban regeneration projects in the UK. We consider that landscape architecture has a crucial pivotal role within the design and development process in creating sustainable environments that will be respected and appreciated for those who will occupy them.

1 Achieving Best Practice

A vital part of achieving quality landscape is to benchmark projects in relation to national and international precedents. To this end, PRP has organised a variety of study tours aimed at raising expectations amongst consultants, client groups and planning authorities. This has included visits to the following new settlements:

Vauban & Rieselfeld, Freiburg, Germany
Kronsberg, Hanover, Germany
HafenCity, Hamburg, Germany
Hammarby Sjöstad, Stockholm, Sweden (Fig. 1)

Fig. 1 Hammarby Sjöstad-High quality landscape utilising surface water drainage for amenity and biodiversity

Cathorst, Amersfoort, The Netherlands

Adamstown, Dublin, Eire

In conjunction with these study tours, PRP in collaboration with URBED and Design for Homes has recently completed a research project entitled Beyond Eco-towns-Applying the Lessons from Europe. This document is now available to download from the PRP website.

Intrinsic to the success of the above projects is the role of landscape design. We are of the opinion that designing successful places is far more challenging and elusive than delivering iconic architecture.

With respect to our recent European research and ongoing project experience, we have identified the following aspects of planning and design that give credence to sustainable landscape and place making:

2 Landscape and Urban Coalescence

The legacy of mid 20th Century town and country planning policy in the UK has assisted in conserving green spaces close to city centres and preventing the worst effects of urban sprawl which is characterised in a majority of developed countries.

The positive effect of this "Green Belt" Policy[3] (Planning Policy Guidance 2) apart from kerbing unbridled economic led city development has prevented urban coalescence and effectively maintained spatial distinctiveness of neighbouring towns and cites. The principals of maintaining openness between nucleated urban developments is still a major consideration on virtually all of the current urban extension and new communities being developed in the United Kingdom.

3 Landscape & Visual Impact Assessment (LVIA)

The European Commission Directive 85/337/EEC requires that an Environmental Impact Assessment (EIA)[4] is undertaken on projects over a particular size and character. This invariably includes new settlements and urban extensions. As part of this exercise a Landscape and Visual Impact Assessment (LVIA) would be carried out.

The LVIA is vital in unravelling the complex nature of existing landscapes, how robust they are to accommodate change and how their character can be reinterpreted through the consequent design and development process. Understanding the three dimensional qualities and the visual impact of a new settlement is paramount to benchmarking its character and determining the best environmental "fit".

4 Landscape as Infrastructure

New planning policy in the UK is providing the impetus at strategic and local level for the designation of "green corridors". This is primarily to connect up isolated ecological areas which once were part of a wider wilderness environment but have become fragmented through agricultural and urban development.

The East London Green Grid Framework[5] (Fig. 2) has recently been adopted as Supplementary Planning

Fig. 2　Proposed East London Green Grid

Guidance to the London Plan. A "Green Grid" lattice is proposed in this part of London and within the Thames Gateway providing the context for new development.

A majority of our master planning projects particularly lower density urban extensions are predicated on providing green infrastructure incorporating amenity biodiversity and productive landscape features as a framework for the built environment.

5 Biodiversity & Local Provenance

The UK Government Code for Sustainable Homes (CSH)[6] provides a method for scoring the environmental performance of a project and this includes its biodiversity rating. The biodiversity rating is benchmarked on the existing ecology of the site and the degree of change created by providing more extensive habitat through native planting and ecological enhancements. There is therefore a presumption on all projects to increase the biodiversity value of a project above that which existed before development.

One of the more subtle aspects of ecological place making, is to identify endemic and indigenous plants that currently or/and historically occupied the area, and to re-establish these in the new proposals as part of its intrinsic character.

Some local authority Biodiversity Action Plans (BAP)[7] stipulates that only local provenance native plant material is used in preference to imported plant material of the same species. This is to conserve genetic diversity and local and regional distinctiveness.

6 Landscapes for People

Consistently the most successful master planning projects are developed with the mandate that pedestrian scale and access is the primary driver to site layout and land use. This is fundamental to achieving successful public spaces and landscapes which require surveillance and need to be animated by pedestrian activity.

Fig. 3 Collaborative Design Workshop with members of the local community-carbon neutral master planning project for English Partnerships

Without exception, all of the best European examples of regeneration and urban extensions have involved close collaboration with the local community and municipality in overarching design proposals and also down to the specifics of private and communal open spaces.

Ownership of the master plan by the local community is also paramount. Whatever we design needs to be appropriate for the community and the end user. Big ideas are fundamental for any successful master plan project. The big idea may start with the design team however it may also be given credibility through collaboration with community group or stakeholder (Fig. 3).

PRP puts great emphasis on involving stakeholders and the wider community in developing major master planning proposals so that the cultural, physical and economic needs of the community are met. This is vital to ensure that nothing has been overlooked by the master planning team during the desktop study and appraisal stages of the project.

7 Landscape for Health & Wellbeing

There is significant empirical evidence that the quality of the environment has a major effect on the state of people's health and wellbeing. Consequently there are increasing expectations to provide access to recreational facilities and green spaces within high density neighbourhood areas. This also includes providing access to healthy food. In response to this the Hammarby project in Stockholm has successfully incorporated vegetable growing areas in the heart of the development.

All large UK master plans now require a full play and recreation strategy to identify provision of facilities across different age bands. Play is also now considered to be an intrinsic part of the landscape instead of seen as an add-on; therefore we now have parkland areas and playable streets that aren't totally dependent upon specific items of play. The latter may constitute a "homezone"[8] where pedestrians are given total priority over traffic. The street effectively becomes a space and focus for social activity and play.

8 Landscape for Microclimate & Air Quality

Vegetation can contribute significantly to improving the microclimate of urban areas and streetscapes. This is particularly important with regard to mitigating the adverse effect of tall buildings and urban canyons to reduce down drafts and wind speeds. Deciduous trees can also temper solar gain for adjacent south facing buildings and trees in the UK generally can save as much as 10% of annual energy consumption around buildings by moderating the local climate.

Vegetation is also effective in precipitating dust and harmful particulates this is particularly relevant along the boundary of urban motorways where vegetation buffers can improve air quality as well as contribute to ecological and amenity value.

The post war development of Freiburg in south-west Germany has involved the deliberate retention of an open south east to north-west urban grid structure. This facilitates the natural phenomenon of cool clean air falling at dusk from the wooded hills to the south-east passing through the city and flushing out the polluted air of the previous day.

9 Landscape for Noise Attenuation

Noise levels above 70dBb are considered to have a substantial adverse impact upon amenity. Many high density schemes in London are also located adjacent to large scale transportation infrastructure. An intrinsic part of the design process is the determination of suitable land uses in relation to existing or proposed noise generators. Often the solution may involve land modelling and the use of landscape bunds and noise barriers. This has been achieved in an exemplary way on the Hammarby project in Stockholm Sweden, where a major urban motorway is barely audible and lost to view

Fig. 4 Green track bed solution for new tramway

by the careful design of the adjacent landscape, provision of noise barriers and a green ecological bridge.

In Vauban, Freiburg, the extension to the tram system incorporates a track bed laid to grass which provides visual amenity, infiltration for surface water run off, ecological habitat and also as a baffle to the noise generated by the trams (Fig. 4).

10 Landscape for Rooftops

Fig. 5 Extensive green roof system to communal cycle storage, Wembley

The London Plan is moving towards a mandatory requirement for all roof surfaces on major developments to be exploited for the use of solar arrays or as green roofs. The latter being either as an accessible roof garden amenity space or as an extensive green or "brown" roof[9] (Fig. 5).

Brown roofs, consisting of crushed aggregate, often reused waste from site clearance can provide habitat for invertebrates and certain species of ground nesting birds that would otherwise be vulnerable to predation by domestic animals at street level. This is of particular value on high density developments where there are few opportunities to incorporate wilderness landscapes at street level.

PRP Landscape has recently proposed a very high density scheme in Blackheath south east, London, where each of the roofs reflects the geological characteristics of the surrounding landscape. The lower buildings have calcareous roofs, reflecting the ground level conditions and the taller buildings are provided with acid heath habitat to reflect the historic character of Blackheath on the elevated escarpment nearby. We have proposed to translocate rare plants and species from an isolated remnant habitat on Blackheath to populate the new green roof areas.

11 Landscape for Drainage

In London, extensive areas of front gardens have been paved over to provide "off street parking" for cars. This has had an enormous impact on increasing surface water run-off and also the loss of amenity and biodiversity to many of London's residential streets. The loss of natural habitat in upper water catchment areas and increasing urbanisation has also exacerbated the level of surface water run-off in many parts of the United Kingdom.

On some sites in the United Kingdom, the Government's Environmental Agency is requiring that the equivalent of green field surface water attenuation levels are achieved even if they are fully urban in character. Consequently the water strategy is now a crucial layer of the master plan. Solutions include the provision of swales infiltration ditches and attenuating ponds. In high density projects this may involve underground storage, combined with requirements for irrigation of plant material in restricted urban locations within street sections or above podiums.

12 Landscape from Waste

Increasingly there are proprietary paving systems available in Europe that incorporates various de-

UNIT 15 LANDSCAPE STUDIES AND DEVELOPMENTS

grees of recycled aggregates. Recycled glass is also being used as a blinding layer and also as a wearing course to pedestrian paved surfaces.

Topsoil is generally a precious commodity on most development sites and in the case of previously developed land may be contaminated or polluted. The obvious approach is to remove the contaminated soil to waste. However there are sometimes opportunities for creating a growing medium on site by using on-site waste material and combining this with waste organic matter to create manufactured topsoil. Many ecological regimes rely on specific growing conditions to flourish and the manufacturing of soil on site can effectively be tailor made to suite the vegetation regime. Species rich calcareous wild flower meadows have effectively been grown over crushed concrete with spectacular results.

13 Landscape for Climate Change and Future Proofing

There is a need to make buildings adaptable and flexible so that structures can be converted to respond to changing tenures in the future. This may be, from residential to commercial and vice versa. Landscapes also need to be flexible to accommodate change of use and to respond to the impacts of climate change.

Our Myatt's Field Park proposal for south London is designed so that most of the green areas could be converted to agricultural production given the concerns for food security in the future. Planting is also being proposed that will be less reliant on irrigation.

14 Case Studies

14.1 New Urban Quarter

A one hundred and twelve hectare urban extension to the historic collegiate city of Cambridge

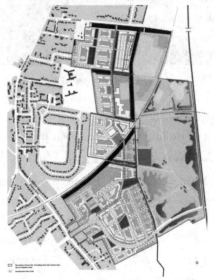

Fig. 6 Urban extension Master Plan to Cambridge

(Fig. 6-Fig. 7). Designing the streets and public spaces is as important as getting the buildings right and the central spine is designed as one entity, incorporating a range of pubic spaces along its route.

Fig. 7 The main spine route and armature to the master plan

(1) The design of the access routes and public spaces came first.

(2) The master plan is "daisy chained[10]" upon the new Cambridge Guided Bus route which runs into the city centre.

(3) The access strategy put pedestrians and cyclists first.

(4) Sustainable Urban Drainage is an intrinsic element of the overall the design.

14.2 New Urban Square

Fig. 8 Night time view of new public square

This space in Earls Court, London was designed as the focus to a new mixed use development, using high quality materials and designed for the day and night time economy (Fig. 8).

(1) Pedestrian movement takes priority over vehicular access.

(2) External pavements and street furniture is co-ordinated with building elevations.

(3) The quality and choice of the new square responded to the elevational quality of the adjacent new and refurbished buildings.

(4) Exotic planting has been provided to provide seasonal visual experience.

14.3 New London Park

Myatt's Field North Estate in south London is earmarked by the local authority for wholesale regeneration. Our proposed solution is driven by the need to create a high quality park that would become the focus for the neighborhood (Fig. 9-Fig. 13).

(1) All pedestrian cycle routes are funnelled through the park spaces.

(2) An "earth sheltered" Community Centre is proposed at the heart of the development and within the park.

(3) Orchard-cum-allotment is provided within secure walled garden.

(4) Playable landscape features have been included in the street spaces as well as the park.

(5) Part of the park can revert to agricultural use.

(6) All surface water run-off from the park is attenuated into wetland strips.

(7) Single aspect development is maintained along all edges of the park to optimise surveillance.

(8) Biodiversity value is optimised by the planting of native species.

(9) Ecological and amenity value. The park also includes opportunities for growing vegetables and fruit by the local community.

14.4 Communal Gardens for High Density

This is part of the first phase of the Wembley Master Plan which surrounds the stadium. We have provided this secluded private communal garden within this high density perimeter block and over a basement car park slab (Fig. 14-Fig. 15). Success of these hemmed-in spaces is dependent on careful collaborative design between landscape architect, engineer and architect and ongoing management strategy to ensure that the landscape is properly nurtured and successfully matures.

(1) A key consideration in delivering this project was to maintain appropriate level of daylight within the courtyard spaces for amenity and plant establishment.

UNIT 15 LANDSCAPE STUDIES AND DEVELOPMENTS

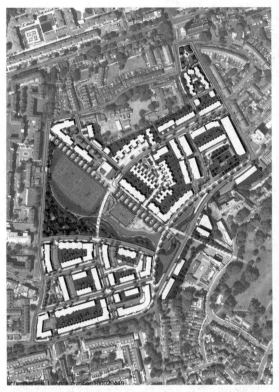

Fig. 9 Myatt's Field North regeneration master plan incorporating major new park.

Fig. 10 Spring

Fig. 11 Summer

Fig. 12 Autumn

Fig. 13 Winter

(2) The design of the supporting slab was paramount with regard to achieving a green solution incorporating significant tree planting. The courtyard incorporates private level "yards" to podium level apartments, the majority of the area however is private communal space.

15 Conclusions

A country's ability to develop in a more sustainable way depends upon the capacity of its people and institutions to understand complex environmental and

Fig. 14 Private communal gardens over podium

development issues so that they can make the right development choices. At the heart of this is the landscape and natural environment. Within the development professions landscape architects therefore have the toughest and most challenging job of all.

Creating great places, where people actually want to live in the long term and which are beautiful as well as deliverable is a major challenge. We consider that the holistic combination of planning, environmental science, design and management skills constituting landscape architecture to be the supreme "Art of Survival".

[NOTES]

[1] 本文经作者同意,转载于《风景园林》2009 年第 1 期,本文属于项目介绍型论文。
[2] David Ellis 戴维·埃利斯:英国风景园林师,PRP 公司副董事,擅长新城镇建设、交通基础设施更新、可持续和环境评估设计。
[3] "Green Belt" Policy 英国绿带政策:是英国城市规划政策最显著的特点之一,并在世界现代城市规划史中占有重要的地位。特别是伦敦城市规划的绿带模式,影响深远,被世界上许多国家的城市效仿。
[4] Environmental Impact Assessment(EIA)环境影响评价:指环保部门对规划和建设项目实施后可能造成的环境影响进行分析、预测和评估,提出预防。
[5] The East London Green Grid Framework 伦敦东部的绿色网络框架:它是最早运用景观和以民为本的绿色基础设施的方法的一种空间框架。这一举措在 2008 年赢得了景观研究所的战略景观规划与总统奖。
[6] Code for Sustainable Homes(CSH)可持续房屋守则:是英国的一种对住房的环境影响评级制度,它为能源效率和不受当前建筑法规强制性规定的持续性制定新的标准。
[7] Biodiversity Action Plans(BAP)生物多样性行动计划:是国际公认的程序,处理受威胁物种的栖息地,旨在保护和恢复生物系统,对于这些计划的原始动力来源于生物多样性公约(CBD),截至 2009 年,已有 191 个国家批准了生物多样性公约。
[8] homezone 家庭区:它是在英国境内用于住宅区街道或主要目的是满足当地社区利益的街区的一个术语,无论是步行、骑自行车或开汽车,都能使主要街道作为社会的公共空间来使用。这些原则类似于共享空间类型的设计,这也适用于更广阔的环境范畴。
[9] green or brown roof 绿色屋顶:是指一栋楼房的屋顶部分或完全被植被与土壤所覆盖,或者是将生长中的中间物种植于防水卷材之上的做法。也可以包括如根部屏障,排水和灌溉系统的附加层。这个术语不是指屋顶的颜色仅仅为绿色,像是绿色的屋面瓦之类的;褐色屋顶,也称为"生物多样化的屋顶",褐色屋顶的建筑技术一般都类似于创建绿化屋顶,主要的区别在于所选用的材质(通常是源于本地的瓦砾、碎石、开掘时所挖起的弃泥等),以满足具体的生物多样性目标。
[10] daisy chained 菊花链:一种阻塞式级联结构,排在前面的优先级高于其后的所有单元。

[GLOSSARY]

crushed aggregate	碎骨料	Landscape and Visual Impact Assessment(LVIA)	景观和视觉影响评估
daisy chain	菊花链	noise attenuation	噪声衰减
deciduous tree	阔叶树,落叶树	site clearance	(施工前的)场地清理
drainage	排水,排水系统	solar gain	吸收日光
hemmed-space	限定的空间	walled garden	墙面花圃;墙园

UNIT 15　LANDSCAPE STUDIES AND DEVELOPMENTS

[**NEW WORDS**]

adjacent	adj.	邻近的,毗连的	lattice	n.	格子,格架,晶格
adverse effect		不利影响,副作用	mandate	n.	命令,指令;委托管理;授权
aggregate	n.	总量,聚合物,合体	mandatory	adj.	命令的;强制的,义务的
allotment	n.	分配,分配物,命运	mitigate	vt.	使缓和,使减轻
amenity	n.	舒适,礼仪,愉快,便利设施	municipality	n.	自治市或区,市政当局,市民
amenity	n.	舒适,礼仪,愉快,便利设施	nucleate	adj.	有核的
animate	vt.	使……有生气,赋予生命	opposition	n.	反对,敌对,反对派
appraisal	n.	评价,估计,估价	overarch	vt.	成拱形
attenuation	n.	衰减,变薄,稀释	particulates	n.	微粒,大气尘,微粒物质
audible	adj.	听得见的	pedestrian	n.	行人,步行者
baffle	n.	挡板,困惑	perimeter	n.	周长,周界,视野计
benchmark	n.	基准	podium	n.	裙房,乐队指挥台,矮墙,墩座墙
buffer	n.	起缓冲作用的人(或物)	precedent	n.	判例,前例,先例
bunds	n.	堤岸工程,河岸,码头	precipitate	n.	沉淀物
calcareous	n.	钙化软骨	predation	n.	掠夺,捕食
canyon	n.	峡谷	presumption	n.	放肆,傲慢,推测
coalescence	n.	联合,接合,合并	proprietary	n.	所有权,所有人
contaminat	vt.	污染	provenance	n.	出处,起源
credence	n.	信任,凭证,祭器台	provision	n.	供应品,准备;条款,规定
cum	prep.	和,连同,附带	refurbish	vt.	翻新,刷新
deliberate	adj.	故意的,从容的,深思熟虑的	regime	n.	政权,政体;社会制度,管理体制
dusk	n.	黄昏,薄暮;幽暗,昏暗	reinterpret	vt.	重新解释
earmark	vt.	在耳朵上做记号,标记	remnant	adj.	剩余的
elusive	adj.	难懂的,易忘的,难捉摸的	retention	n.	留,滞留,保留;记忆力
empirical	adj.	经验的,实验上的	robust	adj.	强健的,健康的;粗野的,粗鲁的
endemic	adj.	风土的,地方性的	secluded	adj.	隐退的,隐居的,隐蔽的
escarpment	n.	悬崖,绝壁,陡坡	skepticism	n.	怀疑,怀疑论,怀疑主义
exacerbate	vt.	使恶化,使加剧,激怒	slab	n.	厚片,厚板,平板,混凝土路面
exemplary	adj.	典范的,可仿效的,惩戒性的	spine	n.	脊柱,脊椎,刺
exotic	adj.	外来的,异国的,异国情调的	stakeholder	n.	利益相关者
flush out	v.	冲掉,排出;驱赶出来	stipulate	vt.	规定,保证
fragmented	adj.	成碎片的,片断的	subtle	adj.	微妙的,敏感的,狡猾的,精细的
funnel	vt.	通过漏斗或烟囱等,使成漏斗形	supreme	adj.	最高的,至高的,最重要的
holistic	adj.	整体的,全盘的	surveillance	n.	监督,监视
impetus	n.	促进,动力,冲力	temper	vt.	调和,锻炼,使缓和
indigenous	adj.	本土的,国产的,固有的,土著的	tenure	n.	占有,任期
indisputable	adj.	明白的,无争论余地的	topsoil	n.	表层土,上层土
infiltration	n.	渗透,渗透物	unbridled	adj.	无拘束的,放肆的,激烈的
intrinsic	adj.	本质的,固有的	unravell	vt.	阐明,解决
invariably	adv.	不变地,总是,一定地	vehicular	adj.	用车辆运载的,车辆的
invertebrate	n.	无脊椎(动)物	vice versa		反之亦然
kerb	vt.	设路缘(或井栏)于……	wholesale	n.	批发

[参考译文]

以可持续方式营造成功景观设计

戴维·埃利斯

摘要： 尽管有不少人怀疑或反对，生态城镇的概念已经引起了人们广泛的关注。毋庸置疑的是，成功的生态城镇都有一个共同的特点，即项目是如何设计和规划，以及这些项目的景观特征和印象是如何从"视觉和知觉"的角度出发形成的，这对塑造项目特征、成功修建宜人的可持续场所起着至关重要的作用。文章结合PRP公司优秀案例进行分析，阐述当今最好的实践案例是如何成功应用的。

关键词： 风景园林；生态城镇；规划设计；可持续发展

在英国的城市人居环境扩展和旧城改造项目领域，PRP在项目开发和营造景观特色方面是最具权威和影响力的有力旗舰。我们认为，可持续景观为使用者所尊重和欣赏，而景观学则在可持续景观的设计发展过程中至关重要。

1 卓越景观的实践之道

要营造有品质的景观，关键是要给项目打上国内或国际前沿的印记。为此，PRP组织了众多不同的旅行考察，以此来达到咨询师、客户群和政府规划部门的期望值。我们考察过的住宅区有德国弗莱堡市的沃班和瑞斯尔弗尔德、德国汉诺威市的康斯伯格、德国汉堡的港口新城、瑞典斯德哥尔摩的哈马比·索斯泰德（图1）、荷兰阿默斯福特的卡索尔斯特以及爱尔兰都柏林的亚当斯敦。

结合这些旅行考察，PRP最近还与URBED公司和霍姆斯设计公司（Design for Homes）合作，完成了一个名为"超越生态城镇——借鉴欧洲实践经验"的研究项目。现在这篇文章可以从PRP网站下载。

景观设计所扮演的角色是上述案例成功的内在因素。我们认为，要成功地设计一个场所，远比实现一个标志性建筑更具挑战性和难以掌控。

结合我们最近对欧洲的研究和正在进行的项目经验，我们确定了以下规划和设计的方向，为可持续性景观的设计和项目开发提供参考。

2 景观和城市结合

在英国，受惠于20世纪中期城镇和乡村规划政策，城市周边地区的绿地得到保护，从而避免了城市扩展所带来的常见不利影响。

这个《规划政策导则2》中"城市绿带（Green Belt）"不仅控制了在经济利益驱使对城市的无止境开发，防止城市的聚集发展，还有效地维持了城镇和城市邻里之间的独特空间景观。在核心城市开发中维持城市开放性的原则，仍是当前所有英国城市扩展和新社区建设过程中需要考虑的主要因素。

3 景观及视觉影响评估（LVIA）

欧盟委员会法例85/337/EEC要求，当项目超过特定规模和特征时，需要进行环境影响评估。新居住区建设和城市扩建常常包含在这些项目中，而景观和视觉影响评估（LVIV）通常作为环境评估的一部分开展。

LVIA是了解景观现状复杂特性、景观适应变化的能力以及如何将它们的特征通过一系列的设计发展过程进行改善的重要方法。而了解新居住区的三维品质和视觉影响对设定其景观特征和环境的适应性是至关重要的。

4 景观作为基础设施

新的英国规划政策正为"绿色走廊"设计提供一个策略性和本土层面的推动力。首要的是将分离的生态区域即因农业和城市的发展而变得支离破碎的广袤原野连接起来。

东伦敦绿色栅格框架（图2）最近被纳入《伦敦规划》的补充规划导则。框架建议以泰晤士河口为背景，在相应的伦敦区域进行"绿色栅格"规划。

我们大部分的总平面规划项目，尤其是低密度城市扩展项目为结合游憩功能、生态多样化和生产景观特征的城市环境框架提供了绿色架构。

5 生物多样化与本土化

英国政府的可持续住宅标准(CSH)为工程的环境特性提供了打分方法,其中包括生物多样性的评级。生物多样性的评级以场地生态现状为基准,分析经过种植本地植物和改善环境以丰富生境所带来的变化。因此,在项目开发之前,所有项目都有一个提高生态多样化价值的提议。

其中创造生态场地更为微妙的一个方法是辨别场地历史上曾覆盖的本地物种,在新提案当中重新种植这些植物,成为本土特征的一部分。

一些地方政府《生物多样性实施计划(BAP)》中规定:同一物种,本土植物应比外来衍生植物优先采用,以保护基因的多样性以及本土和区域的特异性。

6 景观关注民生

最成功的总体规划项目在进行场地划分和土地使用时,对人类尺度的步行道和通道的考虑总是优先于车行道,这是获得安全及可控制人性化的公共空间和景观的成功基础。

无一例外,所有的欧洲城市更新和扩展范例在整个方案设计过程中都涉及与当地社区及政府的合作,包括私人和社区开放空间。

地方社区对总体规划的参与也是至关重要的。无论我们设计什么,都需要与社区和最终使用者相宜。好创意对任何成功的规划来说都是最基本的,有可能从设计小组中产生,也有可能通过与社区小组或股东的合作中产生(图3)。

PRP公司非常重视股东和大社区居住者在总体规划过程中的参与,确保社区的文化、物质和经济需求得到满足以及在项目的研究和评估阶段,没有因素被设计组忽略。

7 景观带来健康与幸福

实践证明环境质量强烈地影响着人们的健康和幸福,因此,在高密度住宅区修建康乐设施和公共绿地的呼声不断增强,同时还包括健康食品的供给。与此相呼应的斯德哥尔摩市哈马比·索斯泰德社区,成功地将蔬菜生长地融入开发区域的中心。

如今,英国所有的大型总体规划设计需要全面的休闲娱乐策略来满足不同年龄阶段的需要。嬉戏游玩是景观设计的一部分,而不是附属品,因此,我们现在拥有稀树草原和游憩街道并不属于具体的游乐项目。后者可能组成一个"家庭地带",而在这里,行人比机动车辆拥有优先权,街道有效地成为一个社交和嬉戏的空间和焦点。

8 景观调节微气候,改善空气质量

植物对于改善城市区域和街区的微气候有着至关重要的作用,尤其对减少高层建筑和密集城市的负面影响,减小不利因素和风速起着特殊作用。落叶林还能调整南向建筑的太阳能吸收量,英国建筑周边树木对当地气候的调节,每年大约能节省15%的能源消耗。

植物还可以有效沉淀灰尘和有害颗粒,高速公路边缘的植物缓冲带不仅具有生态价值和休憩价值,还能改善空气质量。

战后德国西南部的弗莱堡开发中,特意保留了一个从南到北的开放式城市网格结构,有助于干净冷空气在黄昏时分从山林往城市东南部方向的下沉流动,使其穿越城市并且排出前一天的污染空气。

9 景观弱化噪音

噪声70分贝以上对环境休憩功能有极大的负面影响,伦敦许多高密度项目坐落在靠近大型交通基础设施的地方。设计过程中的核心部分是根据现存和拟定的噪声总量来确定土地的合理使用。通常,解决的方法可能是采用土地建模、景观堤岸或噪声隔板,这已经在瑞典斯德哥尔摩市的哈马比项目中成功示范,精心设计的相邻景观噪音隔板和绿色生态桥,最大程度地减少了城市高速干道产生的噪音影响。

在弗莱堡沃班,城市列车系统的扩建将轨道设置与草地结合,不仅达到了愉悦的视觉效果,还有助于地表

径流的净化和生境的形成,并为列车引发的噪音提供了屏障(图4)。

10　屋顶景观

《伦敦规划》逐步强制要求大型开发项目的屋顶必须进行充分利用,用于安装太阳能电池阵或者设计成绿色屋顶。绿色屋顶可以是可达性强的屋顶花园游憩空间,也可以是大型绿色屋顶或棕色屋顶(图5)。

棕色屋顶由小碎石组成,通常是场地清理时的可循环垃圾,能为软体动物和某些地方巢穴鸟类提供栖息地。不然,这些小动物很容易在街上被家养动物捕食。这对很少有机会将野外景观与街景结合的、高密度发展的地区来说有着独特的生态价值。

PRP景观部最近为伦敦东南部的布莱克赫什区提议了一个极高密度项目,每一个屋顶都反映了周围景观的地质特征。低层建筑屋顶有石灰质特性,反映了地面特征;高层建筑长有酸性石南丛,反映布莱克赫什区附近的陡峭悬崖这一历史特征。我们还计划从布莱克赫什区里残留的独立生境中移植少量植物种类来丰富新绿色屋顶植被的多样性。

11　景观引导排水

在伦敦,大量的前花园已经铺上地砖,为车辆提供"路边泊车"区域。这对地表水排放产生了巨大的影响,并且也影响了许多伦敦居住街区应有的游憩性和生物多样性。上游集水区自然栖息地的流失和城市化的加剧使得英国许多地方的地表水排放情况进一步恶化。

英国一些地方政府的环保机构要求,即使在完全城市化的地区,绿地也要达到地表水的衰减程度。因此,排水策略是总体规划的重要部分,解决方案包括洼地过滤和滞留地。在高密度的项目中,还可能涉及地下储存以及植物在有限的城市地区部分街道或架空结构的灌溉。

12　回收材料塑造景观

在欧洲,越来越多的特制铺装系统结合添加了不同比例的回收碎石,可再生玻璃也用于基础垫层和磨耗层铺装在人行道上。

绝大多数开发场地的表层土是珍贵物品。早先开发的土地,表层土可能被污染了,以前的处理方法是直接将污染土壤当成垃圾除掉。现在我们常常利用场地现有材料营造生长媒介,将之与有机垃圾结合在一起创造人工土壤。许多生态领域的繁荣依赖于具体的生长条件,在场地上制造土壤能有效适应植物的生长。物种丰富的野生花草也以惊人的速度生长在碎石之上。

13　景观适宜气候变化和未来

建筑必须具有适应性和灵活性,结构可以转换以应对不断变化的未来。以此类推,景观也需要能够灵活适应功能与气候变化所带来的影响。

我们为南伦敦米亚特野外公园做的设计方案,出于未来的食物安全考虑,将绝大多数绿地转化成农业产地,植被也尽量考虑较少依靠灌溉的品种。

14　案例分析

14.1　新城市一角

在占地面积为112 hm² 的古老剑桥大学城的扩展中,街道和公共空间的设计与建筑的设计同等重要,中心主干道沿路结合一系列的公共空间,成为一个整体(图6－图7)。设计遵循以下原则进行:(1)行人通道和公共空间优先的设计理念;(2)"菊花链"式总体规划沿着新剑桥区到城市中心的公共汽车导游路线展开;(3)通道设计中步行道和自行车道优先设计;(4)可持续城市排水系统是整个设计中最本质的元素。

14.2　新城市广场

这个空间位于伯爵广场,是伦敦多功能综合开发利用中的核心,高质量的材料使用促进了白天和夜间的经济活动(图8)。设计遵循人行道优先车行道的原则,室外地面铺装、街道小品与建筑电梯相协调。新广场的风格与相邻新装修的建筑立面相呼应,而外来植物随着季节变换提供了不同的视觉欣赏。

14.3 新伦敦公园

伦敦南部米亚特的北方野外房产项目由当地政府启动,以刺激批发贸易的发展。方案立意是为街区邻里设计高质量的中心公园(图9—图13):(1)所有的步行路径以漏斗形穿越公园;(2)公园的中心将修建"生土建筑"社区中心;(3)果园与菜地散布在有围墙防护的花园中;(4)游乐景观融合于街区和公园;(5)公园部分空间恢复于农业用途;(6)公园的地表径流浅浅地流进湿地;(7)为了优化监督,公园的所有边缘保留了单项开发项目;(8)本土植被优化了生物多样性价值;(9)公园还具有生态和游憩价值,为当地社区提供种植蔬菜水果的机会。

14.4 高密度的社区花园

这是体育馆周围的威姆布列总规划的第一阶段(图14—图15)。我们将这个社区花园设置在地下停车场之上,并与高密度住区相隔离。这一限定空间的成功规划,取决于风景园林师、工程师和建筑师之间的密切合作,正在实行的管理策略也确保了景观得到健康的培育和成长。工程实施时的主要考虑因素是维持庭院空间适当的日照强度,以便人们进行户外休憩活动和植物的生长。另外,支撑板的设计对绿地与树木的结合来说也极为重要。虽然主要的区域是私人社区空间,但庭院结合了私人的"院"与花坛公寓。因而,这个区域成为住户社交互动的重要场所。

15 结论

一个国家的可持续发展依赖于这个国家的研究机构和人们对复杂的环境和城市发展问题的理解,并做出正确的选择,而景观和自然环境是这项工作的核心之一,因此,风景园林师是最艰巨和最具挑战性的工作之一。

创造人们想长期住居的优美环境是最主要的挑战,全方位的规划、环境科学设计和管理技巧相结合的风景园林才是最伟大的生存艺术。

Reading Material A
Landscape Planning and Design in the Century of the City
Jack Ahern

城市时代下的风景园林规划与设计

Abstract: Urbanization is expected to occur at different rates around the world but will mostly occur in developing countries. Based on the challenges that "the struggle to achieve the millennium development goals will be won or lost in cities", the author discusses four aspects in this paper: urban dynamics typology, using sprawling cities as urban laboratories (taking Shanghai as an example), the significance and measurement of urban ecosystem services, and methods to monitor ecosystem services in greenways and green infrastructures. The paper emphasizes the opportunities and challenges of contemporary urban development that landscape architects and urban planners confront.

Keywords: Urbanization; Ecosystem Services; Urban Sustainability; Green Infrastructure

[1] This article is a summary of the keynote presentation given by the author at the 2013 Fábos Landscape Planning and Greenway Conference[1], University of Massachusetts Amherst.

1 Trends and Challenges of Global Urbanization

[2] At the start of the 20th Century approximately 10 percent of the world's population lived in cities. The 21st Century has already been called the century of the city because, in 2007 for the first time in history, the world's population became more than 50% urban. In the coming decades the

world's total population is expected to rise from 7 billion to 10 billion, with much of that increase occurring in the world's cities. Importantly, this trend towards an urban population is expected to continue throughout the century, reaching 6.3 billion urban inhabitants by 2050 and 8.5 billion by 2100. This urbanization is expected to occur at different rates around the world but will mostly occur in the developing countries. Much of this increase in urban population will occur during the professional careers of the current generation of university students and recent graduates in planning and design. If this new hyper-urban world is to be sustainable, its cities must be sustainable-for urban and landscape planning and design professions.

[3] By the year 2100, the Asian region will see the world's largest increase in urban population from 1.8 billion in 2010 to 3.3 billion in 2050 to 3.8 billion (UN ESA 2012). Through the popular media, China is widely misunderstood as a hyper-urban country. In 2011 China's population was far less urban (50.6%) than the United States (82%), or The Netherlands 83%. However, China is in the process of a massive, unprecedented urban migration. In the next 25 years, China will build more cities with more than one million people than total of current cities with over one million population in United States. In China, this rural-to-urban migration is well underway, with 250 million additional Chinese planned to move into new and existing cities in the next 12 years alone! In this respect China has the potential to become a laboratory from which the rest of the world can learn how to plan and design cities to accommodate this new urban population in a sustainable manner.

[4] The United Nation's Millennium Assessment[2] raised the radical proposition that "the struggle to achieve the Millennium Development Goals will be 'won or lost in cities'". This proposition is based on the link between urbanization and socio-economic development including the effect of urban development on the rural environments around cities, providing engines for rural development and supporting the rural economy. This bold proposition, in a global context, raises a basic challenge for urban planners and designers "How can existing cities be expanded, and new cities created-in a sustainable manner, to meet the crushing demands of the new urban population while managing landscapes and ecosystems to provide the ecosystem services that the cities, and their residents depend on?"

2 A Typology of Urban Dynamics

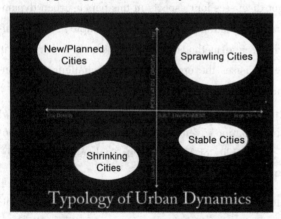

Fig. 1 A typology of Urban Dynamics

[5] The dynamics of urbanization can be understood with a typology based on the level of population growth and the density of the built urban environment (Fig. 1). Sprawling Cities occur where high rates of population growth occur in the context of a high-density built environment. Because these cities continue to grow they expand from the center, typically along transportation routes. In sprawling cities, urban open space becomes scarce due to the intense economic competition for urban space. The effectiveness of planning is often limited by the

exigency for new development, and the existing density that limits change through planning. However, the motivation and opportunity to implement and test innovative urban development and green infrastructure is substantial. The common response to sprawl is containment, or smart growth-allowing the city grow in a more deliberate way crafting a connected, networked open space structure to provide ecosystem services.

[6] Stable cities have a moderate to high density of built environment, but a stable population. These cities often have established urban open space systems, and the opportunity to include new green space in conjunction with new or re-development because the competition for urban space is moderate to low in comparison with sprawling cities. In stable cities urban planning can be intentional and opportunistic, and the need for routine re-development of buildings and infrastructure provides opportunity to implement green building and infrastructure. Because cities need to be rebuilt every few decades, that rebuilding creates the opportunity for sustainability retrofits. Take the example of major roads that generally need repaving/rebuilding every 20 years. During rebuilding a road can become a green street with permeable, high-albedo paving materials, bicycle lanes, tree planting belts, and wildlife underpasses added where appropriate.

[7] New/Planned cities occur in national or regional contexts where population growth is high, but the existing density of the built environment is low. New cities are often located in agricultural regions, requiring resettlement, in "greenfields" where existing development is minimal and the environment is relatively clean, and "brownfields" where prior economic activity left a contaminated environment, but existing population is low. New/Planned cities provide the greatest opportunity for bold, visionary planning and design-employing the best practices for efficient land use and transportation while including an intentional urban open/green space system to deliver ecosystem services. The "eco-city" movement demonstrates the potential for new urbanization to explore innovative models and urbanization concepts.

[8] Shrinking cities are defined by a declining population and eventually produce a low, or lower density of built environment. Shrinking cities occur because of macroeconomic changes and suffer from lack of employment and public revenues. Shrinking cities provide unique opportunities to renew and rebuild through deliberate "editing" of un/underused urban development-by introducing sustainable infill developments, integrated with urban transportation and open space resources. Shrinking cities hold the potential and the challenge to reinvent their identity and economy, and physical form-including their urban/green space structure.

[9] The typology of urban dynamics offers a structured system to classify urban regions based on their population growth and density of urban form. The typology suggests the particular opportunities and challenges that are inherent in a particular type of urban dynamic. In this "Century of the City"[3] perhaps the most challenging and important type are the sprawling cities.

3 Sprawling Cities as Urban Laboratories

[10] Shanghai is a quintessential sprawling city. It has a long urban history and is currently in the midst of a development boom. This mega-regional context of Shanghai and the Yangtze River Delta provides a unique opportunity for research. The Shanghai Key Lab for Urban Ecological Processes and Eco-restoration (SHUES [4]) at East China Normal University has been organized to address the sim-

ultaneous challenge and opportunity.

[11] SHUES is an interdisciplinary research group funded by the Natural Science Foundation of China, among other sources. SHUES conducts research in three areas: 1) complex urban and rural ecosystems, 2) planning managing and design of the urban and regional ecosystem and 3) ecological engineering. SHUES applies its research to solve urban environmental and ecological problems. Under the Direction of Xiang, Wei-ning, SHUES has become a model for using an urban region as a laboratory where monitoring and data base management can document urban conditions over time. In many respects SHUES is similar to the Urban Long Term Research Centers in Phoenix Arizona and Baltimore Maryland in the US. However, because Shanghai is one of the world's largest cities, and is expected to expand into the future, SHUES has a unique opportunity to "learn-by-doing" in understanding complex urban ecological issues, and designing and monitoring innovative and experimental solutions.

4 Measuring Urban Ecosystem Services

[12] The concept of ecosystem services was popularized in the Millennium Ecosystem Assessment. There are four categories of services, supporting, provisioning, regulatory and cultural. These are the services and functions provided by natural ecosystems to the benefit of humans. I've organized them into three broad categories for the sake of illustration: Abiotic, Biotic and Cultural. The abiotic are the non-living services that come from the physical environment, for example hydrology, which is perhaps the most important process in the urban ecology planning and design.

[13] The biotic ecosystem services are related to life and living systems. Ecosystems provide habitat and movement corridors for wildlife species. Air pollution mitigation and remediation can be provided by urban forests. Great philosophers and scientists like Aldo Leopold[5] remind us that it is wise to protect biodiversity, because we don't know how important any species may prove to be in the future. A professor at the University of Massachusetts, Derek Lovley, discovered the geobacter bacteria living in the sediment of the Potomac River [6] in Washington. This bacteria can metabolize and stabilize toxic waste. Here is an example of an organism with no apparent value that has proven to be highly valuable. If our cities are to be sustainable, we need them to be centers of biodiversity-not biodiversity deserts!

[14] Cultural and social ecosystem services are equally important for urban sustainability. Parks are important, of course for human recreation, but also to support human health and to provide places for healthy social interaction. We are creating a new nature in 21st Century cities and we need this new nature to provide a broad suite of ecosystem services, abiotic, biotic and cultural. These, and other concepts both define the term and affirm that ecosystem services are essential for all aspects of human health and well-being. Ecosystem services are what humans depend on for survival in a sustainable world. If the challenge for sustainability will be won or lost in 21st Century cities, it will depend on the ecosystem services that cities provide. Ecosystem services can therefore be understood as the metrics of urban sustainability.

[15] Ecosystem services can serve as assessment metrics to link urban form (pattern) with urban process (ecosystem services). Once understood, measured, and mapped, ecosystem services can become the goals and benchmarks of planning for urban sustainability. In planning and designing sus-

tainable cities, landscape architects have learned from ecologists of the importance of protecting and restoring connectivity between large, and small patches of urban green spaces and habitats. Greenways are understood for their potential to provide a broad suite of ecosystem services in urban areas. Green infrastructure is another contemporary concept for providing urban ecosystem services. Ecological design concepts including greenways and green infrastructure are increasingly understood and valued. As urbanization advances globally, designers will be increasingly challenged to measure and document the ecosystem services that their plans actually provide.

5 Monitoring Ecosystem Services in Greenways and Green Infrastructure

[16] The concept of ecosystem services is increasingly understood as a useful goal and metric for urban sustainability. The next challenge for design professionals is to advocate for and practice monitoring of projects that aim to provide particular ecosystem services. Design professions do not have a tradition of supporting monitoring. If monitoring can be conducted on small-scale, safe-to-fail design experiments, the risk of failure can be minimized and the potential to earn success can be maximized.

[17] One of the expected benefits of urban forests is to increase biodiversity. Measuring of bird diversity can be conducted with the "point count" method where one or more trained/expert observers visit an urban forest or neighborhood and record the number and type(s) of birds seen and heard in a specific observation time(Fig. 2). The results between multiple observers, observation dates, and locations can be compared and averaged. This method is well-known in biology but is rarely practiced by design professionals. These data serve to document the biodiversity benefits of particular types of urban forest plantings and configurations. If landscape architects can learn to promote biodiversity monitoring they can become partners with scientists in "design experiments" to learn how to provide ecosystem services in cities.

FIGURELI:USING POINT COUNT METHOD FOR COUNTING BIRDS IN GI FOREST

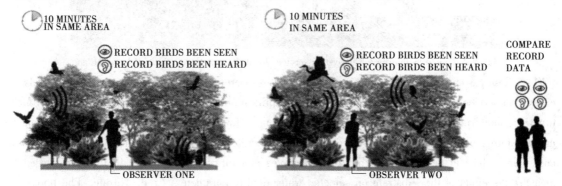

Fig. 2 The point count method for Counting Birds in a Green Infrastructure/Urban Forest. Credit: Shu Liu.

[18] Green roofs are advocated for the multiple ecosystem services they provide: energy savings, stormwater retention, water quality improvement, extending building roof replacement, and supporting biodiversity-particularly of beneficial insects. In highly urbanized environments where green space and habitat is limited, green roofs may become valuable habitats for pollinating insects. The bees that pollinate urban vegetables, trees and flowers can find habitat on green roofs. The presence and diversity of bee species can be measured by several types of traps. The "pan-trap"[7] deploys yellow pans filled with soapy water that attracts and captures the bees. The traps can be collected and monitored on a

regular schedule to learn how many individuals and which species of bees are present on the green roof-and providing pollination services to the surrounding neighborhood. Malaise traps are larger tent-like structures that can capture a wide diversity of flying insects. As with pan traps, malaise traps [8] can be sampled and recorded on a regular basis to learn which insects are present on the green roof.

[19] Created wetlands are commonly used to provide ecosystem services. Floating wetlands are a unique type of green infrastructure that constructs artificial floating habitats on urban waters to provide habitat and water quality improvements. Because floating wetlands can move across large water surfaces, they are difficult to monitor by conventional methods. Figure 3 illustrates a monitoring method that employs GPS [9]-linked sensors to report data to satellites on the floating wetland location, water temperature and water quality parameters. Because this system works with remote sensing it can report data continuously on water quality at different locations with known coordinates (Fig. 3).

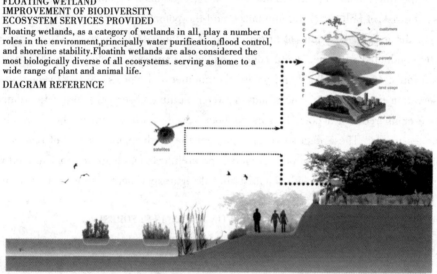

Fig. 3 Remote monitoring of ecosystem services for floating wetlands. Credit: Yiwei Huang.

[20] Rain gardens are one of the more common types of green infrastructure to provide water-related ecosystem services including stormwater retention, infiltration and water quality improvement. Unfortunately monitoring is rarely conducted in rain gardens. However, if a commitment to monitoring is made before the rain garden is constructed, simple monitoring wells can be easily installed at low cost. The wells can be used to gather water samples at multiple depths beneath the rain garden to measure the effect of the substrate on selected water quality parameters (i.e. Nitrate, Phosphorous) through laboratory analysis of the samples (Fig. 4). If these wells are not installed during installation they become very disruptive and expensive to install.

[21] These examples demonstrate how the ecosystem services intended and expected from greenways and green infrastructure can be accurately monitored. If the data are collected with standard accepted methods, it can be scientifically analyzed to learn the differential effects of alternative greenway or green infrastructure configurations or treatments on ecosystem services. When monitoring becomes a regular practice, every urban construction can be understood as an experiment, in an adaptive design process where plans and designs are conceived as opportunities to explore and develop new

UNIT 15　LANDSCAPE STUDIES AND DEVELOPMENTS

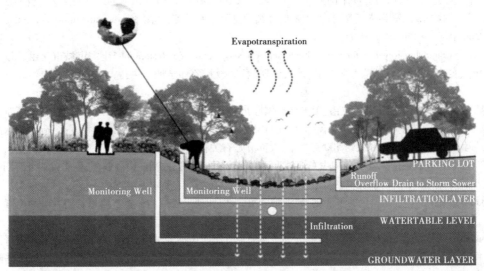

Fig. 4　Monitoring Ecosystem Services: Water quality in rain gardens and bioswales. Credit: Zhangkan Zhou.

research, and to test ideas. It's a different way of thinking about professional planning and design practice than conventional work that tends to apply standard solutions.

6　Summary

[22]　In this century of the city, sustainability will be won-or-lost in existing and future cities. Urban sustainability can be understood, and measured, as the specific ecosystem services provided in cities to support human populations and the biodiversity on which humans depend. Design professionals can contribute to urban sustainability by integrating ecosystem services into plans and designs for infrastructure, settlements and green systems in cities. Because this is a new challenge, and is specific and different in every city, designers must develop and test new ideas and "learn-by-doing" how these ideas perform over time under specific conditions and locations. If planners and designers can effectively practice adaptive design, they may realize the possibility to learn from, and benefit by the process of urbanization. In this way urbanization can change from a perceived human problem-to the source of new ideas and practices that can be important to the solution for urban sustainability.

Biography

　　Jack AHERN, FASLA, Ph. D, is a Professor of Landscape Architecture in the University Of Massachusetts Amherst, USA. He is also the Vice Provost for International Programs.

　　About the translator: HUANG Yi-wei is a graduate student and teaching assistance of Landscape Architecture in the University of Massachusetts Amherst, USA.

[NOTES]

[1]　Fábos Landscape Planning and Greenway Conference 法布士风景园林与绿道规划国际会议：以"美国绿道之父"朱利叶斯·法布士教授（Julius Fábos）之名来命名，每3年举行1次，是以风景园林和绿道规划为主题，以齐聚来自本地以及全球各国专家研讨为目的的聚会。

[2]　The United Nation's Millennium Assessment 联合国新千年生态系统评价：该项目是迄今对全世界生态系统

健康状况最大的评价项目。它在2001年由联合国秘书长安南发起,旨在成为决策者和公众的一个手段。评价于2005年3月完成,该评价由95个国家的1 360名专家参加,有一个80人组成的评议团。文件资料来源于科学文献、已有的数据库、科学模式等,有关MA的详细信息可查阅网址:http://www.millenniumassment.org。

[3] Century of the City 城市的世纪。

[4] SHUES(Shanghai Key Lab for Urban Ecological Processes and Eco-Restoration)上海市城市化生态过程与生态恢复重点实验室:由华东师范大学国家"千人计划"特聘教授象伟宁博士主持,主要从事城市生态方面相关研究。

[5] Aldo Leopold 奥尔多·利奥波德:美国享有国际声望的科学家和环境保护主义者,被称作美国新保护活动的"先知","美国新环境理论的创造者"。他同时又是一个观察家,一个敏锐的思想家,一个造诣极深的文学巨匠。一生共出版了三本书和大约500篇文章,大部分是有关科学和技术的题目。

[6] Potomac River 波托马克河:美国中东部最重要的河流,源出阿巴拉契亚山脉西麓,由北布朗奇河同南布朗奇河汇合而成,注入大西洋的切萨皮克湾,是全美第21大河流。

[7] pan-trap 陷阱锅:是用黄色的平底锅,装满肥皂水,吸引和捕捉蜜蜂的装置。通过定期收集和调度监测陷阱装置。

[8] malaise trap 马氏网诱捕器:用于捕捉昆虫。该采集方法能诱捕多种昆虫,其工作原理是许多飞行昆虫遇到障碍物时(此处为帐幕朝向最高点),在盛有酒精的采集瓶瓶口活动,沿中央幕布底边设置液槽或陷阱收集碰击幕布后跌落的昆虫。

[9] GPS(Global Positioning System)全球定位系统。

[GLOSSARY]

brownfields	棕地	Green roof	绿色屋顶	hyper-urban	超级城市
Global Urbanization	全球城市化	greenfield	绿地	Rain garden	雨水花园
green infrastructure	绿色基础设施	greenways	绿道		

[NEW WORDS]

abiotic	adj.	非生物的,无生命的	Nitrate	n.	硝酸盐
biotic	adj.	生物的;有关生命的	permeable	adj.	能透过的;有渗透性的
configurations	n.	配置;结构;外形	Phosphorous	adj.	磷的,含磷的
coordinates	n.	坐标	pollinate	vt.	对……授粉
disruptive	adj.	破坏的;分裂性的;制造混乱的	remediation	n.	补救;矫正;补习
exigency	n.	紧急,紧迫(性);危急(性)	retrofit	vt.	改进
high-albedo	adj.	高反照率	safe-to-fail	adj.	失败安全
implement	vt.	实施,执行;实现,使生效	sediment	n.	沉积;沉淀物
infiltration	n.	渗透;渗透物	sprawling	adj.	蔓生的
learn-by-doing	v.	做中学	substrate	n.	基质;基片;底层
macroeconomic	n.	微观经济	typology	n.	类型学
mitigation	n.	减轻;缓和;平静	won-or-lost	adj.	成功或失败

UNIT 15　LANDSCAPE STUDIES AND DEVELOPMENTS

[参考译文]

城市时代下的风景园林规划与设计

杰克·埃亨

摘要：城市化进程将会在世界各个领域以不同的速度发生，而且主要将发生在发展中国家。作者基于"城市的成功或失败将决定千年发展目标的实现"，在文章中讨论了以下4个方面：城市动态的类型学；将蔓生的城市作为城市实验室（以上海为例）；城市生态系统服务的重要性和衡量标准；以及在绿道和绿色基础设施中实现监控生态系统服务的方法。文章强调了当代城市发展中，风景园林设计师和城市规划师所面临的机遇和挑战。

关键词：城市化；生态系统服务；城市可持续性；绿色基础设施

[1]　本文是作者在举办于马萨诸塞大学阿莫斯特分校的2013年法布士风景园林与绿道规划国际会议上发表的演讲的要点概述。

1　全球城市化进程的趋势及挑战

[2]　在20世纪初，大约10%的世界人口居住在城市中。而21世纪之所以被人们称为城市的时代，是因为在2007年，城市人口在历史上第一次达到了世界人口的50%。在接下来的数十年中，世界总人口将从70亿增加到100亿，其中大部分增长都将在城市中发生。重要的是，城市人口增长的趋势将在本世纪持续，并预计在2050年达到63亿，在2100年则达85亿。城市化进程会在世界各个领域以不同的速度发生，但主要将发生在发展中国家。城市人口增长这一现象将主要发生在当代大学生以及规划与设计专业的应届毕业生的职业生涯中。如果这个拥有超级城市的新世界想要实现可持续发展，那么它的城市就必须实现可持续——这是一个巨大的挑战，尤其是对城市与风景园林规划和设计领域的挑战。

[3]　到2100年，亚洲地区将发生城镇人口的最大增幅，从2010年的18亿，到2050年的33亿，最后增加到38亿（联合国人类发展报告，2012）。通过大众媒体的传播，中国被广泛误解为一个超级城市国家。事实上，在2011年，中国的城市人口比例（50.6%）远低于美国（82%）和荷兰（83%）。然而，中国目前正在经历一个巨大的、前所未有的城市迁移过程。在未来的25年里，中国将会建造更多人口超百万的城市，甚至超过美国目前百万人口城市数目的总和。在中国，这种农村向城市的人口迁移正在有条不紊地进行，仅仅在接下来的12年中，就有2.5亿人口涌入中国现有的或者新建的城市。从这一点来说，中国将可以成为一个实验室，让世界其他国家学习如何以可持续发展的方式来规划和设计城市，适应新的城市人口。

[4]　此前，联合国千年系统评估提出了一个激进的论点，那就是："'城市的成功或者失败'将决定千年发展目标的实现"。这一主张是基于城市化和社会经济发展之间的联系，包括城市发展对于城市周边农村环境的影响、提供为农村发展和支持农村经济的源动力。这个大胆的论点，在全球范围内掀起了对于城市规划者和设计者的基本挑战，即"如何以可持续的方式，扩张现有城市规模、建设新城市，以满足新城市人口的巨大需求，并同时管理景观和生态，让城市和居民赖以生存的生态系统正常服务？"

2　城市动态的类型学

[5]　城市化的动态可以理解为基于人口增长和建成城市环境的密度水平的类型学（图1）。蔓生的城市会在人造环境密度高、人口增长迅速的情况下出现。随着城市的持续发展，它们会从城市中心开始，通常沿着交通干线，向外扩张。在蔓生的城市，由于城市空间的激烈经济竞争，城市开放空间变得格外稀缺。规划的有效性也往往由于紧急的新发展以及现有的难以更改的密度而受到限制。然而，实践和实验创新的城市发展和绿色基础设施的动力和机会是巨大显著的。人们对于城市扩张的普遍反应是遏制，或者精明增长——让城市在一个可以提供生态系统功能的、相互链接的网络开放空间结构下，从容发展。

[6]　稳定的城市拥有中度到高度的建成环境，并且有一个稳定的人口。这些城市往往已经建立了城市开放空间系统。而这些建成的城市开放空间，在城市空间竞争激烈的背景下，与新的或者重新开发的绿地相链接的概率，比正在增长的城市要低。在稳定的城市中，城市规划会是目的性与机会性并存，而对于常规二次开发的建筑和基础设施的需求，为推行绿色建筑和绿色基础设施提供了机会。因为城市每隔几十年都要更新和重建，这样的重建提供了可持续改造的机会。比如说主要的道路一般在每20年需要重铺/重建。而在这一过程中，重建的道路可改造成一条绿色街道，适当地添加可渗透性、高反照率的铺装，配上自行车道、绿化带以及野生动

物地下通道。

［7］ 新的/计划的城市发生在人口增长速度快,但建成环境密度低的国家或区域范围内。新的城市通常建立在农业区域,需要重新安置时,将现有的建设少、环境相对干净的区域设为"绿地",而将之前经济活动留下来的污染土地但拥有较少人口的地方设为"棕地"。新的/计划的城市为大胆的、富有远见的规划和设计提供了最大机会——在落实高效的土地利用和交通的同时,包含了人为设计的、可提供生态系统服务的城市开放/绿地空间系统。"生态城市"运动展现了探索创新模式和城市化概念的新型城镇化的潜力。

［8］ 萎缩的城市有以下特征:人口不断下降,最终形成密度越来越低的建成环境。城市萎缩发生在宏观经济变化和缺乏就业和公共收入的背景下。在萎缩的城市中蕴含着独特的更新和重建机会,通过谨慎"编辑"未被利用/未被充分利用城市发展——引入可持续的填充式发展,结合城市交通和开放空间资源。萎缩的城市具有重塑自己的身份、经济以及物理形态——包括其城市/绿地结构——的潜力,这同时也是一种挑战。城市动态的类型学提供了一个结构化的系统,根据城市地区的人口增长和城市形态密度进行分类。

［9］ 类型学暗示了内在的一种特定类型的城市动态的特定机会和挑战。在这个"城市的时代",最具挑战性和最重要的类型也许是蔓生的城市。

3 蔓生城市作为城市实验室

［10］ 上海是一个典型的蔓生城市。它有着悠久的城市历史,目前正处于开发热潮之中。这个包含上海和长三角在内的特大型区域为研究提供了难得的机会。上海市城市化生态过程与生态恢复重点实验室(以下简称为SHUES)已经由华东师范大学建立,以期解决共生的挑战和机遇问题。

［11］ SHUES是由包括中国国家自然科学基金委员会在内的多源资助的一个跨学科研究小组。它在以下3个方面开展研究:1)复杂的城市和农村生态系统;2)规划管理和设计城市与区域生态系统;3)生态工程设计。SHUES运用其研究来解决城市环境和生态问题。在象伟宁教授的领导下,SHUES已成为一个使用城市区域作为实验室的典范,他们监测和管理数据库,随时间的推移记录城市状况。在许多方面上,SHUES都类似于美国亚利桑那州凤凰城和马里兰州巴尔的摩的城市长期研究中心。但是,因为上海是世界上最大的城市之一,并有望在未来继续扩张,因此SHUES可以使用"从实践中学习"的方法来理解复杂的城市生态环境问题,并设计和监测创新和实验性的解决方案。

4 衡量城市生态系统服务

［12］ 生态系统服务的概念在千年生态系统评估报告中被广泛推广。它有4类服务类型:支持、供应、监管和文化。这些都是为人类利益而提供的自然生态系统的服务和功能。为便于阐述,可以将它们分为3大类:非生物类、生物类以及文化类。非生物类型是无生命体的服务,它们来自于物理环境,例如水文学,这也许是在城市生态规划与设计中最重要的一环。

［13］ 生物生态系统则是与生活和生命系统相关。生态系统为野生动物物种提供栖息地和行动走廊。空气污染减排及污染防治可以通过城市森林实现。伟大的哲学家和科学家,比如利奥波德(Aldo Leopold)提醒我们,保护生物多样性是明智的,因为我们无法知道在未来,哪一些物种将会被证明有多么重要。马萨诸塞大学教授德里克·洛夫利发现了生活在华盛顿特区波托马克河的沉积物中的地杆菌。这种细菌能代谢和稳定有毒废物。这就是一个表面看似毫无价值,但最终被证明价值非凡的生物体的例子。如果我们想要实现城市的可持续发展,我们就要使它们成为生物多样性的中心——而不是生物多样性的沙漠!

［14］ 文化和社会生态系统服务在城市可持续发展中同样重要。公园是重要的,当然有对人类娱乐的方面,但它也支持着人类健康,为健康的社会交往提供场所。我们正在创造21世纪城市的新的性质,以提供一整套的生态系统服务:非生物类、生物类以及文化类。上述以及其他概念共同定义和肯定着一项,那就是,生态系统服务对人类健康和福祉的各个方面都至关重要。生态系统功能是人类赖以生存在一个可持续发展的世界的保障。可持续性的挑战在21世纪的城市中成功与否,就取决于城市所提供的生态系统服务。因此,我们可以认为,生态系统服务是城市可持续发展的指标。

［15］ 生态系统服务可以作为联系城市形态(模式)和城市过程(生态系统服务)的考核指标。一旦对其进行了了解、测量和映射,生态系统服务便成为城市可持续发展规划的目标和基准。在规划和设计可持续发展的

城市中,风景园林师已经从生态学家那里学习到保护及恢复城市绿地和生境的大、小斑块之间连通性的重要性。绿道具有提供广阔的城市区域生态系统服务套件的潜能。绿色基础设施是另一种为城市提供生态系统服务的现代的理念。生态设计的概念,包括绿道和绿色基础设施,得到人们的日益理解和重视。随着全球城市化进程推进,设计师在测量和记录其所设计的项目所能提供的生态系统服务时,将面临更大的挑战。

5 绿道和绿色基础设施中的生态系统功能监测

[16] 生态系统功能的概念被越来越多地理解为城市可持续性的重要目标和指标。设计领域的下一个挑战将会是提倡以提供特定生态系统功能为目标的项目,并对其实施监测。在设计领域中,并没有支持监测系统的传统。如果监测可以在小尺度的、安全失败的设计实验中实施,那么失败的风险将能被降到最低,而创造成功的潜力将达到最大化。

[17] 城市森林的其中一项预期效益是增加生物多样性。测量鸟类的多样性可以用"点测法"实现(图2),通过一个或多个专业培训的/专家观测者进入一个城市森林,并记录下特定时间段所观察到的或听到的鸟类数量和种类,然后对不同观测者的结果,包括观测时间、观测的地点进行比较和计算。这一方法在生物领域被广泛认知,却很少在设计领域中运用。这些数据将用于记录特定城市森林的种植和组态所产生的生物多样性效益。如果风景园林师可以增加生物多样性的监测,他们就可以与"设计实验"科学家们合作,学习如何提供城市所需的生态系统服务。

[18] 屋顶绿化能提供多元的生态系统服务:能源节约、雨水蓄洪、改善水质、延长建筑物屋顶的更换,并对生物多样性——尤其是益虫的支持。在高度城市化的环境下,绿色空间和栖息地有限,屋顶绿化可以成为传粉昆虫宝贵的栖息地。为城市蔬菜、树木和花卉授粉的蜜蜂可以在绿色屋顶中找到栖息地。而蜂种的存在性和多样性可以通过多种陷阱装置来测量。"陷阱锅"是用黄色的平底锅,装满肥皂水,吸引和捕捉蜜蜂的装置。通过定期收集和调度监测陷阱装置,我们可以了解蜜蜂的种类有多少,哪些是存在于绿色屋顶——为周围社区提供授粉服务。"马莱兹陷阱"是更大一些的帐篷状装置,可以捕捉更广泛多样的飞虫。和"陷阱锅"一样,"马莱兹陷阱"可以进行定期采样和记录,从而了解哪些昆虫存在于绿色屋顶。

[19] 人造湿地通常被广泛用于提供生态系统功能。浮动湿地是一种独特的绿色基础设施,它可以在城市水域建设人工浮动栖息地,以提供栖息地并改善水质。由于浮动湿地可以在大型水面移动,因此很难用常规方法对其进行监测。图3描述了传统的监测方法,它采用GPS相联的传感器将湿地位置、水温以及水质等参数数据报告到卫星上。因为这个系统与遥感相连,它可以在已知坐标的不同位置,连续地报告水质的数据(图3)。

[20] 雨水花园是较常见的一种绿色基础设施,它可以提供与水有关的生态系统服务,包括雨水滞留、渗透以及水质改善。不幸的是我们很少对雨水花园进行监测。但是,如果在建造雨水花园前就决定对其进行监测,那么安装一个简易的监测井就十分容易,所需成本也十分低廉。监测井可以用来收集雨水花园下不同深度穿过的水样品,并通过实验室对样品进行分析,以测量选定样本的水质参数(即硝酸盐、磷)的效果(图4)。如果在建设雨水花园时没有安装监测井,那么后期再进行安装将会变得非常具有破坏性,而且成本昂贵。

[21] 这些示例演示了绿道和绿色基础设施的未来生态系统服务,以及如何对其进行准确的监测。如果以标准认可的方法进行数据收集,就可以科学地对数据进行分析,并了解不同的绿道或绿色基础设施的配置或整改对生态系统服务的不同影响。当监测成为一种惯例,每个城市建设都可以看成一个实验,在一个自适应的设计过程中,计划和设计都可被视为一种机会去探索和开发新研究,并对这些想法进行测试。这是一种不同于采用标准解决方案的专业规划和设计实践的新的方式。

6 总结

[22] 在这个城市的世纪,可持续发展将在现有和未来的城市中成功或失败。城市可持续发展可以被理解为,甚至计量为,城市中特定的支持人口增长的生态系统功能,和人类赖以生存的生物多样性。设计专业领域可以通过在城市规划和设计整合生态系统服务,纳入基础设施、环境和绿色系统来促进城市可持续发展。由于这是一个全新的、具体的挑战,且在每个城市都不尽相同,设计师们必须找到和测试新的想法,以"从实践中学习"的方式,测试这些想法在特定条件和位置下是如何随时间推移而作用的。如果规划者和设计者可以有效地实践适应性设计,他们会意识到从城市化的过程中学习与获益的可能性。这样的话,城市化进程将能够从一个

已察觉的人类问题开始转变——成为新思路和新实践的源头,而这个源头将可能会对城市可持续发展的解决方案产生重大意义。

Reading Material B
Landscape Infrastructure: in plain view[1]
Ying-Yu HUNG[2]
景观基础设施探析

Abstract: This essay examines the underpinnings of infrastructure works as a way of understanding a future medium for open space systems within an urban fabric. The traditional understanding of infrastructure must look beyond the obvious typological definitions (roads, utilities, bridges, etc.) and begin to engage other networked and systematic programs such as the movement of pedestrians and bicyclists, flora, fauna and climate in an attempt to further maximize the diversity and distribution embedded in both. Studies in Landscape Urbanism and international case studies illuminate the progress made in recent years towards a more unified approach to planning and design, while Landscape Infrastructure is framed as a transition for the profession and the principles guiding its approach/application to real world issues and projects.

Key words: Landscape Architecture; Landscape Infrastructure; Research

1 Background

The practice of landscape architecture is a discipline of diverse interests, scales, and territories. The word "landscape" alone can be simultaneously interpreted as a kitchen garden or the entire planet Earth. In this sense, the field of landscape architecture is somewhat vague and requires further clarification. At this writing, two general areas of concentration have emerged: ecological planning to protect and maximize natural resources, and landscape design as a cultural, economic construct serving people's needs. These two broadly defined topics are at times practiced separately due to scale, complexities of the site and personal bias, leading to a severed landscape practice defined by the role of "the designer" or "the planner".

In the last twenty years, new trends in landscape architecture seek to define the practice from a more holistic point of view, one that is not limited by what we create, but the practice itself is an integral part of our philosophy-our way of life. This new world view stems from our realization that we, as a society, have contributed to the degradation of our environment. Landscape architects and urbanists can help reverse the process, cognize that even with our best intentions, the landscape we have created has unpredictable consequences, and that the word "change" is the underlying factor in everything we do.

This philosophical understanding suggests a new trajectory in which to think about landscape architecture. A trajectory meant to further the dialog between ecological process and design. To that end, landscape architecture is crossing disciplines-the physical framework from which landscape architecture operates has no boundaries; and the purpose in which it serves is becoming more infrastructural, socio-political, economic, and environmental. In addition, the practice of landscape architecture today is more closely aligned with architecture, urban design and planning than ever before. Among many leading practitioners, the convergence of these practices share one common outlook: the global landscape is mosaic-based, where edges are permeable, the boundaries between cities and countryside are in flux. (Fig. 1) Within this "mosaic" landscape, there exists a complex set of networks or systems that are highly interconnected and interdependent. The systems cannot be approached in isolation, as even the smallest intervention effects the larger whole. Landscape architecture today offers the means to analyze, synthesize, and provide an organizational framework toward an integrated urban design strategy. At a finer scale, there is distinction between objects and field, the richness of the landscape phenomenon, and the intimate, tactile quality of things which create identity and memory to the place itself (Fig. 2-Fig. 3).

Fig. 1 Qingdao (aerial photo), China, 2009

Fig. 2 Yuyuan, Shanghai, China 2004

Fig. 3 Amsterdam, Netherlands, 2008

2 Identifying the trend: Landscape Urbanism

"Landscape Urbanism"[3] is a term coined by Charles Waldheim[4] twelve years ago, which stated that "landscape has become a lens through which the contemporary city is represented and a medium through which it is constructed." Further, Waldheim contends that landscape architecture has in fact replaced architecture and urban design as the primary discipline that establishes the framework for contemporary city making. The premise that architecture and urban design has become a commodity used to further the economic needs of the city through branding and re-branding has put the professions at risk of becoming irrelevant. By contrast, landscape architecture seeks to take on the context itself, the infrastructure and the "spaces in between" within urban environments, to instill purpose, legibility, and cohesiveness, so that the city as a whole is healthy and robust (Fig. 4).

Fig. 4 SWA Group, Buffalo Bayou, Houston, Texas, 2006

3 In focus: Landscape Infrastructure

To further the thinking beyond landscape urbanism, one can look to the inherent systems that operate beyond the city fabric. The notion of landscape as infrastructure becomes a fresh medium in which to explore the potential for regional connectivity and distribution. As an example, the U. S. interstate highway system was initially developed for national defense to distribute ammunition and war time vessels and dispatch military personnel throughout the United States. Over the last 50 years, the entire system has fallen into disrepair and exceeded its capacity. This situation sets forth a critical juncture in which to re-evaluate the 46,876 miles of contiguous freeway system, to take on multi-modal transportation strategies as ecological conveyance, and as a resource and energy re-distribution mechanism throughout the United States.

Similarly, the South-North Water Transfer project[5] in China was first conceived by Mao Zedong to balance the water needs of the northern region with the abundant water resources available in southern China. Consists of three routes, east, central, and west, the project diverts water from major rivers, tributaries, and lakes across 10 provinces, totaling 2,700 km in length. The Eastern Route requires transferring water uphill through a series of pumping stations and tunnel under the Yellow River. The route also utilizes part of the historic Grand Canal, which has been a major transportation artery since 5th century B. C. The technical and environmental challenge associated with this project presents unprecedented landscape opportunities that could potentially convert a controversial mega-infrastructure into a valuable trans-regional landscape corridor.

4 What is Infrastructure?

Expanding on the topic of infrastructure brings to light that infrastructure, classically defined, is the basic facilities, services, and installations needed for the functioning of a community or society, such as transportation and communications systems, water and power lines, and public institutions including schools, post offices, and prisons. Our society has traditionally placed a high value on the design of mono-functional infrastructural systems, engineered to maximize efficiency at a given time to fulfill one single purpose, but failing to provide a consistent level of efficiency throughout their life spans. The consequence of such a singular approach poses serious impacts on the way infrastructure contributes to urban life. Parking lots, transportation corridors, transit hubs, channelized water ways are left idle between peak hours, thus creating voids and barriers in the city. Industrial sites and landfills are public nuisance that people know they exist but rather forget. We are reminded of their existence when the health and safety of the people's lives are at stake (Fig.5).

Fig. 5 Ballona Creek (Existing photo), Los Angeles, California, 2008

With the rapid growth of our metropolis and the shortage of available open space, one has discovered that the infrastructure is an untapped resource with the capacity to effect positive change.

Through the employment of ecological and social principles, the urban infrastructural systems can play a multi-faceted role that actively contributes to the betterment of urban life.

5 What is Landscape Infrastructure?

Mossop contends that infrastructure such as roads "are required to perform multiple functions: they must fulfill the requirements of public space and must be connected to other functioning urban systems of public transit, pedestrian movement, water management, economic development, public facilities, and ecological systems." The multi-functional aspect of infrastructure also speaks to the importance of diversification as a general principle in city making, leading to an optimized condition in which the city and its infrastructure are one and the same, where infrastructure informs how the city is organized and built. A classic example to support this argument is the Back Bay Fens in Boston designed and engineered by Frederick Law Olmsted. The site was formerly a saltwater marshland tainted with untreated raw sewage from the city's growing settlement. The land reclamation projects in the 1820's marked a series of long dedicated efforts in improving the water quality, controlling flood, and allowing a tidal ecosystem to be re-established. Today, the Back Bay Fens[6] is part of in the 1,100-acre chain of parks, parkways and waterways forming the Emerald Necklace, bringing air quality, urban runoff retention and remediation, urban wildlife habitats, trails, sports venues, and a 265 acre arboretum to the Boston residents (Fig. 6-Fig. 7).

Fig. 6 Olmsted Archive, Emerald Necklace, Boston, Massachusetts

Fig. 7 The Fens, Boston, Massachusetts

Landscape infrastructure is comprised of a specific set of characteristics, expectations and performance standards, encapsulated as follow:

(1) As a non-isolated system, infrastructure has the ability to adhere to a set of requirements and achieve measurable results which if implemented at a large scale; the benefits could be quantified and exhibited to the general public.

Across Los Angeles, the city's alleyways account for more than 900 linear miles of pavement, currently function as service corridors. Putting them all together, the city's alleys would make up about 3 square miles-about half the size of L. A.'s Griffith Park[7], twice the size of New York's Central Park, or the equivalent of about 400 Wal-Marts (not including parking). These alleyways could be retro-fitted with bioswales, exploratory bicycle trails, and pedestrian greenways and pocket parks, vegetable plots and many more uses in addition to being service corridors. (Fig. 8)

Fig. 8　SWA Group, Borderline Neighborhood Improvement Project Proposal, City of Santa Monica, California, 2006

(2) As a collection of aggregates, infrastructure has the ability to improve or cause adverse effect on the end result. The opportunity to "tail or fit" the infrastructure, allows it to remediate and reverse negative impacts toward positive gain.

The city of Chicago has the world's largest square footage of green roof for an urban center. As a collective system, the green roofs operate at a city-scale and can have a tremendous impact on reversing heat-island effect, collecting storm water and creating more habitats.

(3) Infrastructure is a connective tissue that brings together disparate elements, instilling cohesion and purpose between the elements.

Fig. 9　SWA Group, Academy of Sciences at Golden Gate Park, San Francisco, California, 2008

Increasingly old cities that depend on freeways and automobiles as the primary means of transporting goods and services are being retrofitted with public transit. Our carbon heavy freeways will soon become relics of the past. Along these transit corridors, neighborhoods are becoming more identifiable due to their exposure to the public, more face to face interaction between people spur spontaneous conversations and mutual understanding, making the city once again a humane place to live. (Fig. 9)

(4) Infrastructure is a catalytic mechanism by which localized centers, nodes located along the infrastructural spine may share direct benefits due to easy access and high visibility.

Central Park in New York City was envisioned by Andrew Jackson Downing and William Cullen Bryant as a calling to solve the ills of the society: crowded streets, poor immigrants and crime. For a rapidly growing city such as New York in the 1830's, it was necessary that the park be large enough to anticipate the needs of its populace. As a significant landscape infrastructural project, the completions of the 843-acre park in 1860 brought forth unanticipated benefits to its local economy through tourism, increase in property and land values, as well as increased revenues for the government (Fig. 10).

(5) The sheer scale and resources spent on infrastructure presents tremendous opportunities to leverage unrealized potential in the urban environment.

The City of Los Angeles has 511 miles of freeways, 51 miles of channelized waterways, 7,000 miles of power lines, 6,400 miles of streets, and much more infrastructure hidden all over the city that has not been accounted for. The latest survey conducted by the City Council shows that infrastructure improvements

Fig. 10 SWA Group, Bringing Back Broadway, City of Los Angeles, California, 2008

Fig. 11 Central Park Aerial View, New York, New York, 2008

rank top of the list to positively improve Los Angeles' neighborhoods. If any of these systems of infrastructure can reduce the amount of carbon footprint, retain storm runoff and recharge aquifer, use renewable energy generated on site, and promote pedestrian and bicycle use linking the diverse neighborhoods which make the city unique, more people will be inclined to move into the city, the empty lofts of downtown will be brought back to life again (Fig. 11).

6 Future

On February, 2009, the United States Congress passed the American Recovery and Reinvestment Act of 2009[8] in an effort to stimulate the U.S. economy, which has been facing its deepest recession since the Great Depression[9]. The stimulus package provides nominally $787 billion dollars in appropriations towards crisis investments, including $7.2 billion on the environment (wastewater infrastructure, drinking water infrastructure, hazardous waste cleanup, emission reduction from diesel engines, and brownfields); $80.9 billion towards infrastructure investment (including roads, bridges, railways, sewers, high-performance green buildings, wastewater treatment infrastructure improvements, drinking water infrastructure improvements, electric vehicle), and supplemental investments (including Bureau of Reclamation[10], National Park Service[11], Forest Service[12], National Wildlife Refuges[13]). In the energy sector, $81.3 billion provide funding to include an electric smart grid, investments in energy efficiency and renewable energy, cleanup of radioactive waste, training of green collar workers.

Toward off the effects of the global financial crisis, China has also approved a multi-billion dollar package for infrastructure projects, many aimed at rural infrastructure, water, electricity, transport, the environment, and technical innovation. China's Eleventh Five-Year Plan (2006-2010) also places focus on infrastructure with investments diverted into central and western regions, including road networks, railways, power grids, and irrigation system, as the rising middle class in these areas demand an improved standard of living as the rest of the country. We live in a historical moment in

Fig. 12　SWA Group, Jefferson Open Space, Lrvine, California, 2006

which our law makers and our highest government officials share the same vision for a sustainable global development. For individuals who convert their diesel cars into biofuel, gourmet chefs who convert front yard lawns into organic vegetable farms, and academics who teach that the simple technique toward carbon sequestration lies in the preservation of our forest habitats, the future of global sustainability is now! The future of landscape infrastructure projects is in plain view. Our cities need this kind of infrastructural approach that extends beyond perceivable boundaries and simultaneously connects various sites to other sites, people to places, communities to communities, people to people, nature to city and city to nature (Fig. 12).

[NOTES]

[1]　本文经作者同意,转载于《风景园林》2009年第3期,本文属于研究型论文。

[2]　Ying-Yu HUNG 洪盈玉:SWA 洛杉矶董事,美国南加州大学建筑学院景观建筑系客座助理教授。

[3]　Landscape Urbanism 景观都市主义:它是一个城市规划理论,其主要内容为景观比建筑更有能力组织城市和提高城市的经验。

[4]　Charles Waldheim 查尔斯·瓦尔德海姆:现任哈佛大学设计学院景观设计学系主任。曾任加拿大多伦多大学景观设计学院副院长、景观系主任、教授;景观都市主义(Landscape Urbanism)的提出者,编著有 The Landscape Urbanism Reader(《景观都市主义读本》纽约州普林斯顿建筑出版社,2006)。

[5]　The South-North Water Transfer project 南水北调工程:从 20 世纪 50 年代提出"南水北调"的设想后,经过几十年研究,南水北调的总体布局确定为:分别从长江上、中、下游调水,以适应西北、华北各地的发展需要,即南水北调西线工程、南水北调中线工程和南水北调东线工程。

[6]　Back Bay Fens 巴克贝沼泽:是美国马萨诸塞州波士顿的一处荒野,也是城市公共用地。

[7]　Griffith Park 洛杉矶格里菲斯公园:该公园是全美最大的城市公园。位于好莱坞以北,面积 1 600 万平方米。内有高尔夫球场及网球场、骑马道、天文台、动物园等。

[8]　American Recovery and Reinvestment Act of 2009;2009 年美国复苏与再投资法案,计划在 2009 年到 2019 年投入 7 872 亿美元,主要针对减税、政府财政纾困、健康医疗和教育科研投入,以及交通运输和房屋城市发展。其中,在 2009 年将投入 1 849 亿美元,占 2009 年名义 GDP(根据美国国会预算办公室的预计)的 1.3%。

[9]　The Great Depression 大萧条:是指 1929 年至 1933 年全球性的经济大衰退。大萧条的影响比历史上任何一次经济衰退都要来得深远。

[10]　Bureau of Reclamation 美国垦务局,是美国水利设施建设,最重要的研究、制造与规划单位,尤其在水资源、坝工技术与河川水库淤沙的移运分析处理上,具有非常丰富的经验。

[11]　National Park Service (美国)国家公园管理局:于 1916 年 8 月 25 日根据美国国会的相关法案成立,隶属于美国内政部,主要负责美国境内的国家公园、国家历史遗迹、历史公园等自然及历史保护遗产。

[12]　Forest Service (美国农业部)林务局:成立于 1905 年,是一个美国农业部的机构。

[13]　National Wildlife Refuges 美国国家野生动物保护区:它是世界上第一个把公共土地和水域预留出来以保护美国的鱼类、野生动物和植物的系统。自 1903 年罗斯福总统指定佛罗里达州的鹈鹕岛国家野生动物保护区为第一个野生动物保护以来,该系统已发展到 150 多万英亩,584 个国家野生动物保护和庇护系统和其他单位,以及 37 个湿地管理区。

UNIT 15 LANDSCAPE STUDIES AND DEVELOPMENTS

[GLOSSARY]

arboretum	植物园,(供科研等的)树木园	mega-infrastructure	巨大的基础设施,巨型构架
bioswale	生态走廊	metropolis	大都市,大城市,重要中心
brownfield	棕地,污染场地,废弃场地	node	节点
fauna	动物群,动物区系	square footage	建筑面积
flora	植物群,植物区系	transit hubs	交通枢纽
grand canal	大运河		

[NEW WORDS]

align with	v.	与……结盟	fabric	n.	织物,构造,建筑物,组织
ammunition	n.	弹药,军火	gourmet	n.	美食家
aquifer	n.	蓄水层,含水土层	illuminate	vt.	照亮,阐明,说明
artery	n.	动脉,干道,主流	in flux		在变化,不定
at stake		处于危险中,在紧要关头	inherent	adj.	固有的,内在的,与生俱来的
bias	n.	偏见,偏爱,斜纹,乖离率	instill	vt.	徐徐滴入,逐渐灌输
biofuel	n.	生物燃料	integral part		整数部分,主要的部分
carbon sequestration		碳封存,碳固定,碳回收	interstate	n.	(美)州际公路
catalytic	adj.	接触反应的,起催化作用的	intervention		介入,妨碍,调停
chef	n.	大厨,主厨	intimate		至交,知己
cognizant	adj.	审理的,已认知的	juncture		连接,接合,接缝
cohesion	n.	内聚力,结合,凝聚	landfill		垃圾堆,垃圾填埋地
cohesiveness	n.	黏结性,内聚力,凝聚力	legibility		易读性,易辨认
coin	vt.	杜撰,创造,铸造(货币)	lens	n.	透镜,镜头,晶状体
connective tissue	n.	结缔组织	leverage	n.	杠杆作用,手段,影响力
contemporary	n.	同时代的人,同时期的东西	life span		寿命,使用期限
contiguous	adj.	接触的,邻近的,连续的	marshland	n.	沼泽地,沼泽地区
convergence	n.	会聚,集合	mono-functional	adj.	功能单一的
diesel engine		柴油引擎,柴油发动机	mosaic-based		镶嵌式的
disparate	n.	无法相比的东西	multi-faceted	adj.	多方面的,多才多艺的
divert	vt.	转移,使转向	perceivable	adj.	可知觉的,可感知的
embed	vt.	栽种,使嵌入,使插入	permeable	adj.	能透过的,有渗透性的
encapsulated	adj.	密封的,包在荚膜内的	populace	n.	平民,大众,人口
envision	vt.	预想,想象	practitioner	n.	开业者,实践者,实习者

· 297 ·

continued

premise	n.	前提,上述各项,房屋连地基	tailor fit	v.	定制适合的
public nuisance	n.	妨害公众利益的人或事物	taint with	v.	沾染,使受不良影响
radioactive waste		放射性废弃物	tidal	adj.	潮汐的,定时涨落的
recession	n.	衰退,凹处,后退,不景气	trajectory	n.	弹道,轨道,轨线
relic	n.	遗迹,遗骸,纪念物	trans-regional	adj.	跨区域的
remediation	n.	补救,矫正,补习	tributary	n.	支流
retention	n.	扣留,滞留,保留,记忆力	underpinning	n.	支柱,支承结构,基础
sheer	adj.	绝对的,透明的,峻峭的	untapped	adj.	未使用的,未开发的
simultaneously	adv.	同时地	utility	n.	效用,实用,功用,公共设施
spontaneous	adj.	自发的,自然的,无意识的	vague	adj.	不明确的,含糊的,模糊的
spur	vt.	激励,鞭策,给……装踢马刺	venue	n.	聚集地点,场馆,发生地点
synthesize	vt.	合成,综合	vessel	n.	容器,船舶,血管

[参考译文]

景观基础设施探析

洪盈玉

摘要:探寻了如何将基础设施架构视为一种综合的理解方式,并以此理解城市肌理中开放空间系统未来的载体。我们必须超越对基础设施的传统理解和单一的类型定义如道路、水电、桥梁等,并开始结合其他的网络系统工程,例如行人和自行车的流动、动植物和气候的关系,力图进一步拓展其多样性和分布范围,而这两者又是紧密结合的。通过分析景观都市主义和国内外范例项目,阐释了近年来日趋将城市规划和景观设计合为一体的设计实践,同时景观基础设施这一定义和设计导则将引领 LA 的转型,并成为解决现实问题和现实项目的指导原则。

关键字:景观建筑;景观基础设施;研究

1 背景

LA 是一门融合不同兴趣、尺度和领域的学科。就"景观"一词而言,它可以同时被理解为非常不同的景物,比如一个菜园可称为景观,而整个地球也可以被统称为一个大的景观。从这个角度来看,景观建筑学的范围并不是很清晰,还需要进一步界定。本文集中在两方面对此进行阐述:以保护并充分利用自然资源为目的的生态规划和以满足人们文化、经济需求为目的的景观设计。由于项目的规模、场地的复杂性以及个人理解的不同,这两种定义广泛的主题有时会被区分开来,并由此而产生了由"设计师"和"规划师"所进行的不同的景观设计实践。

在过去的 20 年里,人们开始试图从加全面的观念来定义景观建筑学实践,这种观点不拘泥于我们具体创造的是什么,而认为景观建筑学实践的本身就是我们的生活哲学和生活方式中不可分割的一部分。这种新的世界观来源于我们的以下认识:我们每一个社会成员都对环境退化负有责任,而景观建筑师和城市规划师能够帮助扭转这一进程。我们也认识到在设计过程中尽管我们抱有最良好的愿望,但是我们所创造的景观给环境带来影响是很难以预测的,而"变化"是我们所有努力中唯一的永恒。

这种哲学层面上的理解暗示着一种对景观建筑学的新的思考,它意味着生态过程与设计过程之间更积极和广泛的对话。因此,景观建筑学是一门交叉学科——它的实践不受实体框架的限制;它的范围变得更加接近基础设施,并与社会政治、经济以及环境更为密切地互相关联。此外,与以前相比,当代的景观建筑学与建筑学、城市设计和规划联系得更加紧密。很多业内专家认为,这些领域的交汇区有一个特征,这就是整个世界的景观就像一幅马赛克图画,而这幅图画的各个边缘是互相渗透的,城市与乡村的边界在不断变化之中(图1)。在这种"镶嵌式"的景观的内部存在着一套复杂的网络和系统,它们互相联系、互相依赖。对于这些系统不能孤立分析,因为即使最小的干预也会对最大的整体有所影响。当今的景观建筑学为整合式的城市设计策略提供了分析与合成的方法,并为城市设计战略提供整体的组织架构。从一个细化的尺度来讲,个别的物体与整个场地之间、丰富的景观现象本身和由景观中的物体带给人们的亲密触感所形成的人们对于这一场所的独特记忆之间存在明显区别(图2、图3)。这种类似于在"花蜜"与人造甜味素之间、鸟鸣与"甲壳虫"乐队的音乐之间、春天的潮声与自来水的滴答声之间、长满青苔的荒地与发烫的柏油路面之间、人为创造的空间与广袤的自然空间之间、世界大事与发生在某时某地的小事之间所存在的诗意转换正是人类的多样性以及创造性的丰富来源。[1]

2 趋势确定:景观都市主义

"景观都市主义"是 Charles Waldheim 在 12 年前首次提出的,其内涵是:"景观已成为展示当代都市的一个透镜,也是城市本身得以构成的重要组成因素。"此外,Waldheim 认为景观建筑学事实上已代替了建筑学和城市设计,成为一门为城市建设提供框架结构的主要学科。建筑学与城市设计已作为一种通过品牌塑造及品牌再塑造来促进经济需要的商品,这种行业行为将这些专业带入了被孤立的险境。[2]相反,景观建筑学则设法考虑周边环境、基础设施以及城市环境内"场所之间的空间",并灌输设计和规划的目的性、可读性和内聚性,从而使得城市成为一个健康的、充满活力的整体(图4)。

Waldheim 所描述的景观都市主义的原则阐明了一系列当今景观建筑学实践所具有的特质[3]:

(1) 景观建筑学的实践包括认可和接受城市环境中生态和社会变迁的不可测定性。在设计过程中的战略性阶段,适度的投入和控制可能带来更加丰富并具有多样性的结果。

(2) 作为帮助公众了解自然在城市环境中的作用的一种方式,自然生态在极大程度上承担着表述功能,它是一种大众可以很容易理解和接受的公共景观和公共展示。同样,景观建筑设计的实践也可以被视作环境艺术,一种被艺术家和设计师所使用的媒介,设计师们通过这种媒介来展示自然界的微妙美景,也通过它来详细地记录自然的力量。

(3) 引导和激发公众对于以生态可持续性为主旨的开发方式的兴趣(生态认知力),能够给当今的景观建筑师带来创造新的生态开发范例的机会。这些范例在生态上具有可行性,并具有文化认同力,而且能够带来经济效益。

(4) 将景观都市主义的理论转化为城市的实体形象越来越多地依赖于"参数"处理过程。在这种过程中,一个设计问题会涉及众多的变量或参数。对变量(可能是生态价值驱使的)的处理会产生多个可供选择的解决方案,比如建筑密度或者是建筑布局的决策最终有可能取决于最优化的开放空间网络的形式等。

Waldheim 对景观都市主义的这些关键要点的阐述,设立了景观建筑学新的焦点,也重申了景观建筑学这一学科的重要意义。很显然,景观建筑学的作用开始于政策制定之前,而且是在公众利益与个人利益就某一共同目标而达成共识的情况下而产生作用的。这种共同目标形成一份蓝图,由之而制定的相关政策被纳入规划方针,并被用来指导特定地区的所有未来的开发项目。SWA 集团在中国四川省米易县所制定的安宁河总体规划项目很好地例证了这一过程。这个 330 hm² 的项目原本是一个设计招标,目的是寻求超越传统的生态规划方案。作为中标规划方案,SWA 的方案将该地区现有的宝贵资源作为未来基础设施架构的重要组成部分,其中包括 56 hm² 农田、现有灌溉渠道及一座水电站,并将这些资源作为公众生态意识的教育工具。这一项目还进一步发展了用来提高安宁河水质的创新技术。它采用生物修复的手段提高水质,并将清洁后的河水用于城市湖泊、湿地及森林,以此来提高生物多样性。这个行之有效的生态规划方案帮助当地政府获得了公共发展基金,赋予了城市独特的个性,并为其后的开发项目拟定了指导方针(图5)。

3 聚焦:景观基础设施

为了拓宽景观都市主义的思考范围,我们可以审视一下在城市肌理以外运行的、现有的基础设施系统。这种将景观视作基础设施的概念成为一种新鲜的媒介,使我们能以此来探讨地区之间的连通性及配送能力。例如,最初开发美国洲际高速公路系统是出于国防目的:以便在全美范围内分布弹药、战舰并高效率地调动军事人员。在过去50年中,整个系统年久失修并且超负荷运行。这种现象成为一种契机,促使美国重新评估它46 876英里连绵不断的高速公路系统,并启用多种形式的对生态较为友好的运输策略,同时促成了遍布全美的资源和能源的再分配机制。

类似的例子有中国的南水北调工程。南水北调工程最早是由毛泽东所构思的,调运中国南部丰富的水资源来平衡中国北部的水源缺乏。这个项目计划由东部、中部和西部的三条水道将南部河流的干、支流和湖泊中的水资源穿过10个省,跨越2 700 km的距离调往北方。其中东部水道需要一系列的泵站将水提升到地势较高的地域,并需要打通地下河道来穿越黄河。这条水道利用了自公元5世纪起就作为中国运输动脉的京杭大运河的部分已有河道。这个项目所带来的环境方面和技术方面的挑战给景观基础设施带来空前的机遇,这种机遇有可能把这个富有争议性、尺度巨大的基础设施项目变为一个跨区域的生态和景观长廊。

4 什么是基础设施?

要拓展有关基础设施的话题,首先让我们来看一看它的传统定义:基础设施是"一个社区或社会正常运作所需要的基本设施、服务以及机构,如运输及通信系统、供水及供电线路、公共机构如学校、邮局及监狱等。"[4] 我们的社会通常重视单一功能的基础设施系统的设计,以工程化的设计来保证这些系统在特定时间内可以最高效地完成某个单一目的,但这并不能使这些设施在其整个生命周期内提供连续性的使用效益。这种单一效益的思考方式严重影响了基础设施对城市生活的贡献。停车场、交通走廊、交通枢纽、沟渠水道在非高峰时段则闲置不用,从而给城市造成空置和空间阻隔。工业场地及垃圾填埋场是环境公害,人们虽然了解但更愿意忽视它们的存在。只有当人类生活的健康与安全受到威胁时,人们才会重新记起它们(图6)。

随着城市的急速增长和开放空间的缺乏,我们发现基础设施是一个能够产生积极影响确却未被充分开发的资源。通过运用生态规律和社会原则,城市基础设施系统能够在改善城市生活上贡献出多方面的积极作用。

5 什么是景观基础设施?

Mossop认为基础设施,例如道路,"应具有多种功能:必须满足公共空间的要求,必须串连其他功能性的城市系统,诸如公共交通、人行系统、水处理系统、经济开发、公共设施以及生态系统。"[5] 基础设施的多功能性同时也指出了多样化作为城市建设的总体原则的重要性,从而形成一种优化状态,即城市及其基础设施融为一体,而基础设施向人们展示该城市的组织及建设方式。体现这一观点的典型范例是由弗雷盐沼(Back Bay Fens)。这一场地原为一处因城市人口增长而导致的直排污水所污染的盐水沼泽。受益于19世纪20年代的土地开垦项目所带来的改善水质、控制洪水并重新建立潮汐生态系统等一系列卓见成效的持久努力,现在后海湾盐沼已成为波士顿著名的占地1 100英亩开放空间网络"翡翠项链"(Emerald Necklace)中的重要组成部分。"翡翠项链"开发空间网络包括公园、公园大道及水路,调节和净化城市径地表流的场地、城市野生动物栖息地、散布道、体育活动场所以及一处占地265英亩的植物园,并为波士顿居民带来了清新空气(图7)。

景观基础设施包含一套具体的特征、期望值及性能标准,在这里总结为:

(1)作为非孤立系统,景观基础设施能够达到一系列要求并取得可测量的成果。通过大规模普及,这种成效是可以量化并向大众展示的。以洛杉矶的窄巷系统为例。洛杉矶有长度超过900英里的窄巷,位于城市住宅的后墙与后墙之间,全部为硬质铺装,并且仅用于收集垃圾等市政服务。这些窄巷的总面积累计达3平方英里——相当于洛杉矶格里菲斯公园面积的一半,纽约中央公园面积的3倍,或约400个沃尔玛超市的面积总和(不包括停车场)。除了用于单纯的市政服务,这些窄巷完全可以被改建为生态走廊、自行车道、林荫步行道、袖珍公园、小菜园等其他用途(图8)。

(2)作为一个众多元素的集合体,基础设施既能提高,也能削弱城市功能。如果我们能够给基础设施以积极的导向,它们就能够修复城市发展中的负面因素,甚至产生积极的贡献。如芝加哥拥有世界上面积最大的位于城市中心的绿色屋顶。作为一个整体系统,绿色屋顶在整个城市范围内发挥生态作用,对逆转热岛效应、收集雨水及创建更多动植物栖息地具有重要意义。

(3)作为将众多因素连接在一起的结缔组织,基础设施能够将具有不同特质的元素融合在一起,并逐渐加强这些元素之间的凝聚力和统一的目的性。如越来越多地以高速公路和汽车作为货物与服务运输的主要途径的老城市开始选择以公共交通体系作为未来的发展方向,导致严重排放的高速公路交通方式不久将成为历史遗迹。更多地面向公众,沿着交通走廊分布的众多邻里社区将给人们留下越来越深刻的印象,而人与人之间面对面的交流将促进更多自发的交谈,增进人们互相之间的了解和理解,促使城市再次成为人类的理想栖息地(图9)。

(4) 基础设施是一种催化机制,沿线的区域中心和节点将会因它们的可达性和高度的可见性而直接受益。

以纽约中央公园为例,Andrew Jackson Downing 和 William Cullen Bryant 曾经预见纽约的中央公园将成为医治包括拥挤街道、贫困移民以及犯罪等社会弊病的良药。19世纪30年代的纽约发展迅速,因此中央公园必须有足够的面积来满足密集人口的需求。作为一项重要的景观基础工程,占地843英亩并于1860年完成的中央公园通过旅游业给当地经济带来了未曾预料的收益,它使得周边房地产与地价大幅升值,并因此而增加了政府税收(图10)。

(5) 基础设施的规模及为此花费的大量资源预示了人们充分使用其潜力的无可限量的机会。

举例说明,洛杉矶市拥有511英里的高速公路、51英里的沟渠河道、7 000英里的输电线路、6 400英里的街道,以及更多的遍布城市各处、未经统计的基础设施。市议会最近进行的一项调查显示,改善基础设施被认为是改善洛杉矶地邻里区的所有措施中最重要的一条。如果这些基础设施系统能够减少碳排放量、收集并保持地表径流并恢复向地下蓄水层的回渗功能,使用场地上产出的再生能源,促使人们更多地使用连接不同社区的该城市独特的人行道及自行车道系统,将会有更多的人愿意移居市内,而市中心商业区内目前空置的楼宇也将重新焕发生机(图11)。

6 未来

2009年2月,美国国会通过了2009美国复苏与再投资法案以刺激美国经济,而此次危机是美国自大萧条以来面临的最严重的经济衰退。该经济刺激一揽子计划提供了约7 870亿美元用于适当的危机投资,其中有72亿美元的环境投资,包括废水处理设施、饮用水基础设施、有害废物清除、柴油发动机的废气减排,以及治理受污染土地等方面;809亿美元的基础设施投资,包括道路、桥梁、铁路、污水系统、高效环保建筑、改善废水处理基础设施、改善饮用水基础设施、加大电动车辆的研发和使用等方面,以及面向垦务局、国家公园管理局、林务局、美国国家野生动物保护区等的补充投资。在能源领域,813亿美元的资金被投放在智能电网、能源节约及可再生能源、放射性废物清理、环保人员培训方面。

为了规避全球金融危机的影响,中国已经通过了资金高达数十亿美元的一系列基础设施项目,许多项目针对农村基础设施、水电、交通、环境以及技术改造等方面(新华网,2008年11月)。中国的"十一五计划"(2006—2010年)同样集中在基础设施的投资,其中较多的投入到中西部地区,包括道路网络、铁路、电网及灌溉系统,因为这些地区的居民如同发达地区的居民一样需要提高和改善生活条件。我们生活在一个立法者和最高政府官员对于全球的可持续性发展与我们持有相同的愿望的历史性时刻。对于那些将自己的汽车由传统燃油行转换为生物燃料汽车的个人、那些将自己前院草坪变为种植有机蔬菜的菜园的厨师,以及传授用保存森林资源这种简单易行的方法来降低人类的碳排量的学者来说,全球可持续性发展的未来就在我们当下的点滴行动当中! 景观基础设施的重要性和不可阻挡的发展趋势清晰可见。我们的城市需要这种基础设施的设计导向,这种导向超越了视觉边界,连接着场地与场地、人群与场所,连接着社区与社区、个体与个体,还连接着自然与城市以及城市与自然(图12)。

[思考]

(1) 可持续景观设计主要体现在哪些方面?
(2) 如何衡量绿色基础设施的生态服务功能?
(3) 景观都市主义和景观基础设施的概念和内涵是什么?

附 表

附表一 欧美地区设置风景园林专业的院校及网址一览(部分)

国家	学校名称	网 址
美国	哈佛大学设计研究生院 (Harvard University-Graduate School of Design)	www.gsd.harvard.edu
	宾夕法尼亚大学 (University of Pennsylvania)	www.upenn.edu
	伊利诺伊大学香槟分校 (University of Illinois at Urbana-Champaign)	illinois.edu
	宾夕法尼亚州立大学 (Pennsylvania State University)	www.psu.edu
	加利福尼亚大学伯克利分校 (University of California at Berkeley)	berkeley.edu
	康奈尔大学 (Cornell University)	www.cornell.edu
	马萨诸塞州大学 (University of Massachusetts)	www.umass.edu
	明尼苏达大学 (University of Minnesota)	www.umn.edu
	加州理工大学 (California Polytechnic State University)	www.calpoly.edu
	南加州大学 (University of Southern California)	www.usc.edu
	弗吉尼亚大学 (University of Virginia)	www.virginia.edu
	德州农工大学卡城分校 (Texas A&M University)	www.tamu.edu
	密西根大学 (University of Michigan)	www.umich.edu
	乔治亚大学 (University of Georgia)	www.uga.edu

续表

国家	学校名称	网址
美国	华盛顿大学西雅图分校 (University of Washington, Seattle)	www. washington. edu
	俄亥俄州立大学 (Ohio State Univeristy)	www. osu. edu
	得克萨斯大学奥斯汀分校 (University of Texas, Austin)	www. utexas. edu
	罗德岛设计学校 (Rhode Island School of Design)	www. risd. edu
	北卡罗来纳州州立大学 (North Carolina State University)	www. ncsu. edu
	佛罗里达大学 (University of Florida)	www. ufl. edu
	堪萨斯州立大学 (Kansas State University)	www. k-state. edu
	鲍尔州立大学 (Ball State University)	cms. bsu. edu
	纽约州立大学 (State University of New York)	www. suny. edu
	路易斯安那州立大学 (Louisiana State University)	www. lsu. edu
英国	剑桥大学 (University of Cambridge)	www. cam. ac. uk
	谢菲尔德大学 (University of Sheffield)	www. shef. ac. uk
	伦敦格林威治大学 (London university of Greenwich)	www. gre. ac. uk
	利兹城市大学 (Leeds Metropolitan University)	www. lmu. ac. uk
	爱丁堡艺术学院 (Edinburgh College of Art)	www. eca. ac. uk
	纽卡斯尔大学 (University of Newcastle)	www. ncl. ac. uk
	曼彻斯特大学 (University of Manchester)	www2. man. ac. uk
	瑞丁大学 (University of Reading)	www. rdg. ac. uk

续表

国家	学校名称	网　址
英国	曼彻斯特城市大学 （Manchester Metropolitan University）	www. mman. ac. uk
	卡迪夫大学 （University of Cardiff）	www. cardiff. ac. uk
	伯明翰中英格兰大学 （Birmingham of Central England）	www. uce. ac. uk
	利物浦大学 （University of Liverpool）	www. liv. ac. uk
	金斯顿大学 （University of Kingston）	www. kingston. ac. uk
德国	柏林工业大学 （Technische Universität Berlin）	www. tu-berlin. de
	德累斯顿工业大学 （Technische Universität Dresden）	tu-dresden. de
	汉诺威大学 （Leibniz Universität Hannover）	www. uni-hannover. de
	慕尼黑工业大学 （Technische Universität München）	www. tu-muenchen. de
	卡塞尔大学 （Universität Kassel）	www. uni-kassel. de/uni/
	安哈尔特应用技术大学 （Hochschule Anhalt）（Anhalt University of Applied Sciences）	www. hs-anhalt. de masterla. de
	柏林工业大学 （Technische Universität Berlin）	www. tu-berlin. de
	达姆施塔特工业大学 （Technische Universität Darmstadt）	www. tu-darmstadt. de
	汉堡-哈尔堡工业大学 （Technische Universität Hamburg-Harburg）	www. tu-harburg. de
法国	凡尔赛国立高等风景园林学院 （École nationale supérieure de paysage）	www. ecole-paysage. fr
	国立高等自然与景观学校 （L'école nationale supérieure de la nature et du paysage）	www. ensnp. fr
	国立波尔多建筑与风景园林学院 （Ecole Nationale Superieure D'architecture et de Paysage de Bordeaux）	www. bordeaux. archi. fr

续表

国家	学校名称	网 址
荷兰	瓦戈宁根大学及研究中心 (Wageningen University and Research Center)	www.wur.nl/en.htm
	阿姆斯特丹艺术学校 (Amsterdamse Hogeschool voor de Kunsten)	www.ahk.nl
	阿姆斯特丹建筑学院 (Amsterdam Academie van Bouwkunst)	www.academievanbouwkunst.nl
丹麦	奥胡斯建筑学院 (Aarhus School of Architecture)	en.aarch.dk/en/
	丹麦皇家艺术学院 (Royal Danish Academy of Fine Arts)	www.kadk.dk/en/
	哥本哈根大学 (University of Copenhagen)	www.ku.dk
比利时	布鲁塞尔自由大学 (Vrije Universiteit Brussel)	www.vub.ac.be
	鲁汶大学 (Katholieke Universiteit Leuven)	www.kuleuven.be
爱尔兰	都柏林大学 (University College Dublin)	www.ucd.ie
	都柏林理工学院 (Dublin Institute of Technology)	www.dit.ie
俄罗斯	莫斯科建筑学院 (Moscow Architecture Institute)	www.marhi.ru
	莫斯科国立林业大学 (Moscow State Forest University)	www.msu.ru/en/
	马里埃尔州立工业大学 (Mari-El State Technical University)	www.marstu.net
	奥廖尔国立农业大学 (Orel State Agrarian University)	www.orelsau.ru
西班牙	巴塞罗那大学 (Universitat de Barcelona)	www.ub.edu
	马德里理工大学 (Universidad Politecnica de Madrid)	www.upm.es
瑞典	瑞典农业科学大学 (Sveriges Lantbruks Universitet)	www.slu.se
瑞士	苏黎世大学 (Universität Zürich)	www.uzh.ch

续表

国家	学校名称	网址
加拿大	不列颠哥伦比亚大学 (University of British Columbia)	www.ubc.ca
	圭尔夫大学 (University of Guelph)	www.uoguelph.ca
	曼尼托巴大学 (University of Manitoba)	www.umanitoba.ca
	多伦多大学 (University of Toronto)	www.utoronto.ca
	蒙特利尔大学 (University of Montreal)	www.umontreal.ca
	滑铁卢大学 (University of Waterloo)	www.uwaterloo.ca
	卡尔顿大学 (University of Carleton)	www.carleton.ca
	麦吉尔大学 (University of McGill)	www.mcgill.ca
澳大利亚	悉尼大学 (University of Sydney)	www.sydney.edu.au
	墨尔本大学 (University of Melbourne)	www.unimelb.edu.au
	皇家墨尔本理工大学 (RMIT University)	www.rmit.edu.au
	新南威尔士大学 (University of New South Wales)	www.unsw.edu.au
	阿德莱德大学 (University of adelaide)	www.adelaide.edu.au
	西澳大学 (University of Western Australia)	www.uwa.edu.au

附表二 国际风景园林组织机构一览(部分)

机构名称	Institution name	网址
联合国教育、科学及文化组织世界遗产委员会(UNESCO)	United Nations Educational, Scientific and Cultural Organization, World Heritage Committee	http://whc.unesco.org/
国际自然资源保护联盟(IUCN)	International Union for Conservation of Nature and Natural Resources	http://www.iucn.org/
国际风景园林师联合会(IFLA)	International Federation of Landscape Architects	http://www.ifla.org/
欧洲景观设计协会(EFLA)	European Foundation for Landscape Architecture	
欧洲景观教育大学联合会(ECLAS)	European Council of Landscape Architecture School	http://www.eclas.org/
美国风景园林师协会(ASLA)	American Society of Landscape Architecture	http://www.asla.org/
美国风景园林设计教育理事会(CELA)	Council of Educators in Landscape Architecture	http://www.thecela.org/
美国风景园林专业资格评估委员会(LAAB)	Landscape Architectural Accreditation Board	http://www.asla.org/accreditationlaab.aspx
美国建筑师协会(AIA)	American Institute of Architects	http://architectfinder.aia.org/
美国规划师协会(APA)	American Planning Association	http://www.planning.org/
英国风景园林协会(LI)	The Landscape Institute	http://www.landscapeinstitute.org
澳大利亚景观设计师协会(AILA)	Australian Institute of Landscape Architects	http://www.aila.org.au/
中国风景园林学会(CHSLA)	Chinese Society of Landscape Architecture	http://www.chsla.org.cn/
台湾造园景观学会	—	http://www.clasit.org.tw/
台湾景观学会(中华民国景观学会)(TILA)	Taiwan Institute of Landscape Architects	http://www.landscape.org.tw/
香港园境师学会(HKILA)	Hong Kong Institute of Landscape Architects	http://www.hkila.com/

附表三　国外风景园林期刊一览（部分）

国家 Country	名　称 Name	网　址 Website
美国	Landscape Architecture	http：//www. asla. org/
	Landscape Journal	http：//lj. uwpress. org/
	Harvard design magazine	http：//www. harvarddesignmagazine. org/
	Garden Design	http：//www. gardendesign. com/
英国	Landscape Research	http：//www. landscaperesearch. net/
	Landscape	https：//www. landscapeinstitute. org/journal-issues/
荷兰	JoLA	http：//www. jola-lab. eu/
	Landscape and urban planning	http：//www. journals. elsevier. com/landscape-and-urban-planning/
	Scape	http：//www. scapemagazine. com/
	Urban Forestry & Urban Greening	https：//www. journals. elsevier. com/urban-forestry-and-urban-greening/
德国	Topos	http：//www. topos. de
	Garten + Landschafts	www. garten-landschaft. de
日本	Landscape design	
澳大利亚	Landscape Architecture Australia	http：//www. australianmagazinesubscriptions. com/art_and_design_magazines/landscape_architecture_australia. html
加拿大	Landscape Paysages	https：//www. bcsla. org/initiatives/landscapes-paysages
	Landscaping & Groundskeeping magazine	
新西兰	Landscape New Zealand	https：//worldarchitecture. org/amp/？q = pzfnz
丹麦	LANDSKAB	http：//arkfo. dk/
意大利	Gardenia	http：//www. cairocommunication. it/Mensili/Gardenia. html

附表四 国际风景园林学术会议

会议名称 Conference Name	简 介 Brief Introduction	其他相关信息 Website
国际风景园林师联合会 International Federation of Landscape Architects, IFLA	IFLA 是"The International Federation of Landscape Architects"的英文简称,中文译为"国际风景园林师联合会"。IFLA 于 1948 年 9 月在英国剑桥成立,总部设在法国凡尔赛,是受联合国教科文组织指导的国际风景园林行业的最高组织,现有 52 个国家的风景园林学会会员。2005 年中国风景园林学会正式加入了 IFLA,成为代表中国的国家会员。IFLA 每年召开一次全球性年会,轮流在三大区(亚太区、美洲区、欧洲区)举行。	网址:https://www.ifla.org 召开频次:每年一次 召开时间:国际竞赛报名时间(8—11月) 既往主题: 2017年:改变世界的10天(10 Days to Change the World) 2016年:品味景观(Tasting the Landscape) 2015年:未来的历史(History of the Future) 2014年:景观的思考与行动(Thinking and Action on Landscape) 会议其他构成:IFLA 国际竞赛
国际风景园林师联合会亚太区年会 International Federation of Landscape Architects, IFLA ASIA-PACIFIC	IFLA 亚太地区(IFLA APR)是国际风景园林师联合会(IFLA)的一个子组织。该组织由位于亚太地区的各专业风景园林组织提名的代表组成,这些机构都是 IFLA(目前为14人)的成员。代表们代表他们各自在 IFLA 的国家,作为一个合作小组,共同致力于亚太地区的景观架构问题和项目。	网址:http://iflaapr.org/ 召开频次:每年一次 召开时间:国际竞赛报名时间(5—10月) 既往主题: 2015年:未来的山和火山-创造力的繁荣(The Future Mountain And Volcanoscape-Creativity To Prosperity) 2014年:绿色明天(A Greener Tomorrow) 2013年:救赎的景观(Redeeming landscape) 2012年:风景园林让生活更美好(Better Landscape, Better life) 会议其他构成:主题发言、分论坛、竞赛
世界公园大会 World Parks Congress, IUCN	代表人类最高追求目标的世界自然保护联盟,每 10 年召开一次公园大会,总结过去 10 年的经验与差距,评估现在面临的问题,共商未来发展大计与对策。	网址:https://www.iucn.org/theme/protected-areas/about/congresses/world-parks-congress 召开频次:10年一次 召开时间:11月 既往主题: 2014年:公园、人类、地球,启发解决方案

续表

会议名称 Conference Name	简介 Brief Introduction	其他相关信息 Website
国际城市与区域规划师学会 International Society of City and Regional Planner Joint Conference, ISOCARP	国际城市与区域规划师年会由ISOCARP国际城市与区域规划师学会主办,该学会成立于1965年,是联合国人居署和欧洲理事会认证的国际专业组织,同时也是联合国教科文的官方顾问机构。	网址:https://isocarp.org/ 召开频次:每年一次 召开时间:10月 既往主题: 2017年:智慧社区(Smart Community) 2016年:我们拥有的城市 vs 我们需要的城市(Cities We Have VS. Cities We Need) 2015年:城市拯救世界,让我们重塑规划(Cities Save the World, Let's Reinvent Planning)
世界景观生态学大会 International Association for Landscape Ecology, IALE	四年一度的国际景观生态学大会代表了当前国际景观生态学领域研究的最高水平,旨在搭建景观生态学多尺度主题讨论和交流的国际平台,开展当前人类社会面临的环境挑战和可能的解决途径的深入研讨,是探讨景观生态学发展方向和前沿领域、推动不同地区之间学术交流的重要平台。	网址:http://www.ialeworldcongress.org/ 召开频次:每4年一次 召开时间:7月 既往主题: 2015年:跨越规模、跨越国界:全球应对复杂挑战的方法(Crossing Scales, Crossing Borders: Global Approaches to Complex Challenges) 2011年:变化的世界中的生态学(Ecology in a Changing World) 2007年:25年的景观生态学:实践中的科学原理 会议其他构成:主题报告,专题研讨会,技术性专题研讨会,会间和会后考察
国际恢复生态学大会 Society for Ecological Restoration, SER	国际恢复生态学学会1987年在美国成立,拥有60多个国家成员和合作伙伴,并在14个国家或地区设有办事处。促进生态修复作为一种维持地球上的生命的多样性和重建生态健康的自然和文化之间的关系。生态恢复成为一个基本组件的保护和可持续发展项目在世界各地由于其固有的能力,不仅为人们提供机会修复生态破坏,同时也改善人类的生活条件。	网址:http://www.ser.org/ 召开频次:每年一次 召开时间:8—11月 既往主题: 2015年:恢复城市、乡村和原野(Restore the city, countryside and wilderness) 2013年:回顾过去,引领未来(Look back and lead the future) 2009年:在变化的世界中创造变化(Create change in a changing world) 2007年:生态恢复的全球性挑战(The global challenges of ecological restoration) 会议其他构成:全体会议、分组报告、考察

续表

会议名称 Conference Name	简 介 Brief Introduction	其他相关信息 Website
国际湿地大会 International Wetlands Conference, INTECOL	国际湿地大会是在国际生态学协会及湿地工作组组织下举办的国际盛会,是湿地科学与应用领域中最大的、最有影响力的国际性会议,参会人员均为世界湿地方面的顶级专家及管理人员。大会宗旨是推动全球湿地保护,湿地资源管理经验及湿地科学领域的学术交流。国际湿地大会已经成功举办10届,2016年是大会首次在亚洲召开,地点是江苏常熟。	网址:http://www.conference.ifas.ufl.edu/INTECOL/ 召开频次:每4年一次 召开时间:6—10月 既往主题: 2016年:全球气候变化下生物多样性和生态系统的热点问题 2012年:气候变化对湿地生物区系、生物地球化学循环、水文、固碳、温室气体、盐度、水质,以及长期存在的营养物质和污染物的影响 2008年:大湿地大关注 2004年:1.湿地在综合水资源管理中的角色;2.湿地科学在环境管理中的地位;3.湿地的生物地球化学循环;4.植被在湿地环境中的作用;5.湿地的保护和管理;6.湿地恢复和重建;7.在世界范围内湿地和气候变化的关系;8.湿地在水环境改善中的作用;9.热带湿地道德功能和合理利用
国际景观数字大会 Digital Landscape Architecture Conference, DLA	国际数字景观大会是由德国安哈尔特应用技术大学(Anhalt University of Applied Sciences)主办的国际性会议,自2000年开始,每年举办一届。到目前为止,会议已成功举办14届。会议由该大学发起,会议地点大部分在德国德绍(Dessau)的安哈尔特大学,近几年开始每2年选择其他国家主办。每年5—6月,世界各地的风景园林及相关专业学者相聚在一起,共同交流和探讨数字景观发展的最新成果和趋势。	网址:http://www.digital-la.de/ 召开频次:每年一次 召开时间:5—6月 既往主题: 2017年:响应的风景(Responsive Landscapes) 2016年:代表、评估和设计景观:数字方法(Representing, Evaluating and Designing Landscapes:Digital Approaches) 2015年:景观规划和设计中的系统思维(Systems Thinking in Landscape Planning and Design) 2014年:风景园林和规划:在地理设计中发展数字方法(Landscape Architecture and Planning: Developing Digital Methods in GeoDesign) 会议其他构成:会议发言、分论坛
欧洲景观教育大会联合会年会 European Council of Landscape Architecture Schools, ECLAS	风景园林学校的欧洲理事会的开端可以追溯到欧洲景观教育的开端。旨在促进欧洲风景园林学科信息的交流、经验和想法,讨论刺激和鼓励欧洲园林学校之间的合作。	网址:http://www.eclas.org/ 召开频次:每年一次 召开时间:9月

续表

会议名称 Conference Name	简介 Brief Introduction	其他相关信息 Website
英国景观行业协会年会 British Association of Landscape Industries, BALI	英国景观行业协会（BALI）成立于1972年，是英国景观行业最具代表性的贸易协会，一直致力于提高行业标准，追求卓越。一年一度的英国BALI国家景观奖评选始于1976年，目的是评选在广泛的景观设计领域展现了杰出的专业性和技术性的BALI会员，已成为英国最大、最富盛名的景观奖项。每年的颁奖盛会都会吸引近千名专业人士参加，成为行业交流的最佳机遇。	网址：http://www.bali.org.uk/about-bali/ 召开频次：每年一次 召开时间：12月 会议其他构成：年度国家景观奖颁奖
澳大利亚园林大会 Australian Landscape Conference, ALC	澳大利亚园林大会在墨尔本召开，本会议由国际景观设计师进行十个左右的专业演讲，促进专业知识的交流和最新进展。	网址：http://www.landscapeconference.com/ 召开频次：每三年一次 召开时间：3—4月 既往主题： 2018年：设计与自然：重新连接人和空间（Design with Nature: Reconnecting People and Place） 2015年：艺术与自然：景观中的冲突与和谐（Art and Nature: Conflict and Harmony in the Landscape）
美国风景园林师协会年会 American Society of Landscape Architects, ASLA	美国风景园林师协会ASLA（AMERICAN SOCIETY OF LANDSCAPE ARCHITECTS）主办，美国景观设计师协会年会和展览是景观设计专业人士最看重的活动，该年会是全球最大的景观设计专业人士和学生聚会。	网址：http://www.asla.org/ 召开频次：每年一次 召开时间：10月 既往主题： 分为专业组和学生组，奖项设置由卓越奖和荣誉奖组成，分为综合设计类、住区设计类、分析与规划类、研究范畴、信息类、地标奖六大类，奖励在水、交通、公园等领域出色的景观设计。 会议其他构成：它是世界上最大的景观设计贸易展览，每年有5 000位与会者和500个参展商，每年组织竞赛。

续表

会议名称 Conference Name	简介 Brief Introduction	其他相关信息 Website
美国景观教育年会 Council of Educators in Landscape Architecture, CELA	CELA 的使命是鼓励、支持和继续教育风景园林领域的相关教学、研究、奖学金和公共服务。CELA 年度会议关注最近在风景园林的各个方面的研究和奖学金。	网址:http://thecela.org/ 召开频次:每年一次 召开时间:1—9 月 既往主题: 2017 年:沟通(Bridging) 2016 年:困境:辩论(Dilemma:debate) 2015 年:鼓励改变|改变洞察力(Incite Change | Change Insight) 2014 年:景观、城市和社区(Landscape, City, and Community) 会议其他构成:主题报告,分论坛
加拿大魁北克国际花园节 The International Garden Festival, Grand-Métis, Quebec, Canada	国际花园节自 2000 开始举办,是一场关于花园创新性和实验性的特殊展览。它是北美地区最重要的当代花园展览活动。花园节的空间设计结合了视觉艺术、建筑、景观、设计和自然等元素,每年夏天这些独特的装置都会给参观者们带来惊喜。	网址:http://www.chla.com.cn 召开频次:每年一次 召开时间:7—8 月
切尔西花展 RHS Chelsea Flower Show	一年一度的"切尔西花展"是英国的传统花卉园艺展会,也是全世界最著名、最盛大的园艺博览会之一。由英国皇家园艺协会主办的切尔西花展创办于 1862 年,最初在肯辛顿(Kingsington)举行,自 1913 年起移至伦敦的切尔西地区举办。创办时间距今已有 150 余年的历史。	网址:http://www.rhs.org.uk/shows-events/rhs-chelsea-flower-show 召开频次:每年一次 召开时间:5 月 会议其他构成:切尔西花展分为室内、室外两个展区。室内展览在一个巨型帐篷内进行,展出多种最新的园艺珍品,包括育种家提供的新优花卉、优秀的花艺作品以及各式各样的园艺产品。室外展区主要是花园展览,包括展示花园(Show Gardens)、时尚花园(Chic Gardens)、城市花园(City Gardens)和庭院花园(Courtyard Gardens)4 类。

续表

会议名称 Conference Name	简介 Brief Introduction	其他相关信息 Website
新加坡国际空中绿化会议 International Skyrise Greenery Conference	会议由国家公园委员会和国际绿色屋顶协会(IGRA)共同组织,来自不同学科的国际城市绿化专家,建筑师、景观设计师、景观承包商、决策者等聚集在一起,讨论这个不断增长的行业的当前和未来趋势。	网址:https://www.nparks.gov.sg/skyrisegreenery/events/conferences 召开频次:两年一次 召开时间:11月 既往主题: 2017年:共生:城市中紧张的自然环境(Symbiosis:Intense Nature Within Cities) 2010年:创意表面快乐之窗(Surfaces of Creativity:Spaces of Delight)